Excel

Get the Results You Want!

AF867485

YEAR 12 MATHEMATICS STANDARD 1

Lyn Baker

PASCAL PRESS

CONTENTS

TOPIC: ALGEBRA

CHAPTER 1 TYPES OF RELATIONSHIPS

Simultaneous linear equations

1 Graphing lines

- **If x and y are related by the equation $y = mx + c$** then they are in a linear relationship.
- **A linear function is a straight line** when graphed.
- **When graphing a line with pencil and paper**, we can draw up a table of values and plot some points to show the position of the line.
- **Linear functions can also be graphed** using graphing software.

EXAMPLE 1

Consider the rule $y = 3x - 1$.

a Complete the table of values.

x	1	2	3	4
y				

b Graph the line $y = 3x - 1$.

a $y = 3x - 1$

x	1	2	3	4
y	2	5	8	11

When $x = 1$, $y = 3 \times 1 - 1$; when $x = 2$, $y = 3 \times 2 - 1$; when $x = 3$, $y = 3 \times 3 - 1$; when $x = 4$, $y = 3 \times 4 - 1$.

b

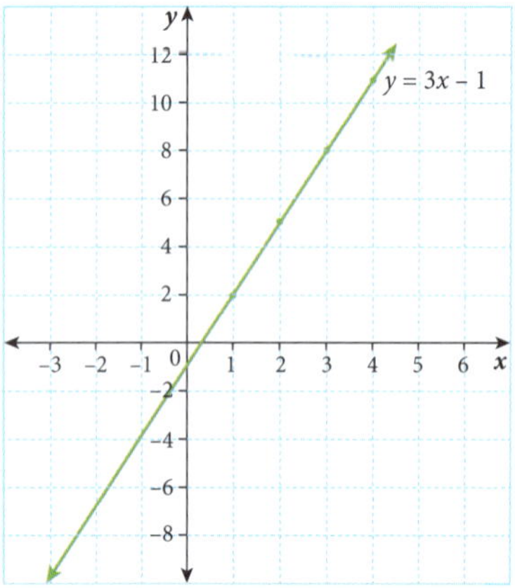

When drawing a line you should show the line going right across the grid.

Don't just join the dots from your table of values because these are just a few of the infinite number of points that lie on the line.

EXAMPLE 2

Complete the table of values for the rule $y = -x - 1$ and graph the straight line on a number plane.

x	−2	−1	0	1	2
y					

x	−2	−1	0	1	2
y	1	0	−1	−2	−3

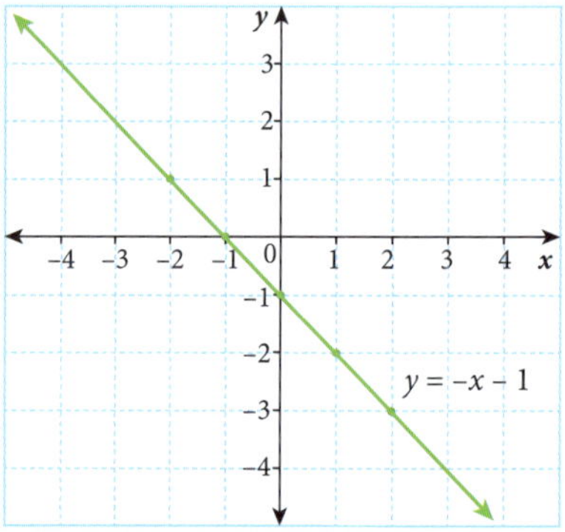

Two points are sufficient for graphing a line but use a couple of extra points in a table of values just as a check.

EXAMPLE 3

Draw the straight line $y = 6 - 2x$.

$y = 6 - 2x$

x	0	1	2	3
y	6	4	2	0

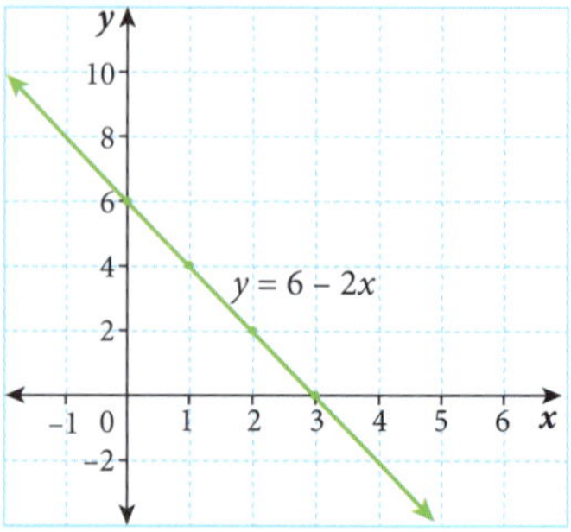

If asked to draw a line we can always draw a table of values even if we haven't been specifically asked to do so.

When choosing values to place in a table of values, you can choose any numbers at all but it makes sense to choose easy numbers, (0, 1, 2 and 3) for example.

EXAMPLE 4

Draw the line $C = \frac{3}{4}n + 40$ for values of n between 0 and 120.

$C = \frac{3}{4}n + 40$

n	0	40	80	120
C	40	70	100	130

When plotting points we normally use the *x* and *y* axes.

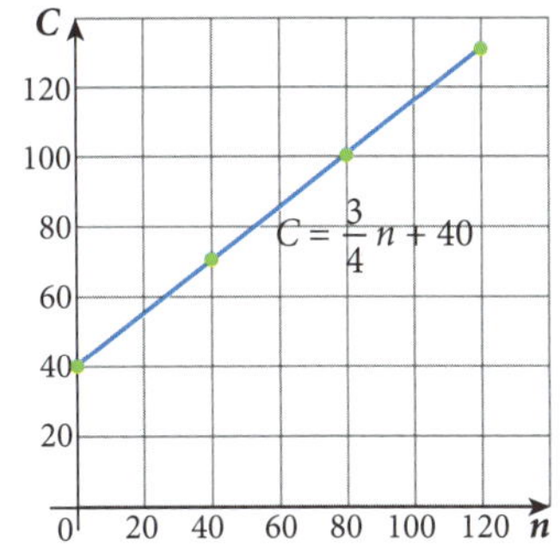

We can plot a line using any variables but we must take care to label the axes correctly.

The independent variable will usually be shown on the horizontal axis and the dependent variable will be on the vertical axis.

→ PRACTICE

1 Consider the rule $y = 4x + 1$.

  a Complete the table of values.

x	0	1	2	3
y				

  b Graph the line $y = 4x + 1$.

2 Complete the table of values for the rule $y = -2x + 1$ and graph the straight line on a number plane.

x	−2	−1	0	1	2
y					

3 Draw the straight line $y = 8 - x$.

4 Draw the line $h = 0.8t + 60$ for values of t between 0 and 100.

5* Use graphing software to draw the line $y = \frac{2}{3}x + \frac{3}{4}$.

Answers ➲ p. 30

2 Points of intersection

→ **Two distinct straight lines will only meet in**, at most, one point. If parallel the lines will not intersect; otherwise there will be one point of intersection.

→ **A pair of simultaneous equations** can be solved graphically by finding the point of intersection of the two lines.

EXAMPLE 1

The diagram shows the lines $y = x + 1$, $y = \frac{x}{2} + 4$ and $y = -x + 7$.

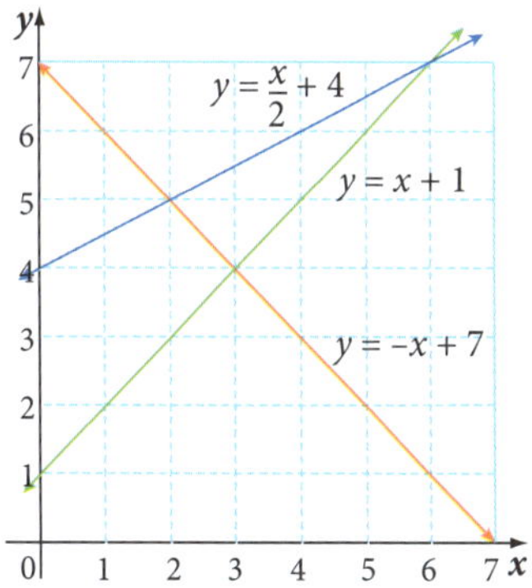

a **Where do the lines $y = x + 1$ and $y = -x + 7$ intersect?**

b **Where does the line $y = \frac{x}{2} + 4$ intersect with the line $y = -x + 7$?**

c **What is the point of intersection of $y = \frac{x}{2} + 4$ and $y = x + 1$?**

a When $x = 3$ and $y = 4$. — or the point (3, 4)

b When $x = 2$ and $y = 5$. — or the point (2, 5)

c The point of intersection is (6, 7). — or where $x = 6$ and $y = 7$

EXAMPLE 2

The diagram shows the lines $y = 3x$ and $y = x + 8$.

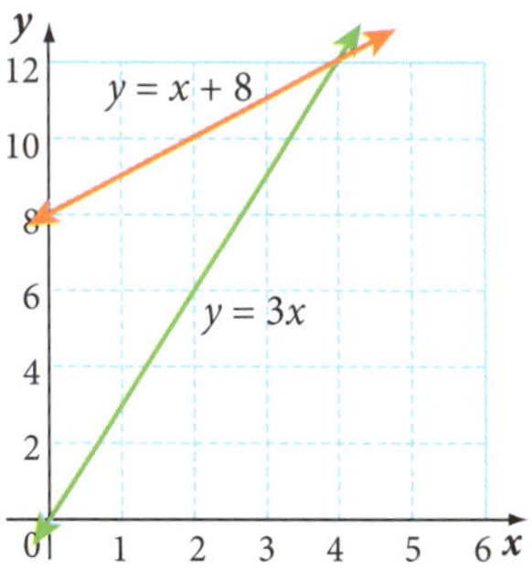

What is the solution when $y = 3x$ and $y = x + 8$ are solved simultaneously?

$x = 4$ and $y = 12$

Read the values from the graph. You can always substitute the values into the equations to check that the solution is correct.

EXAMPLE 3

a Use graphing software to graph the lines $y = 48.5 - 8x$ and $y = 2.5x + 6.5$.

b What is the simultaneous solution of the equations $8x + y = 48.5$ and $y = 2.5x + 6.5$?

a

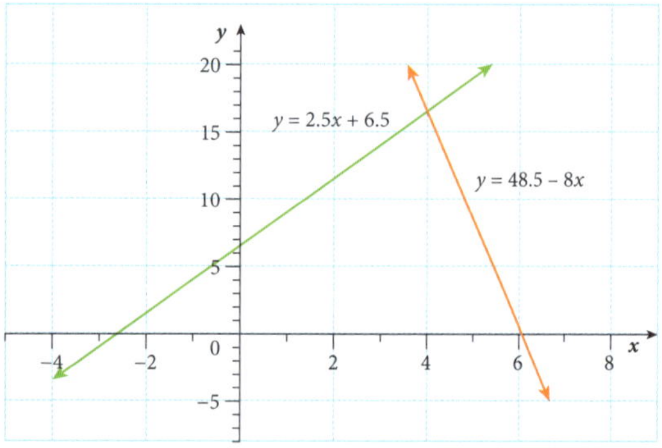

b $8x + y = 48.5$

$y = 48.5 - 8x$

From the graph, the simultaneous solution is when $x = 4$ and $y = 16.5$.

By rearranging we can see that this is the same equation that has been graphed in part **a**.

→ PRACTICE

1 The diagram shows the lines $y = x + 2$, $y = 2x$ and $y = -x + 12$.

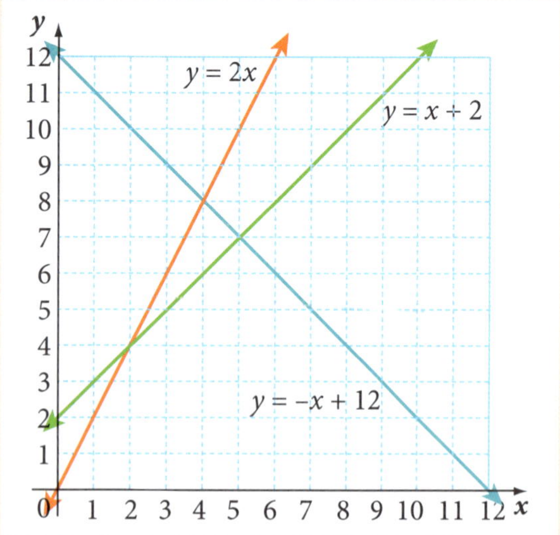

a Where do the lines $y = x + 2$ and $y = -x + 12$ intersect?

b Where does the line $y = 2x$ intersect with the line $y = -x + 12$?

c What is the point of intersection of $y = 2x$ and $y = x + 2$?

2 The diagram shows the lines $y = -x + 9$ and $y = x + 5$.

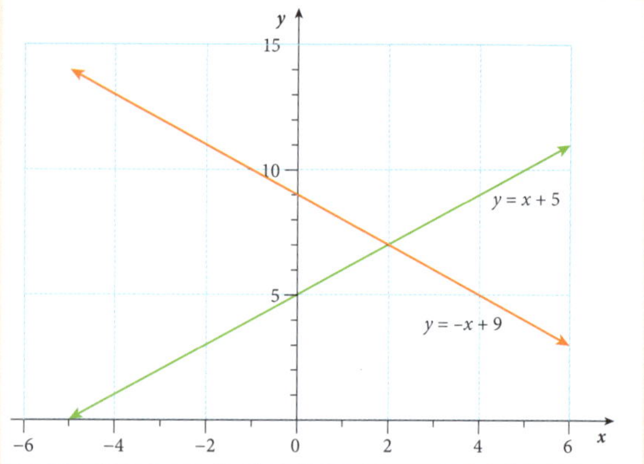

What is the solution when $y = -x + 9$ and $y = x + 5$ are solved simultaneously?

3 a Use graphing software to graph the lines $y = 22 - 5x$ and $y = 3x - 2$.

b What is the simultaneous solution of the equations $5x + y = 22$ and $y = 3x - 2$?

4* Find, graphically, the simultaneous solution of $y = 4x + 9$ and $y = \frac{1}{2}x - 5$.

Answers ➲ pp. 30–31

3 Practical interpretations of points of intersection

→ **When two equations that represent models of physical phenomena are solved simultaneously**, that solution has a practical significance.

→ **The simultaneous solution is the only solution** that works for both equations at the same time.

EXAMPLE 1

Two different hire car companies offer slightly different rates for hiring a car for a day. Each charge is made up of a fixed rate and a cost depending upon the number of kilometres driven.

The graph shows the cost for hiring cars from the two companies for different distances.

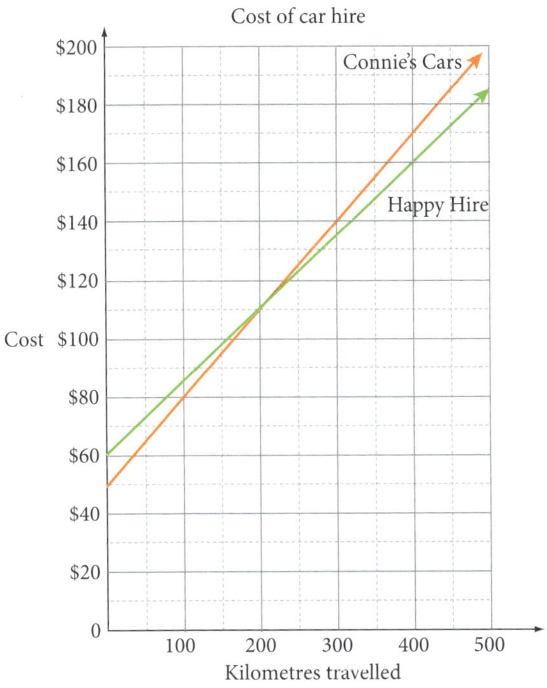

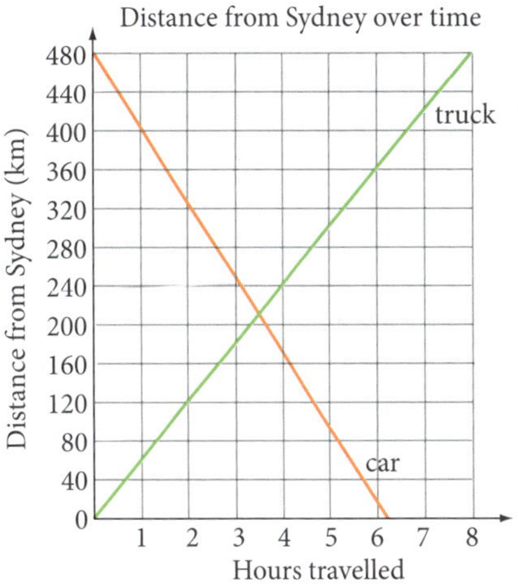

After $3\frac{1}{2}$ hours. Approximately 210 km from Sydney.

Until $3\frac{1}{2}$ hours from the start the truck is closer to Sydney than the car, but after $3\frac{1}{2}$ hours the car is closer to Sydney.

a How much does Connie's Cars charge as a fixed rate?

b How much will it cost to hire a car from Happy Hire if you travel 100 km?

c At what distance travelled are the two rates the same?

d If you needed to travel 300 km, which would be the most economical company to choose and by how much?

a $50

b $85

c 200 km

d Happy Hire is cheaper by $5. Connie's Cars cost $140 but Happy Hire is $135.

The fixed rate is the cost when no distance is travelled.

EXAMPLE 2

A truck leaves Sydney and travels to Gunnedah, a distance of 480 km. At the same time a car leaves Gunnedah and travels to Sydney. The distances from Sydney after a certain amount of time are shown on the graph. After how many hours are the vehicles the same distance from Sydney and what is that distance?

EXAMPLE 3

Two different phone companies offer different rates to make calls to a certain country overseas.

a One company, TelephX, charges a $2 connection fee and 40 cents per minute. Let $\$C$ be the cost of a call that lasts m minutes. Write an equation for the cost of a call using TelephX.

b The second company, PhoneY, charges a connection fee of $3.20 and then 30 cents per minute. Write a similar equation for the cost of a call using PhoneY.

c Draw the graphs of the costs for both companies for the first 20 minutes of a call.

d How much cheaper is it to use TelephX for an 8-minute call?

e After how many minutes is it cheaper to use PhoneY?

a TelephX: Cost = $2 + ($0.40 for each minute)

So $C = 2 + 0.4m$

Because the cost is given as $C, there is no need to include any dollar signs in the equation.

b PhoneY: Cost = $3.20 + ($0.30 for each minute)

So $C = 3.2 + 0.3m$

c

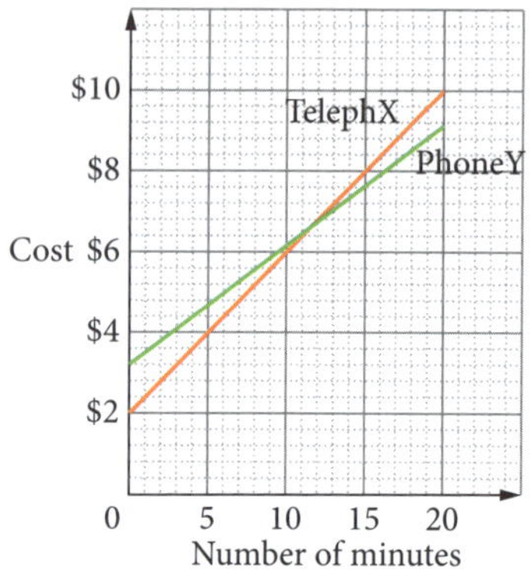

d For an 8-minute call:

TelephX charges $5.20

PhoneY charges $5.60.

It is 40 cents cheaper to use TelephX.

e After 12 minutes it is cheaper to use PhoneY.

For 12 minutes, both companies charge the same amount: $6.80.

→ PRACTICE

1 Two different taxi companies offer slightly different rates for hiring a cab. Each charge is made up of a fixed rate and a cost depending upon the number of kilometres travelled. The graph shows the cost for hiring cabs from the two companies for different distances.

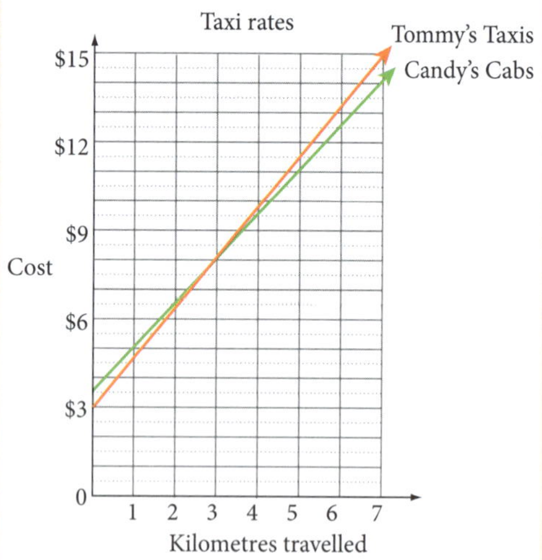

a How much does Candy's Cabs charge as a fixed rate?

b How much will it cost to hire a taxi from Tommy's Taxis if you travel 7 km?

c At what distance travelled are the two rates the same?

d If you needed to travel 5 km, which would be the most economical company to choose and by how much?

2 A bus leaves Sydney and travels to Grafton, a distance of 640 km. At the same time a car leaves Grafton and travels to Sydney. The distances from Sydney after a certain amount of time are shown on the graph.

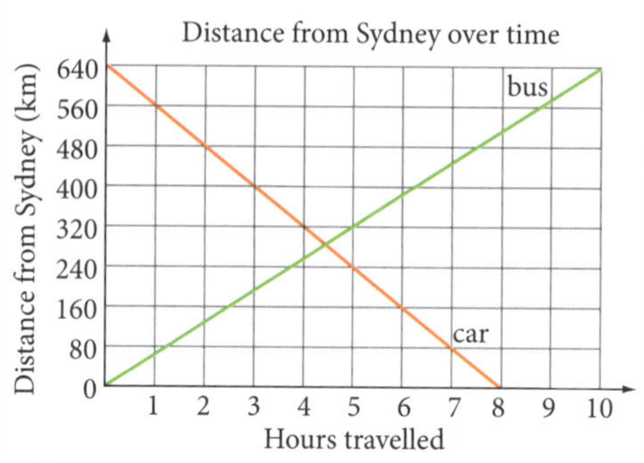

After how many hours are the vehicles the same distance from Sydney and what is that distance?

3 Two different phone companies offer different rates to make calls to certain numbers.

a One company, CallsRUs, charges a $0.50 connection fee and 10 cents per minute. Let $\$C$ be the cost of a call that lasts m minutes. Write an equation for the cost of a call using CallsRUs.

b The second company, Ringsabell, charges a connection fee of $0.20 and then 15 cents per minute. Write a similar equation for the cost of calls using Ringsabell.

c Draw the graphs of the costs for both companies for the first 20 minutes of a call.

d By how much is it cheaper to use Ringsabell for a 5-minute call?

e After how many minutes is it cheaper to use CallsRUs?

4* a A freezer originally set at −19 °C is not working properly and the temperature is increasing at the rate of 1.5 °C per hour. Write an equation that shows the temperature T °C at time h hours.

b A second freezer has been turned on to replace the first one. The temperature in the second freezer was originally 25 °C and it is cooling at the rate of 4 °C per hour. Use a graph to find the optimum time to transfer the contents of the first freezer to the second.

Answers ➲ p. 31

4 Break-even analysis

- **When graphs are used to illustrate the cost of producing goods and the return from selling those goods**, then the point of intersection gives us the 'break-even' point.
- **When the cost is greater than the return**, there will be a loss.
- **When the return is greater than the cost**, there will be a profit.
- **But when the cost and return are equal** there is no profit or loss and the enterprise is said to break even.

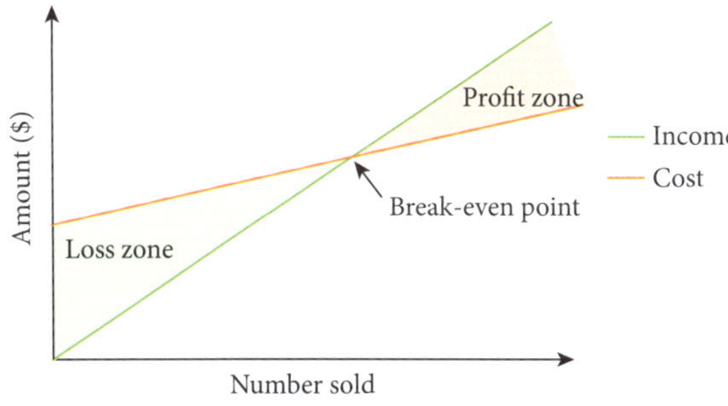

EXAMPLE 1

Guy decides to teach people to train their dogs and charges a fixed fee for each person who comes to the class. The costs involved with holding the class are $60. The graph shows the cost and return to Guy for up to 10 students.

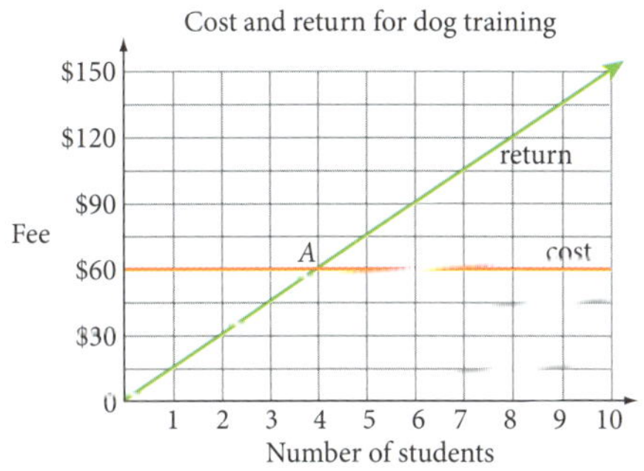

a **What does the point A represent?**

b **How much does Guy charge each person?**

c **How much profit does Guy make if seven people attend?**

d **How much does Guy lose if only two people attend?**

a A is the 'break-even' point. If four people attend, Guy will receive $60 and so cover his costs.

b Each person is charged $15.

c If seven people attend, Guy receives $105.

Profit = $105 − $60
= $45

The return is greater than the cost so a profit is made.

d If two people attend, Guy receives $30.

Loss = $60 − $30
= $30

The return is less than the cost so a loss is made.

EXAMPLE 2

The graph shows the cost to make 'babybugs'.

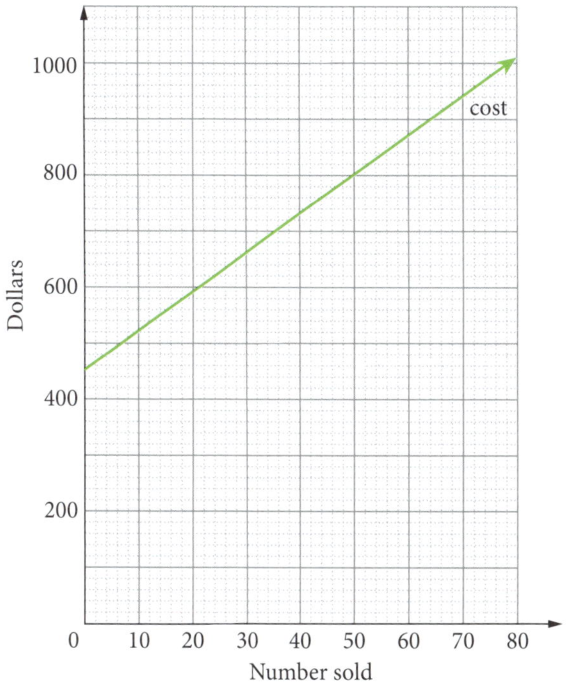

a **What is the gradient and what does it represent?**

b **What is the y-intercept and what does it represent?**

c **Each 'babybug' sells for $22. Copy the graph and show the income on the same graph.**

d **What is the break-even point?**

e **How much is the profit when 50 'babybugs' are sold?**

f **If 20 'babybugs' are sold will there be a profit or a loss? What is the size of this profit or loss?**

a $m = \frac{140}{20}$
$= 7$

The gradient is 7 and it represents the extra cost to make each 'babybug'.

b The y-intercept is 450. It represents the fixed costs.

Fixed costs are the costs for things like machinery which must be paid regardless of how many items are produced.

c

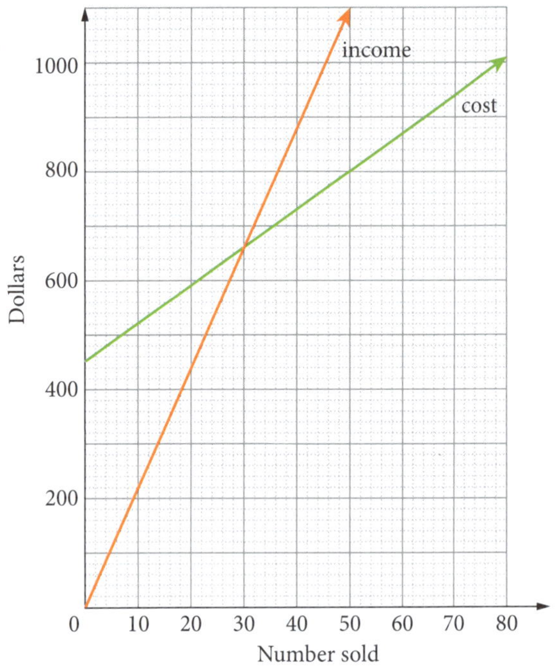

d The break-even point is when 30 'babybugs' are sold. Both the cost and income are $660.

e The profit when 50 babybugs are sold is $300.

f If 20 babybugs are sold there will be a loss of $150.

There is no need to calculate either the cost or the income. The profit can be seen on the graph as the difference between them.

→ PRACTICE

1 Angus decides to run a coaching class and charges a fixed fee for each student who comes to the class. The costs involved with holding the class are $50. The graph shows the cost and return to Angus for up to 12 students.

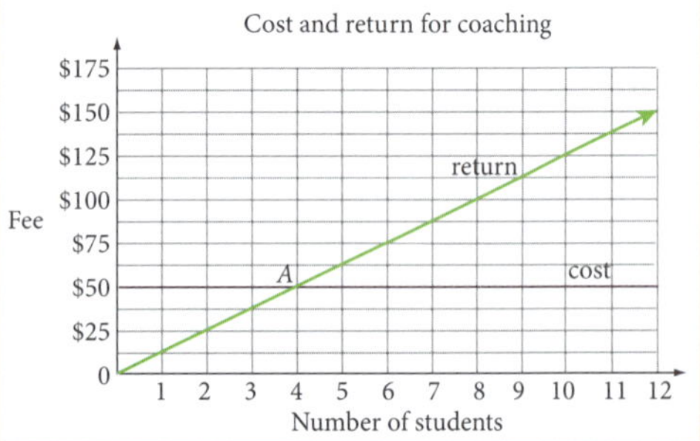

a What does the point A represent?

b What does Angus charge each student?

c How much profit does Angus make if 10 people attend?

d How much does Angus lose if only two people attend?

2 The graph shows the cost to make bags of 'sweetbeans'.

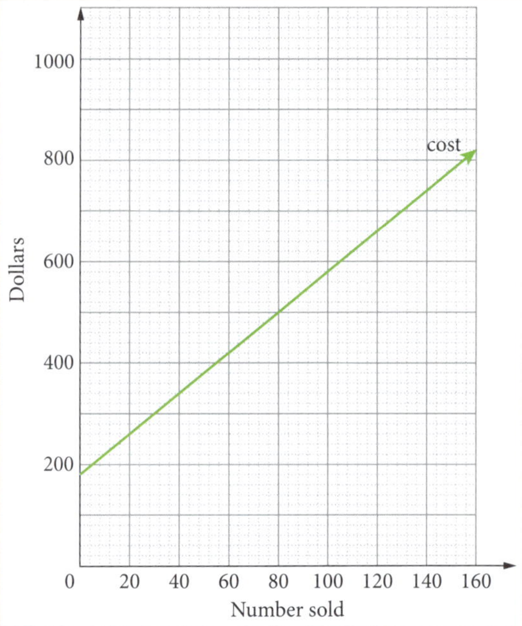

a What is the gradient and what does it represent?

b What is the y-intercept and what does it represent?

c Each bag of 'sweetbeans' sells for $5.50. Copy the graph and show the income on the same graph.

d What is the break-even point?

e How much is the profit when 150 bags of 'sweetbeans' are sold?

f If 100 bags of 'sweetbeans' are sold will there be a profit or a loss? What is the size of this profit or loss?

3* Holly makes packets of cupcakes, which she sells at a market. She sells the packets for $3.50 each. Each packet costs Holly $2.75 to make and the fixed costs are $120.

a Write an equation to show the income, $$I$, Holly receives if she sells x packets.

b Write an equation to show the cost, $$C$, to Holly of producing x packets.

c Show the equations from parts a and b on a graph.

d Find the number of packets Holly must sell to make a profit.

Answers ➲ p. 31

Graphs of practical situations

1 Constructing graphs

- **In many practical situations graphs representing various phenomena** might be linear or non-linear.
- **Plotting points from a table of values** gives a visual representation.

EXAMPLE 1

a Complete the table of values for $y = 10x - x^2$.

x	0	1	2	3	4	5	6	7	8	9	10
y											

b Sketch the graph of $y = 10x - x^2$.

a

x	0	1	2	3	4	5	6	7	8	9	10
y	0	9	16	21	24	25	24	21	16	9	0

b

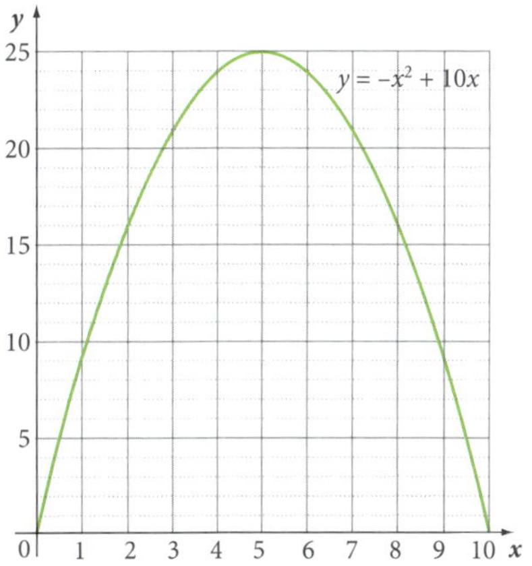

The points do not form a linear graph so don't use straight lines to join the points.

EXAMPLE 2

Liam recorded these figures, showing the height of the river at the boundary of his property during a recent flood.

a Draw a graph of the figures.

b If the height of the bridge into Liam's property is 5 m, between roughly what times will Liam not be able to cross?

Friday		Saturday	
Time	Water level (m)	Time	Water level (m)
0000	1.2	0000	5.8
0300	1.2	0300	6.2
0600	1.5	0600	6.0
0900	1.9	0900	5.6
1200	2.1	1200	5.2
1500	4.4	1500	4.9
1800	4.6	1800	4.5
2100	5.0	2100	4.3

a

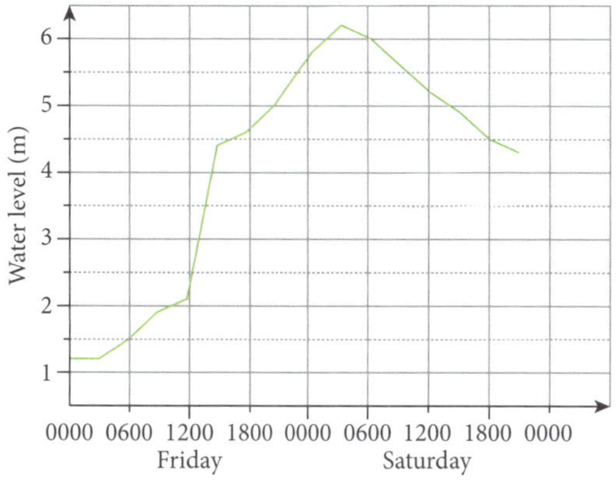

Often expressions of practical situations use 24-hour time as is done here.

b Between 9 pm on Friday and 3 pm on Saturday

→ PRACTICE

1 a Complete the table of values for $y = x^2 - 8x + 20$.

x	0	1	2	3	4	5	6	7	8
y									

b Sketch the graph of $y = x^2 - 8x + 20$.

2 Leah recorded these figures showing the height of the creek that flows through her property during a recent flood.

Sunday		Monday	
Time	Water level (m)	Time	Water level (m)
0000	0.8	0000	4.0
0300	0.9	0300	3.9
0600	1.1	0600	3.5
0900	1.4	0900	3.6
1200	2.0	1200	3.0
1500	3.5	1500	2.5
1800	5.1	1800	2.0
2100	4.2	2100	1.8

a Draw a graph of the figures.

b If the height of the bridge into Leah's property is 3.5 m, between roughly what times will Leah not be able to cross?

3* On a tidal river, high tide is at 7 am at a height of 2.35 m. The depth of water over the following 11 hours is shown in the table.

Time	Depth (m)
8 am	2.24
9 am	1.96
10 am	1.55
11 am	1.14
12 noon	0.81
1 pm	0.66
2 pm	0.71
3 pm	0.96
4 pm	1.34
5 pm	1.76
6 pm	2.12
7 pm	2.32

a Sketch a graph of the river height over that time.

b A boat needs a depth of 1.2 m. Between what times would it not be able to travel across that section of the river?

Answers ➲ p. 32

2 Using descriptions

- **From descriptions of a practical situation** it may be possible to draw a graph to illustrate it.
- **The graph might be linear** (a straight line) when there is direct variation between variables, or non-linear (exponential for example).

EXAMPLE 1

A person leaves home and drives for $2\frac{1}{2}$ hours at an average speed of 70 km/h. He then stops for $\frac{1}{2}$ hour before continuing, reaching his destination 240 km from home after another hour and a half. After a stop of an hour, he heads straight home arriving there $8\frac{1}{2}$ hours after he left.

a Show this information on a travel graph.

b About how far did he travel on the second leg?

c What was his average speed on the journey home?

a

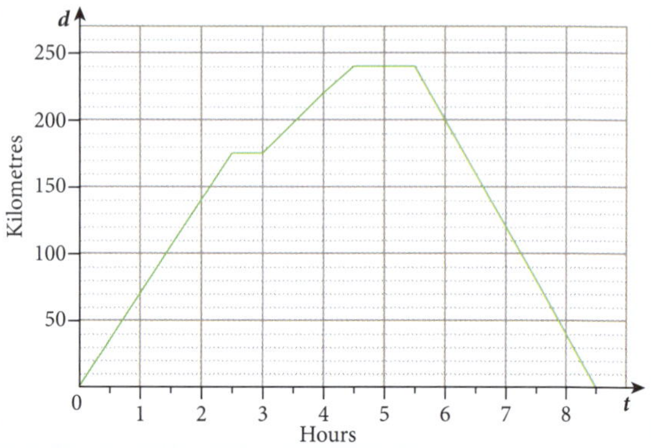

b After 3 hours he had travelled 175 km.

Distance on second leg = (240 – 175) km
= 65 km

c Distance travelled = 240 km

Time taken = 3 hours

Speed = (240 ÷ 3) km/h
= 80 km/h

The average speed can be calculated but not the actual speed at any time.

EXAMPLE 2

The diagram shows a vase. Water is poured into the vase at a constant rate.

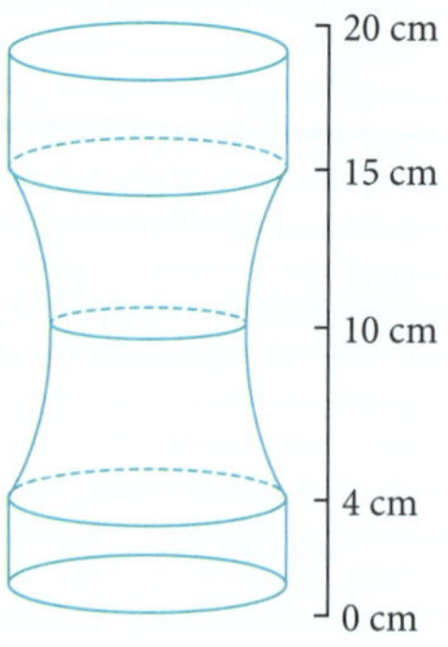

a As water is poured into the bottom 4 cm of the vase is the relationship between the length of time and the height of the water linear or non-linear?

b Is that relationship the same as the water fills the next 6 cm of the vase? Justify your answer.

c Sketch a possible graph of the height of the water in the vase over the time it takes to fill to a height of 17 cm.

a Linear

The depth is changing at a constant rate.

b No, the rate at which the depth is changing increases over the time. The depth will increase more quickly as the vase narrows.

c

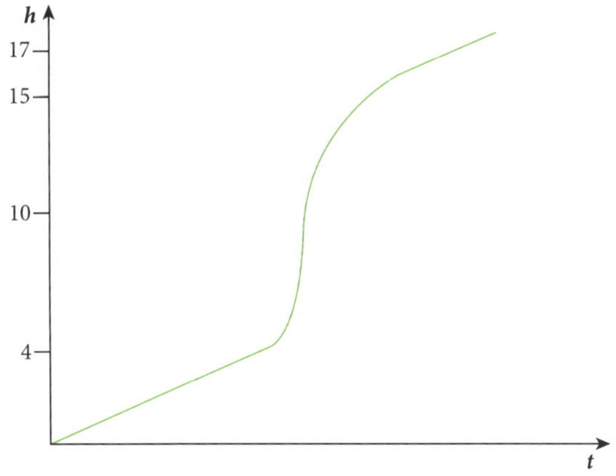

There is no exact answer because we don't know the rate at which the water is flowing.

EXAMPLE 3

The diagrams show the cross-sections of swimming pools. Sketch a graph of the depth of water over time as each is drained at a constant rate.

a

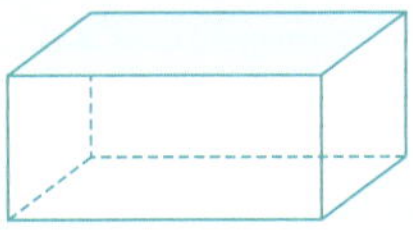

b

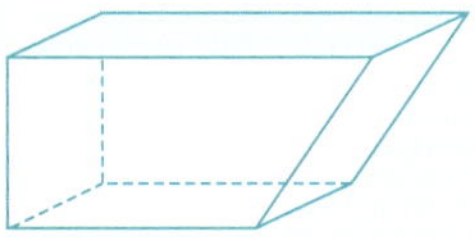

a

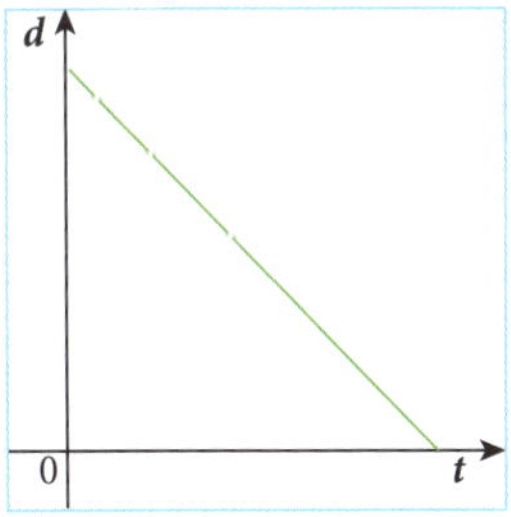

The depth will decrease at a constant rate.

b

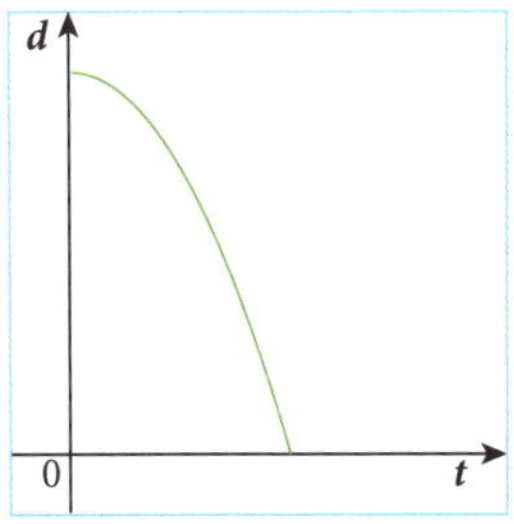

The depth will decrease at an increasing rate.

→ PRACTICE

1 A person leaves home and drives for $1\frac{1}{2}$ hours at an average speed of 80 km/h. She then stops for an hour before continuing, reaching her destination 280 km from home after another $2\frac{1}{2}$ hours. After stopping for another hour she travels straight home, arriving 10 hours after she left.

- a Show this information on a travel graph.
- b About how far did she travel in the last hour before reaching her destination?
- c What was her average speed on the journey home?

2 The diagram shows a bottle. Water is poured into the bottle at a constant rate.

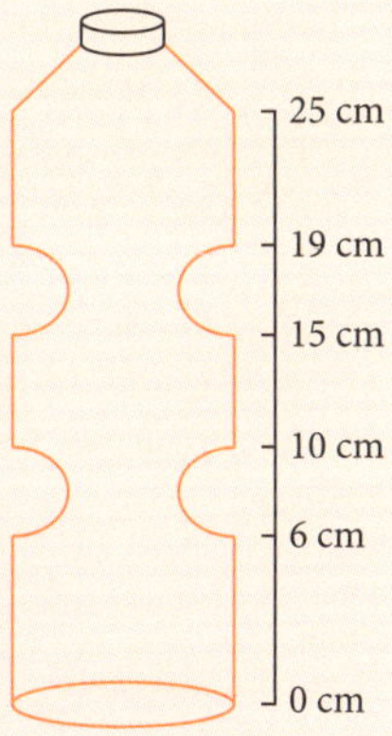

- a As water is poured into the bottom 6 cm of the bottle is the relationship between the length of time and the height of the water linear or non-linear?
- b Is that relationship the same as the water fills the next 4 cm of the bottle? Justify your answer.
- c Sketch a possible graph of the height of the water in the vase over the time it takes to fill to a height of 24 cm.

3 The diagrams show the cross-sections of water tanks. Sketch a graph of the depth of water over time as each is filled with water flowing into them at a constant rate.

a

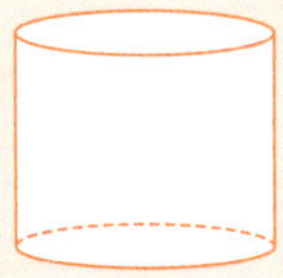

b

4* In a certain country, no tax is payable for the first \$10 000 earned. Between \$10 000 and \$40 000 there is a constant rate of tax payable, with a total of \$2000 tax being paid when \$40 000 is earned. For incomes between \$40 000 and \$90 000 there is a higher constant rate of tax payable. When \$90 000 is earned the amount of tax is \$12 000.

- a Show this information on a graph.
- b What is the gradient of the curve for incomes between \$10 000 and \$40 000?
- c Explain the meaning of the gradient found in part b.

d About how much tax would be paid on an income of $70 000?

e May pays $3200 in tax. What is her income?

Answers ➲ pp. 32–33

3 Further linear graphs of practical situations

➔ **In many practical situations one quantity will vary directly with another**. This variation can be shown by the graph of a straight line.

➔ **In the equation $y = mx + c$, m is the gradient and c** the y-intercept. The meaning of the gradient and of the y-intercept can be explained in the context of the relationship between the quantities.

➔ **If we are given information about the relationship between variables**, then a graph can be easily drawn. Using the gradient and y-intercept the equation of the graph can be found.

EXAMPLE 1

The graph shows the cost, $C, of buying n tickets for a concert.

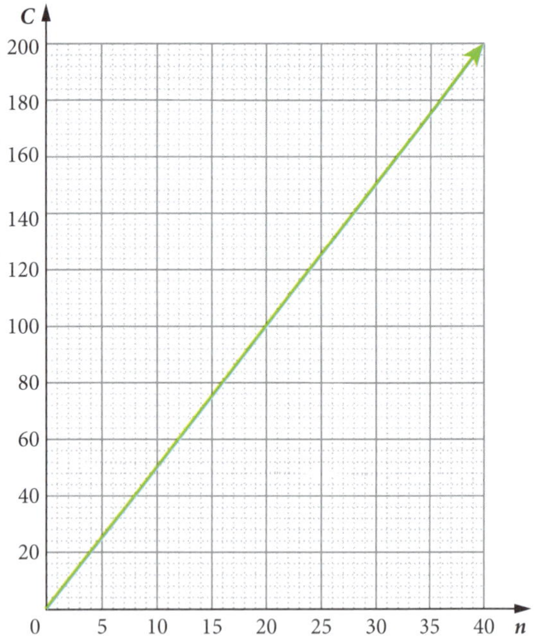

a What is the cost of 12 tickets?

b Sam paid $130 for some tickets. How many tickets did he buy?

c Determine the gradient of the line.

d Explain the meaning of the gradient in this context.

a 12 tickets cost $60.

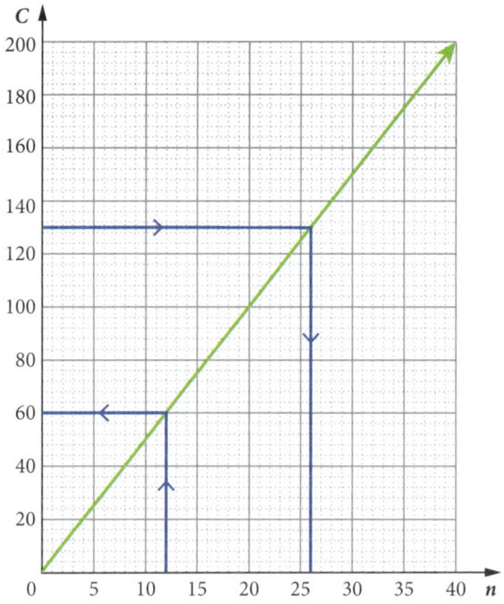

b Sam bought 26 tickets.

c $m = \frac{60}{12}$

$= 5$

The gradient is equal to the vertical rise over the horizontal run.

d The gradient represents the cost of one ticket. Each ticket will cost $5.

EXAMPLE 2

This graph has been drawn to show the value of a boat.

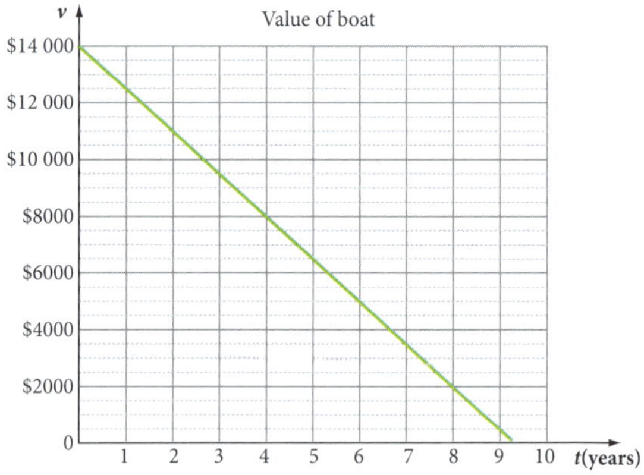

a What is the vertical intercept?

b What does the vertical intercept represent?

c What is the gradient?

d What does the gradient represent?

a The vertical intercept is 14 000.

b The vertical intercept gives the original value of the boat.

c When $t = 1$, $v = 12\,500$

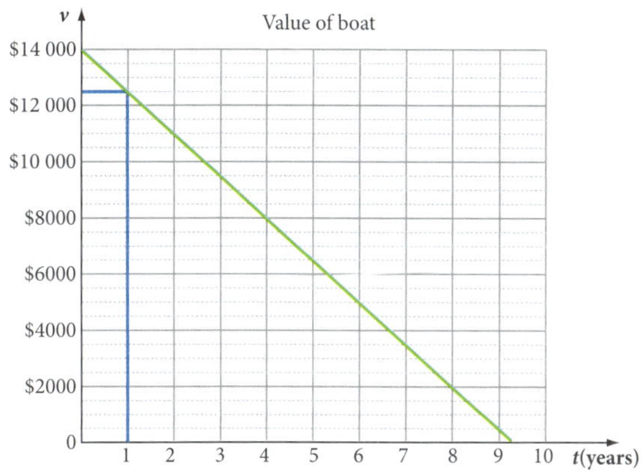

So the vertical change = 12 500 − 14 000

= −1500

$$m = \frac{-1500}{1}$$

$= -1500$

The line leans to the left (\) so the gradient is negative.

d The gradient gives the amount by which the boat is decreasing in value each year (the amount by which it is depreciating).

EXAMPLE 3

Nathan invests $10 000 in an account at a bank. Each year his investment increases by $500. The graph shows the value of the investment for the first 12 years.

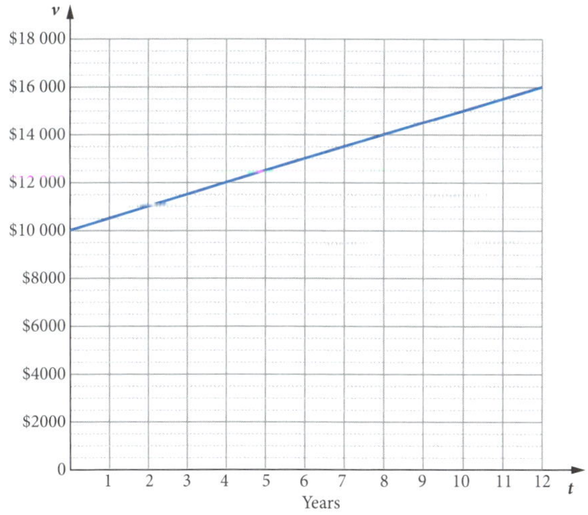

a What is the value after 6 years?

b After how many years is the value $14 500?

c What is the y-intercept and what does it represent?

d What is the gradient and what does it represent?

e What is the equation of the graph?

a After 6 years the value is $13 000.

b The value is $14 500 after 9 years.

c The y-intercept is $10 000. It represents the initial investment.

d The gradient is $500. It represents the amount by which the investment increases each year. (It is the amount of interest earned.)

e $v = 500t + 10\,000$

In this case the independent variable is t and the dependent variable is v.

→ PRACTICE

1 The graph shows the income received, $$I$, when n items are sold.

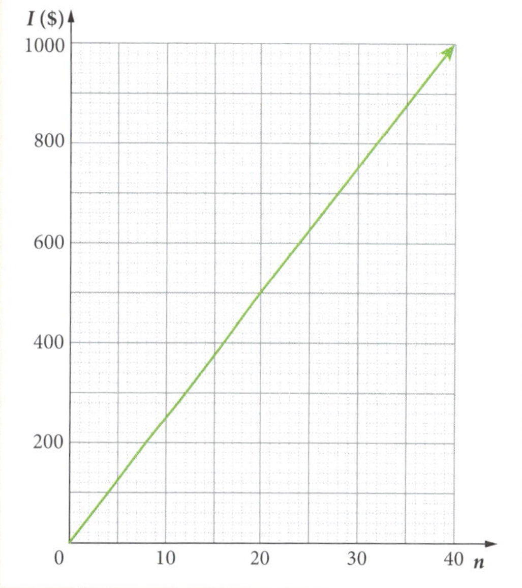

a What is the income when 24 items are sold?

b One day the income was $875. How many items were sold that day?

c Determine the gradient of the line.

d Explain the meaning of the gradient in this context.

2 This graph has been drawn to show the value of a car.

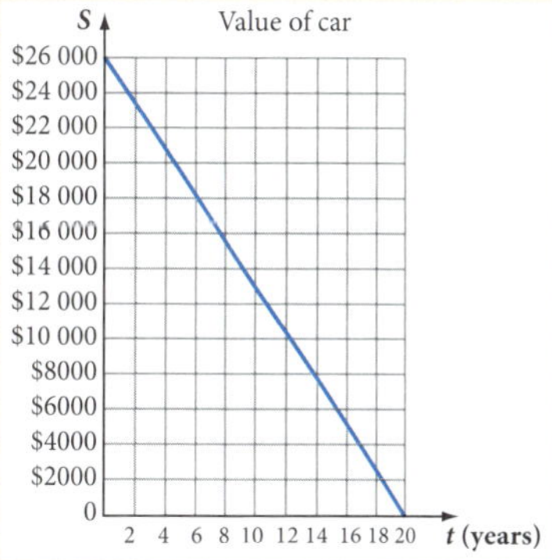

a What is the y-intercept?

b What does the y-intercept represent?

c What is the gradient?

d What does the gradient represent?

3 Natalie buys a car for $20 000. Each year the car decreases in value by $1250. The graph shows the value of the car over the first 8 years.

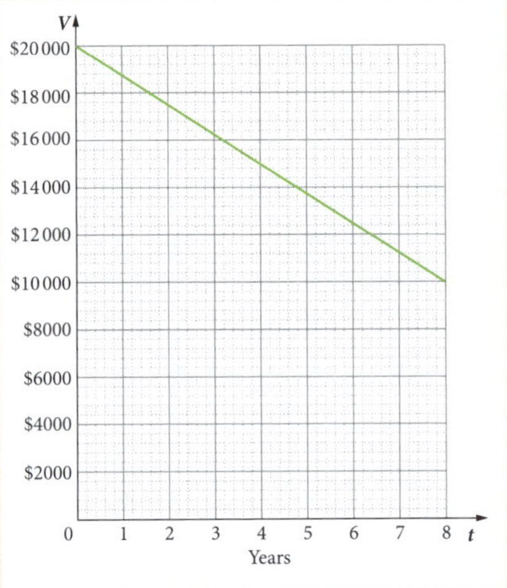

a What is the value after 6 years?

b After how many years is the value $15 000?

c What is the y-intercept and what does it represent?

d What is the gradient and what does it represent?

e What is the equation of the graph?

4* A graph has been drawn to convert temperatures in degrees Fahrenheit to temperatures in degrees Celsius and vice versa.

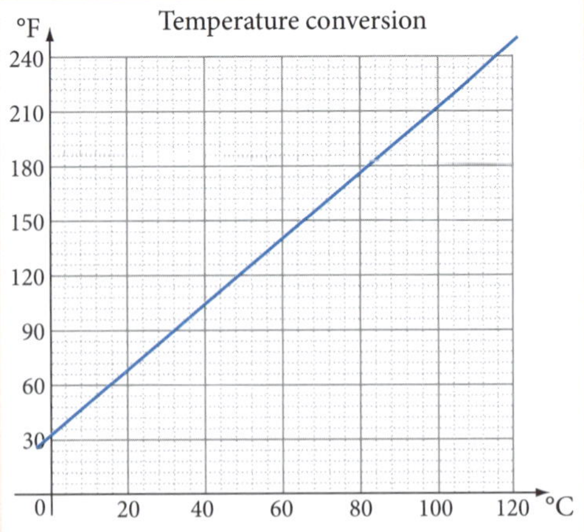

a What is the approximate temperature in degrees Fahrenheit if it is 32 °C?

b What is the approximate temperature in Celsius if it is 60 °F?

c The equation of the line is $F = 1.8C + 32$. What is the intercept on the vertical axis and what does it mean?

d What is the gradient of the line and what does that represent?

e Frederick is cooking a cake and needs to set the oven to 180 °C. Because the graph doesn't show 180 °C he finds that the temperature for 90 °C is 195 °F. He then doubles this amount to get 390 °F. He sets the oven at 390 °F and is surprised when the cake is burnt. Explain what is wrong with Frederick's reasoning.

f At what temperature should Frederick have set his oven?

Answers ➲ pp. 33–34

4 Exponential graphs

→ **The graphs of many practical situations** are examples of exponential graphs.

→ **If the y-values get very large very quickly**, we have exponential growth. The graph is of the form $y = a^x$.

→ **If the y-values get very small very quickly**, we have exponential decay. The graph is of the form $y = a^{-x}$.

EXAMPLE 1

If an amount of $500 is invested at 6% p.a. compound interest, then the balance (A) of the account after n years is given by $A = \$500(1.06)^n$, the graph of which is shown below.

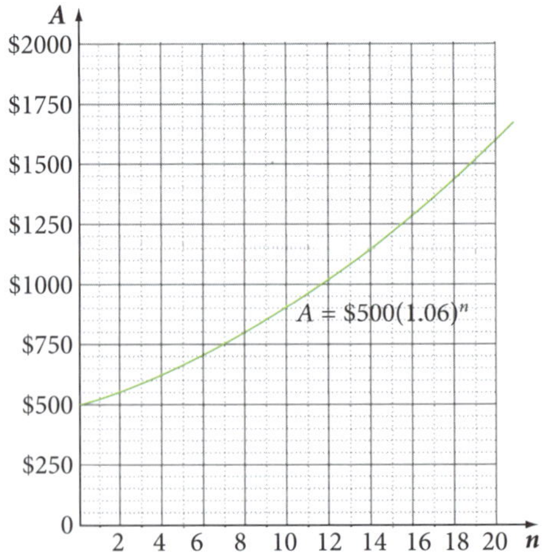

Use the graph to answer the following questions.

a How much would be in the account after eight years?

b After how many years will the balance of the account be $1275?

c **How much interest is earned during the tenth year?**

a After 8 years there is \$800 in the account.

b After 16 years the balance of the account is \$1275.

c When $n = 10$, $A = \$900$

When $n = 9$, $A = \$850$

Difference = \$900 − \$850

= \$50

\$50 interest is earned during the tenth year.

The compound interest formula is an example of exponential growth.

EXAMPLE 2

The population in millions, P, of a city is approximated by the formula $P = 6(1.02)^t$ where t is the time in years since 2010.

a **Use graphing software to draw the graph of $P = 6(1.02)^t$ for values of t between 0 and 50.**

b **After approximately how many years will the population reach 10 million?**

c **Briefly explain the significance of the y-intercept.**

a

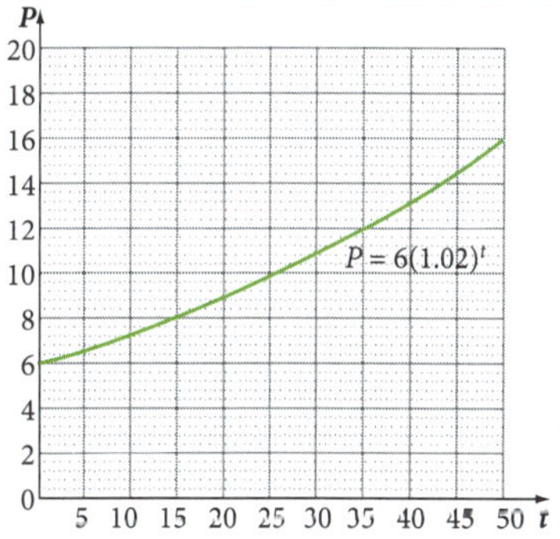

b From the graph, the population would reach 10 million in approximately 26 years.

so in about the year 2036

c The y-intercept is 6. It is the population, in millions, at the beginning of the specified time, i.e. in 2010.

Populations are usually examples of exponential growth.

EXAMPLE 3

If a car originally valued at \$25 000 is depreciated at the rate of 12% per year, then the salvage value S of the car after n years is given by $S = \$25\,000(0.88)^n$. The graph of $S = \$25\,000(0.88)^n$ is shown.

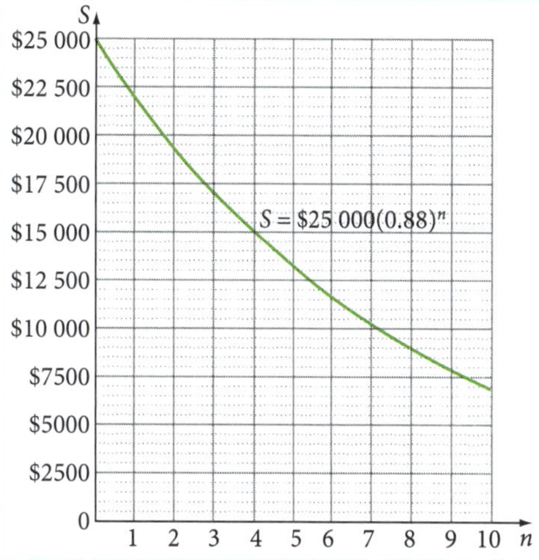

Use the graph to answer the following questions.

a **What is the approximate value of the car after 7 years?**

b **When does the car fall below \$15 000 in value?**

c **What is the intercept on the vertical axis and what does this represent?**

d **How much does the car depreciate in the first year?**

a After 7 years the value of the car is about \$10 200.

b After 4 years the value of the car falls below \$15 000.

c The intercept on the vertical axis is \$25 000. This represents the original value of the car: the price at which the car was originally bought.

d When $n = 1$, $S = \$22\,000$

Depreciation = \$25 000 − \$22 000

= \$3000

The car depreciates by \$3000 in the first year.

This is an example of declining-balance depreciation which is another way that assets can be depreciated other than the straight-line method.

EXAMPLE 4

A radioactive substance disintegrates over time. The mass (in grams) of a particular radioactive substance present after t years is given by $M = 80(0.95)^t$.

a **What is the original amount of this substance?**

b **Use graphing software to graph $M = 80(0.95)^t$.**

Use the graph to answer the following questions:

c **How much of the substance is present after 7 years?**

d **After how many years does the mass present fall below 20 g?**

e **How long does it take for half of the mass to disintegrate?**

a $M = 80(0.95)^t$

When $t = 0$,

$M = 80 \times 0.95^0$

$= 80$

The original amount was 80 g.

b

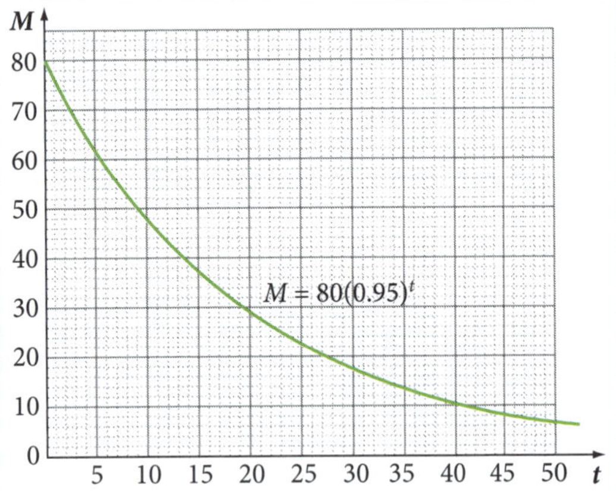

Because we have the equation we can use that to determine particular values, but the graph is quicker and easier to use to find approximate values.

c After 7 years, approximately 56 g remain.

d After 27 years, the mass present falls below 20 g.

e When $t = 13.5$, $M = 40$

It takes approximately $13\frac{1}{2}$ years for half of the mass to disintegrate.

The rate at which a radioactive substance disintegrates is proportional to the amount present.

→ PRACTICE

1 If an amount of $2000 is invested at 7% p.a. compound interest then the balance (A) of the account after n years is given by $A = \$2000(1.07)^n$, the graph of which is shown below.

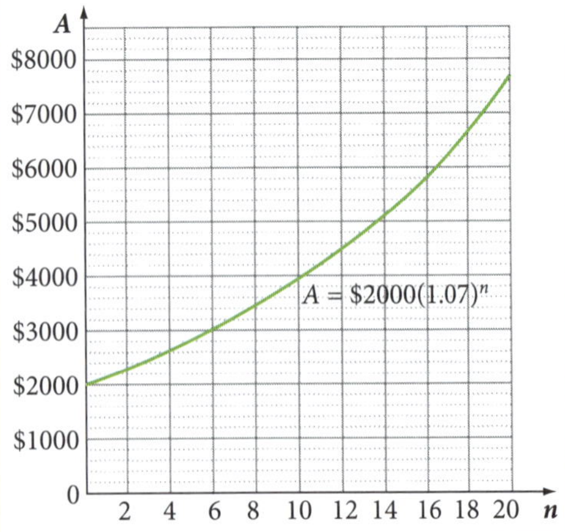

Use the graph to answer the following questions.

a How much would be in the account after 12 years?

b After how many years will the balance of the account be $5150?

c How much interest is earned during the eighth year?

2 The population in millions, P, of a city is approximated by the formula $P = 8(1.03)^t$ where t is the time in years since 1990.

a Use graphing software to draw the graph of $P = 8(1.03)^t$ for values of t between 0 and 50.

b After approximately how many years will the population reach 25 million?

c Briefly explain the significance of the y-intercept.

3 If a car originally valued at $20 000 is depreciated at the rate of 8% per year, then the salvage value S of the car after n years is given by $S = \$20\,000(0.92)^n$. The graph of $S = \$20\,000(0.92)^n$ is shown below. Use the graph to answer the following questions.

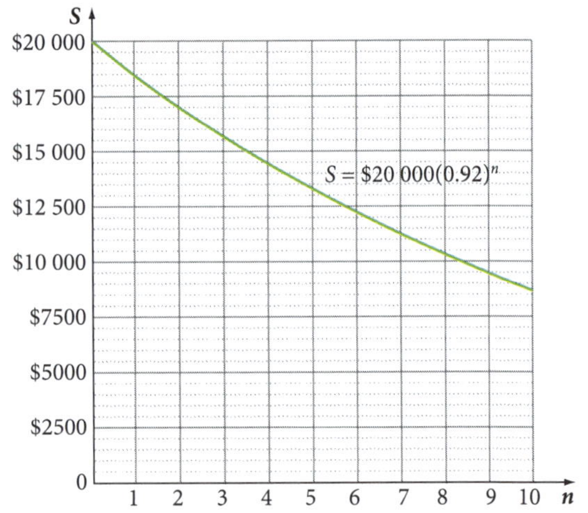

a What is the approximate value of the car after 6 years?

b When does the car fall below $10 000 in value?

c What is the intercept on the vertical axis and what does this represent?

d How much does the car depreciate in the first year?

4 A radioactive substance disintegrates over time. The mass (in grams) of a particular radioactive substance present after t years is given by $M = 60(0.9)^t$.

a What is the original amount of this substance?

b Use graphing software to graph $M = 60(0.9)^t$.

Use the graph to answer the following questions:

c How much of the substance is present after 13 years?

d After how many years does the mass present fall below 10 g?

e How long does it take for half of the mass to disintegrate?

5* The graph shows the population (in millions) of a country over 40 years from 1979 until 2019.

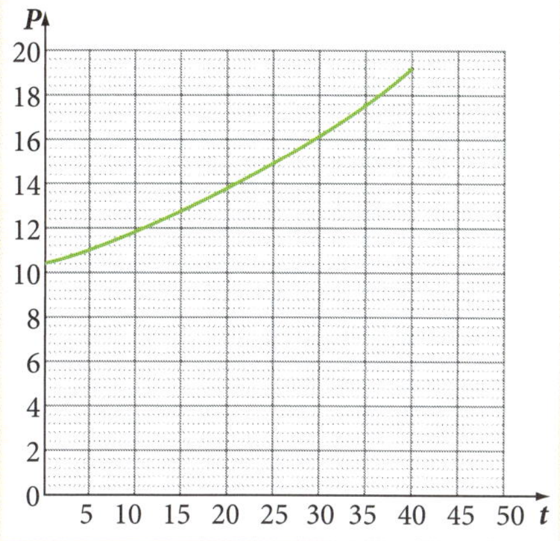

a What was the population in 2019?

b When did the population reach 14 million?

c Nicole looked at the graph and predicted that the population would be 22 million in 2029. Do you agree? Justify your answer.

Answers ⊃ p. 34

5 Inverse variation

➔ **In many practical situations, one variable will vary in proportion to another.** For example, the distance that is travelled at a fixed speed is directly proportional to the time taken to travel that distance.

➔ **If two quantities vary inversely with one another,** then one would get larger as the other gets smaller. For example, the number of jellies each person gets to eat if a packet is shared equally varies inversely with the number of people sharing. If there are 24 jellies in the packet, one person could eat all 24, but if there are twice as many people they get half as many each. Two people would get 12 each, three people eight each and eight people would have three each.

➔ **If two quantities x and y vary inversely** with each other, then $y = \frac{a}{x}$ where a is a constant.

➔ $y = \frac{a}{x}$**, where a is a constant,** is a reciprocal function.

➔ **The graph of $y = \frac{a}{x}$** is a **hyperbola**.

➔ A **hyperbola** has **two branches**. The x and y axes are asymptotes of the curve. The graph of $y = \frac{a}{x}$ $(a > 0)$ is shown below.

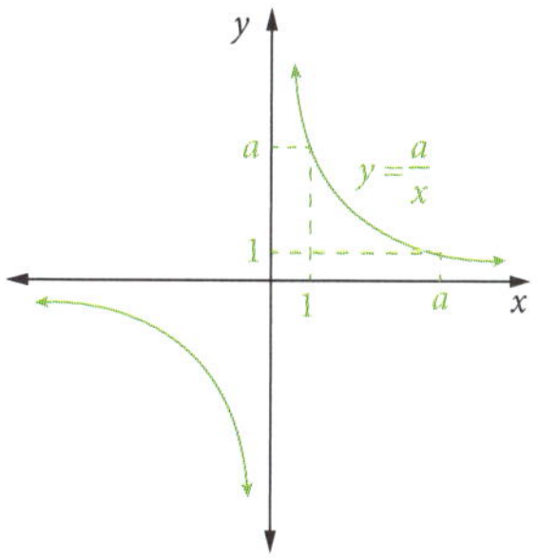

➔ **Many models of physical phenomena** only use the positive branch of the hyperbola.

➔ **In this course, graphs do not involve negative values** of the independent variable.

EXAMPLE 1

Eight friends are trying to decide whether to charter a plane for a joy flight. The cost to charter the plane is $420.

a Determine the cost per person if different numbers of people, from one to eight, share the cost.

b Show the different costs on a grid.

c Louise would like to take the flight but can only afford to pay $100. How many other people would need to take the flight so that Louise could go?

a Let n be the number of people and $\$C$ the cost per person.

n	1	2	3	4	5	6	7	8
C	420	210	140	105	84	70	60	52.50

b

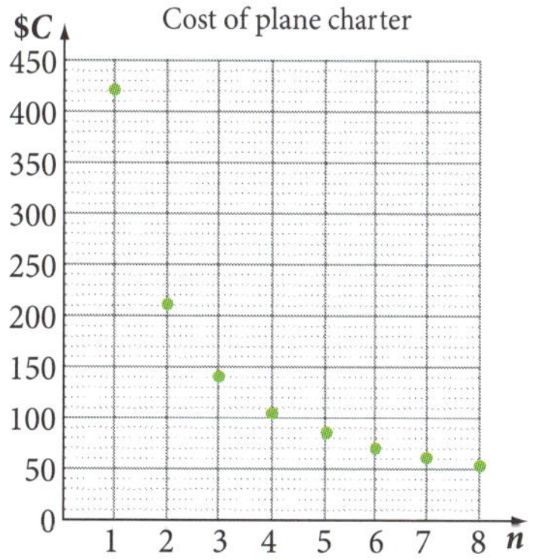

The graph only needed to show the points because it doesn't make sense to have anything other than whole numbers of people.

c At least five people would need to travel for the cost to be less than $100 each.

Louise would need four other people to take the flight.

EXAMPLE 2

The time needed to fill a swimming pool is inversely proportional to the rate at which water is flowing into the pool. It takes 50 hours to fill the pool if water is flowing at 25 L per minute.

a **Show that $T = \dfrac{75\ 000}{r}$ where T is the time in minutes to fill the pool and r the rate (in litres per minute) at which the water is flowing.**

b **Find the time needed to fill the pool, in hours and minutes, if the water is flowing at 60 L per minute.**

c **Stacey fills the pool in 8 hours and 20 minutes. At what rate was the water flowing?**

a 50 hours = 50 × 60 minutes
= 3000 minutes

$T = \dfrac{a}{r}$

When $T = 3000$, $r = 25$

$3000 = \dfrac{a}{25}$

$a = 25 \times 3000$
$= 75\ 000$

$\therefore T = \dfrac{75\ 000}{r}$

The time is inversely proportional to the rate so the time is equal to some constant divided by the rate.

b $T = \dfrac{75\ 000}{r}$

When $r = 60$,

$T = \dfrac{75\ 000}{60}$
$= 1250$

It takes 1250 minutes to fill the pool.

1250 minutes = (1250 ÷ 60) hours
= 20.833 33 ... hours
= 20 hours and 50 minutes

It takes 20 hours and 50 minutes to fill the pool when water flows at 60 L/min.

c 8 hours and 20 minutes
= (8 × 60 + 20) minutes
= 500 minutes

$T = \dfrac{75\,000}{r}$

When $T = 500$,

$500 = \dfrac{75\,000}{r}$

$500r = 75\,000$

$r = \dfrac{75\,000}{500}$
$= 150$

The water was flowing at 150 L per minute.

EXAMPLE 3

A fun run covered a distance of 8 km and participants could run, walk or cycle. The average speed at which a person covered the course is inversely variable to the time taken so that $s = \dfrac{8}{t}$. The graph of $y = \dfrac{8}{x}$ is shown below.

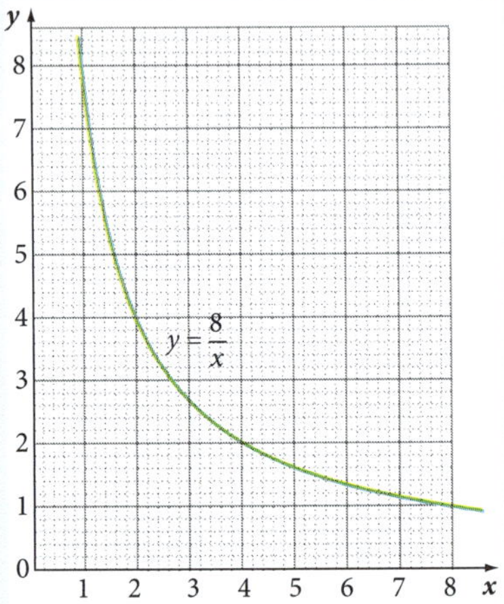

a **Briefly explain why this graph can be used to find the speed given the time taken or the time given the average speed.**

b **If a participant took 3 hours and 12 minutes to complete the fun run, what was their average speed?**

c **If the average speed of a competitor was 5 km/h, how long did he or she take to complete the course?**

a Let x be the time taken and let y be the average speed and the required formula becomes $y = \dfrac{8}{x}$. The graph of $s = \dfrac{8}{t}$ would be exactly the same, just using different variables.

b 3 hours and 12 minutes is 3.2 hours.

When $x = 3.2$, $y = 2.5$

The average speed was 2.5 km/h.

c When $y = 5$, $x = 1.6$

Now 1.6 hours is 1 hour and 36 minutes.

So it would take the competitor an hour and 36 minutes.

Notice that when the speed doubles, the time halves.

→ PRACTICE

1 Nine friends are trying to decide whether to charter a boat for a fishing trip. The cost to charter the boat is \$504.

a Find the cost per person if different numbers, from 1 to 9, share the cost.

b Show the different costs on a grid.

c Samuel would like to go on the trip but can only afford to pay \$70. How many other people would need to take the trip so that Samuel could go?

2 The time needed to travel across a particular region is inversely proportional to the speed. It takes $2\frac{1}{2}$ hours to cross the region if the speed is 8 m per second.

a Show that $T = \frac{72\,000}{s}$ where T is the time in seconds and s the speed (in metres per second).

b Find the time to travel across the region, in hours and minutes, if the speed is 12 m per second.

3 A race was 12 km long. The average speed at which a person ran the race is inversely variable to the time taken so that $s = \frac{12}{t}$. The graph of $y = \frac{12}{x}$ is shown below.

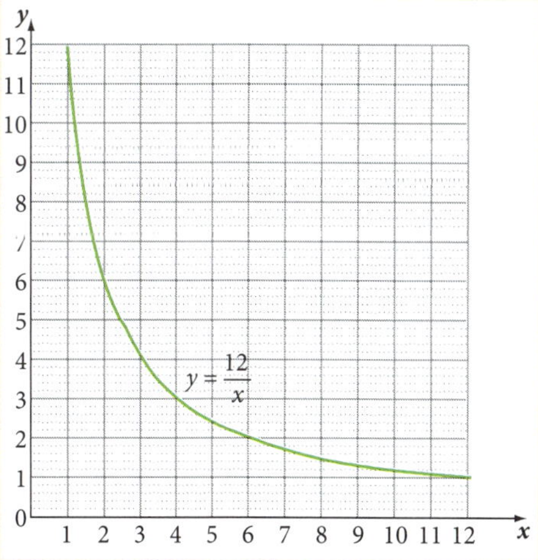

a Briefly explain why this graph can be used to find the speed given the time taken or the time given the average speed.

b If a participant took a total of 2 hours 20 minutes to complete the race, what was their average speed?

c If the average speed for a runner was 1.75 km/h, how long did they take to complete the race?

4* It is known that a particular rectangle has area 64 units^2. If the sides of the rectangle are l cm and b cm, then $b = \frac{64}{l}$.

a Use graphing software to draw the graph of $b = \frac{64}{l}$.

b If the length of the rectangle is 12 cm, use the graph to find the breadth.

c If the breadth is 42 cm, use the graph to find the length of the rectangle.

d If it is also known that the particular rectangle has its length 8 cm longer than its breadth, find the dimensions of the rectangle.

Answers ➲ pp. 34–35

6 Further graphs of practical situations

→ **The graph of an expression that has practical significance** might be of many different types, such as parabolas and cubic curves.

→ **Given an equation that represents something of practical significance**, a graph can be drawn to illustrate. We can answer questions by **reading values from the graph** or by using the equation.

EXAMPLE 1

The area, A m^2, of a rectangle with perimeter 60 m is given by the expression $A = 30x - x^2$ where x is the length of one of the sides. The diagram shows the graph of $A = 30x - x^2$.

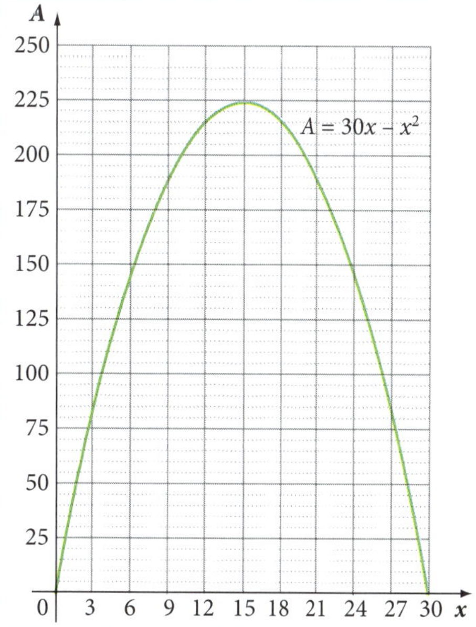

a What is the area of the rectangle if one side is 10 m?

b What are the dimensions of the rectangle if its area is 125 m²?

c What is the maximum possible area of a rectangle with perimeter 60 m?

d What is special about the rectangle with maximum area?

a When $x = 10$, $A = 200$

The area is 200 m².

b When $A = 125$,

$x = 5$ or $x = 25$

If the area of the rectangle is 125 m², the dimensions will be 25 m by 5 m.

The value of both dimensions can be read from the graph. One will be the length and one the width.

c The maximum area is 225 m².

d When $A = 225$, $x = 15$

There is only one possible value for the length of the side.

A square is a rectangle.

The rectangle is a square of side 15 m.

EXAMPLE 2

The volume of a sphere can be approximated by the formula $V = 4r^3$ where r is the radius of the sphere. The graph of $V = 4r^3$ is shown below.

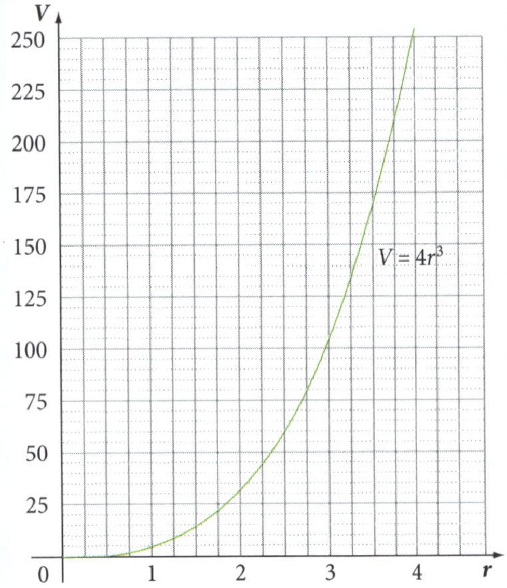

a Using the graph, what is the approximate volume of a sphere of radius 1.25 m?

b Using the graph, what is the approximate radius of a sphere of volume 80 cm³.

c Find the volume of a sphere of radius 3.8 mm using the formula $V = \frac{4}{3}\pi r^3$.

Give the answer to the nearest cubic millimetre.

d Find the difference between the answer to part c and the value from the graph. Express this difference as a percentage of the answer to part c.

a When $r = 1.25$, $V \approx 7.5$

The volume of a sphere of radius 1.25 m is approximately 7.5 m³.

b When $V = 80$, $r \approx 2.7$

A sphere of volume 80 cm³ has radius approximately 2.7 cm.

c $V = \frac{4}{3}\pi r^3$

When $r = 3.8$,

$V = \frac{4}{3} \times \pi \times (3.8)^3$

$= 229.847\,2961 \ldots$

$= 230$ (nearest unit)

$V = \frac{4}{3}\pi r^3$ gives the exact volume.

The volume is 230 mm³ to the nearest cubic millimetre.

d From the graph, when $r = 3.8$, $V = 220$

Difference $= 230 - 220$

$= 10$

Percentage difference $= \frac{10}{230} \times 100\%$

$= 4.347\,826\,087 \ldots\%$

$= 4.3\%$ (1 d.p.)

→ PRACTICE

1 The area, A m², of a rectangle with perimeter 48 m is given by the expression $A = 24x - x^2$ where x is the length of one of the sides. The diagram shows the graph of $A = 24x - x^2$.

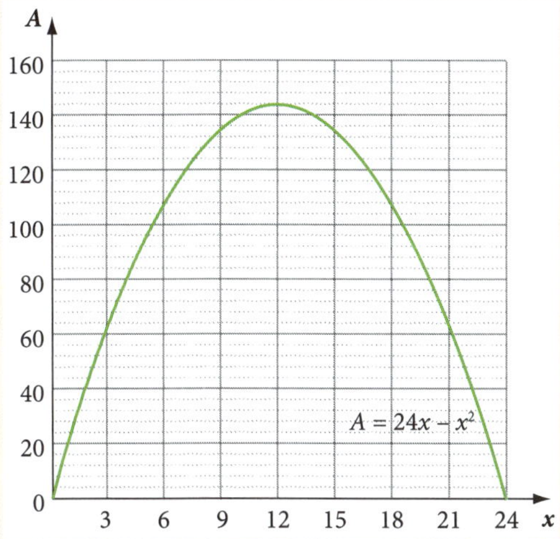

a What is the area of the rectangle if one side is 4 m?

b What are the dimensions of the rectangle if its area is 108 m^2?

c What is the maximum possible area of a rectangle with perimeter 48 m?

d What is special about the rectangle with maximum area?

2 The mass (in grams) of a particular type of ball can be approximated by the formula $M = 6r^3$ where r is the radius of the ball in centimetres. The graph of $M = 6r^3$ is shown below.

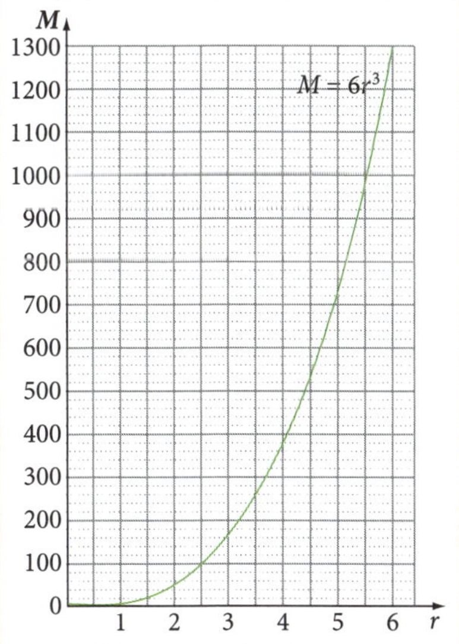

a Using the graph, what is the approximate mass of a ball of radius 4.5 cm?

b Using the graph, what is the radius of a ball of mass 1 kg?

c Find the volume of a sphere of radius 2.4 cm using the formula $V = \frac{4}{3}\pi r^3$. Give the answer to the nearest cubic centimetre.

d Use the graph to find the mass of a ball with radius 2.4 cm and hence find the density in g/cm^3 by dividing the mass by the volume.

3* The equation $y = 0.005x^2 + 0.45x$ can be used to model stopping distances on a dry road, where x is the speed in km/h and y is the stopping distance in metres.

a Use graphing software to draw the graph of $y = 0.005x^2 + 0.45x$ for values of x between 0 and 120.

b About how much further will it take to stop at 120 km/h than at 100 km/h?

c At what maximum speed is it possible to stop in a shorter distance than that in the answer to part b?

Answers ➲ p. 35

7 Limits and restrictions

- ➔ **There are often limits that must be placed on models of physical phenomena** so that they make sense.
- ➔ **For some values of a variable, a formula is not valid.**
- ➔ **Restrictions are placed on the independent variable** so that an equation will work for all the allowable values.
- ➔ **A model might only be linear**, for example, for certain values and may then take on characteristics of a different graph.
- ➔ **We often place limits on the values that occur in an equation** to prevent answers that cannot physically occur.

EXAMPLE 1

Kayla buys a quad bike for \$8400 and depreciates it (using the straight-line method) at \$1200 per year. The salvage value, \$*S*, of the bike after *n* years is given by $S = 8400 - 1200n$.

a What is the salvage value of the quad after seven years?

b What limits must be placed on the formula for giving the value of the bike? Justify your answer.

a $S = 8400 - 1200n$

When $n = 7$,

$S = 8400 - 1200 \times 7$

$= 0$

The salvage value is \$0 after 7 years.

b The formula must be restricted to values of n less than or equal to 7. It makes no sense for the value to be less than zero. Also the number of years cannot be negative. So $0 \le n \le 7$.

From part **a** we know that the bike has no value after 7 years. The formula would give a negative value for values of n greater than 7.

EXAMPLE 2

A special supplement was being fed to certain animals at a zoo. The keeper plotted the average weight (in kg) for the first 3 months and used graphing software to draw this line.

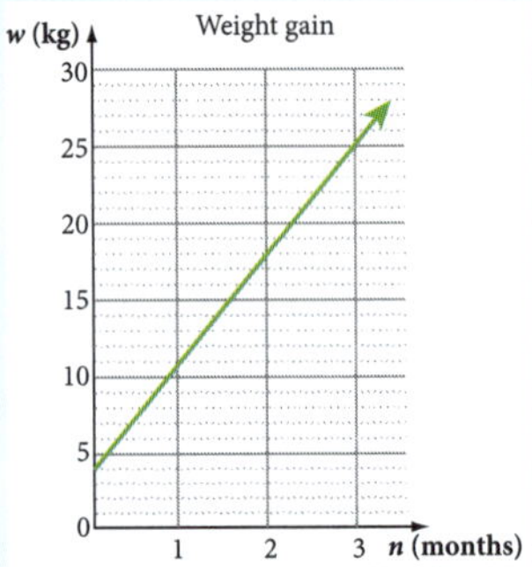

a **If w is the weight and n the number of months, what is the equation of the line?**

b **If the weight gain continues at the same rate, how much would you expect an animal to weigh at the end of 12 months?**

c **After how many months would you expect an animal to weigh 60 kg?**

d **Briefly explain why there must be some sort of limit placed on the equation for finding the weight.**

a vertical change $= 25 - 11$
$= 14$
horizontal change $= 3 - 1$
$= 2$

$$m = \frac{\text{vertical change in position}}{\text{horizontal change in position}}$$
$$= \frac{14}{2}$$
$$= 7$$

$c = 4$

$w = mn + c$

$w = 7n + 4$

Find two points on the graph that the line passes through and use the horizontal and vertical changes in position to determine the gradient.

b $w = 7n + 4$

When $n = 12$,

$w = 7 \times 12 + 4$

$= 88$

You would expect an animal to weigh 88 kg.

c $w = 7n + 4$

When $w = 60$,

$7n + 4 = 60$

$7n = 56$

$n = 8$

You would expect an animal to weigh 60 kg after 8 months.

d The animals cannot continue to gain weight indefinitely. At some stage they must reach their ideal weight and will no longer need the supplement.

EXAMPLE 3

The diagram shows the graph of $y = \dfrac{900}{x}$.

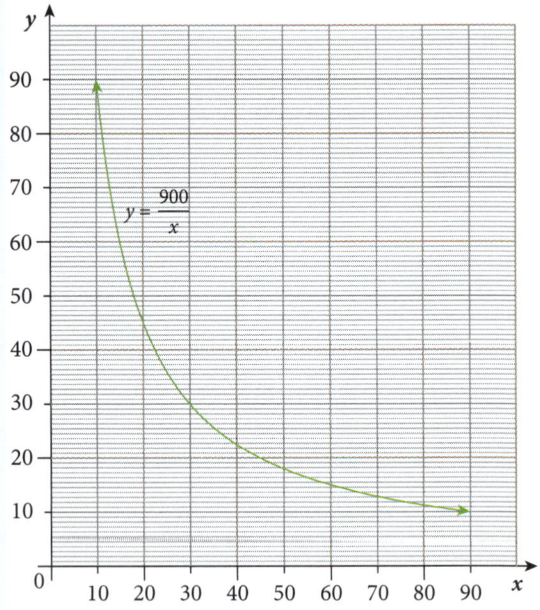

a **What is the value of y when $x = 36$?**

b **When will y first fall below 60?**

c **What will be the value of y when $x = 150$?**

d **What is the minimum value of y? Justify your answer.**

a $y = 25$

b When x is greater than 15

c $y = \dfrac{900}{x}$

When $x = 150$,

$y = \dfrac{900}{150}$

$= 6$

The value of $x = 150$ is not shown on the graph but we have the equation so can use that to find the answer.

d There is no actual minimum value of y. The value will get closer and closer to 0 as x increases but will never actually reach 0.

EXAMPLE 4

When a ball is thrown up and out from the roof of a building, the height (h metres) of the ball at time t seconds can be approximated by the formula $h = 35 + 30t - 5t^2$ for times up to and including 7 seconds.

a Find the value of h when $t = 0$ and briefly explain the significance of the answer.

b What is the value of h when $t = 7$?

c Why is the formula not valid for values of t greater than seven?

a $h = 35 + 30t - 5t^2$

When $t = 0$,

$h = 35 + 30 \times 0 - 5 \times 0^2$

$= 35$

Negative values of t will give answers for h but these make no sense in the context of the question.

The height of the ball initially is 35 m.

This means that the roof of the building is 35 m high.

b $h = 35 + 30t - 5t^2$

When $t = 7$,

$h = 35 + 30 \times 7 - 5 \times 7^2$

$= 35 + 210 - 245$

$= 0$

c After 7 seconds the height of the ball is zero. The ball has reached the ground. It cannot continue to fall (and will probably bounce). The formula gives negative values of h when $t > 7$ and this is impossible.

EXAMPLE 5

Bert is building a yard at his farm to hold goats. He has enough materials to fence 400 m and wants the yard to be rectangular, but he is not sure how long to make each side.

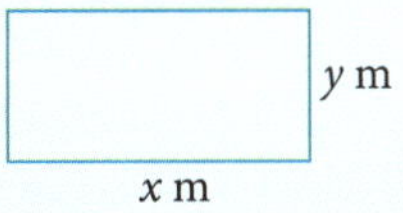

a If the length of the yard is x m and the width is y m, show that $y = 200 - x$.

b Show that the area (A m²) of the yard will be given by $A = 200x - x^2$.

c Why must x always be greater than 0?

d What is the greatest possible value of x? Justify your answer.

e Bert sketched the parabola $A = 200x - x^2$. What will be the value of k? Justify your answer.

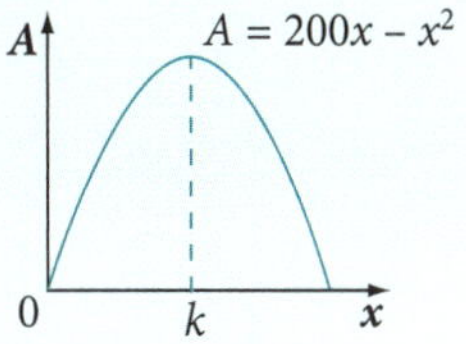

f What is the maximum possible area of the yard and what will be its dimensions?

a Perimeter = 400 m

$2x + 2y = 400$

$x + y = 200$

$y = 200 - x$

b Area: $A = xy$

$= x(200 - x)$

$= 200x - x^2$

c x m is the length of the yard. x must be greater than 0 or the yard will not exist.

d y m is the width of the yard so y must be greater than 0. This means that x must be less than 200.

e A is zero when $x = 0$ and $x = 200$.

Because the parabola is symmetrical, k is halfway between 0 and 200. $\therefore k = 100$

f The maximum area occurs when $x = 100$.

$A = 200x - x^2$

When $x = 100$,

$A = 200 \times 100 - 100^2$

$= 20\,000 - 10\,000$

$= 10\,000$

For all rectangles with the same perimeter, the one with maximum area will be a square.

The maximum area is 10 000 m² (1 hectare) when the length and breadth are both 100 m.

→ PRACTICE

1 **Kelvin buys a van for \$23 000 and depreciates it (using the straight-line method) at \$1150 per year. The salvage value, \$$S$, of the van after n years is given by $S = 23\,000 - 1150n$.**

a What is the salvage value of the van after 20 years?

b What limits must be placed on the formula for giving the value of the van? Justify your answer.

2 **A number of trees were planted and the height (in metres) recorded at the end of each year for the first 3 years. Graphing software was used to draw this line.**

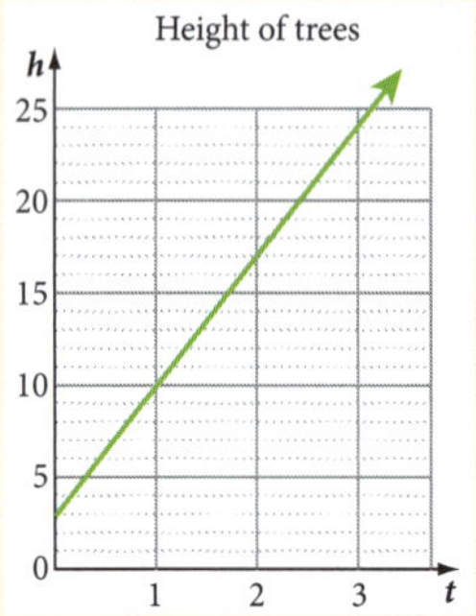

a If h is the height and t the number of years, what is the equation of the line?

b If the trees continue to grow at the same rate, how high would you expect a tree to be after 6 years?

c After how many years would you expect a tree to reach 66 m?

d Briefly explain why there must be some sort of limit placed on the equation for finding the heights of the trees.

3 The diagram shows the graph of $S = 10\,000(0.8)^n$.

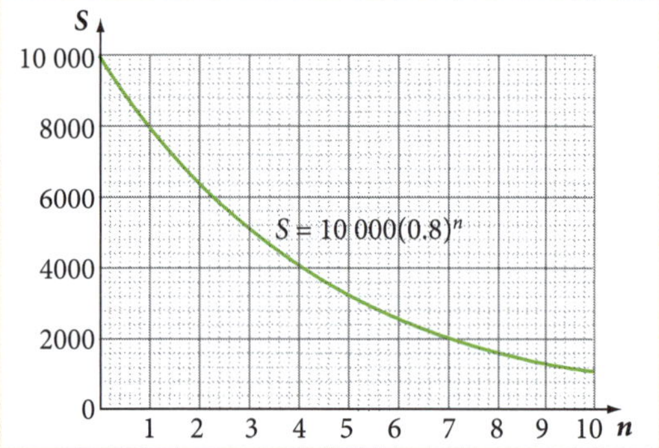

a What is the value of S when $n = 2$?

b When will S first fall below 3000?

c What will be the value of S when $n = 17$?

d What is the minimum value of S? Justify your answer.

4 When a ball is thrown from the roof of a building, the height (h metres) of the ball at time t seconds can be approximated by the formula $h = 96 + 14t - 5t^2$ for times up to and including 6 seconds.

a Find the value of h when $t = 0$ and briefly explain the significance of the answer.

b What is the value of h when $t = 6$?

c Why is the formula not valid for values of t greater than six?

5 Betty is building a pen in her backyard to hold hens. She has plenty of roofing materials but only enough materials to fence 18 m on the sides. She wants the pen to be rectangular but is not sure how long to make each side.

a If the length of the pen is x m and the width is y m, show that $y = 9 - x$.

b Show that the area (A m^2) of the pen will be given by $A = 9x - x^2$.

c Why must x always be greater than 0?

d What is the greatest possible value of x? Justify your answer.

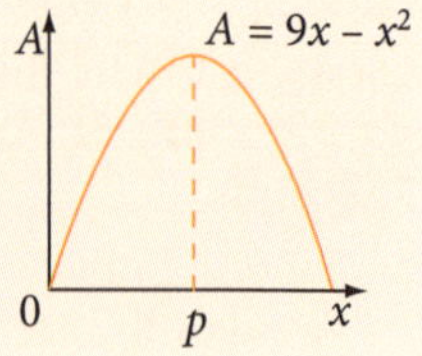

e Betty sketched the parabola $A = 9x - x^2$. What will be the value of p? Justify your answer.

f What is the maximum possible area of the pen and what will be its dimensions?

6* The distance that a train was from Sydney was modelled using the formula $D = 960 - 80h$. The graph of this equation was drawn using graphing software.

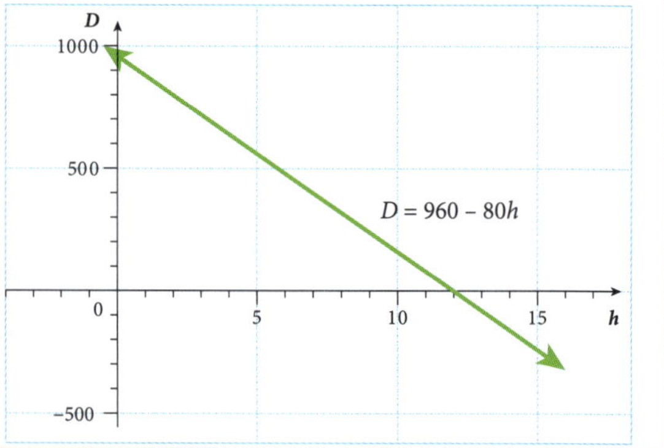

For what values of h is the formula valid? Justify your answer.

Answers ➲ p. 36

HSC EXAM-TYPE QUESTIONS

Now for the real thing! The following questions are modelled on the types of questions you will face in the HSC Examination. If you get extensive practice at answering these sorts of questions, you will be more confident answering them in the HSC Examination.

Remember: The Answer section is located straight after this section so that you can check the answers readily. There are worked solutions and also ticks that indicate the mark allocation for all questions. Both these features will help you to strive for full marks.

Make sure also that when you check your work you highlight any questions you found difficult and earmark these areas for extra study.

Total marks: 30

PART A

(Suggested time: 12 minutes)

(1 mark each)

1 Which of these graphs will pass through the point (4, 9)?

A $y = 2x + 1$ only

B $y = x^2 - 7$ only

C $y = \frac{36}{x}$ only

D all of these

2 Ellen makes homemade chocolates, which she sells at her local market. The diagram shows the cost of producing the packets of chocolates and the return from their sale.

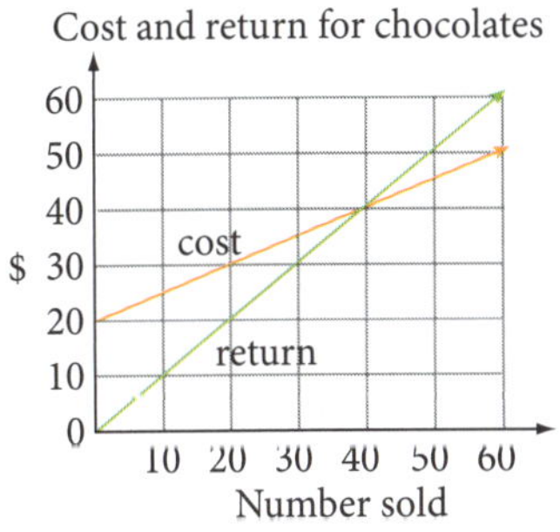

One Saturday, Ellen sells 50 packets. Which statement is correct?

A Ellen makes a profit of \$50.

B Ellen breaks even.

C Ellen loses \$15.

D Ellen makes a profit of \$5.

3 In order to graph $y = 3x^2 - x$, Alex drew a table of values.

x	0	1	2	3
y	0	2	6	24

One of the y-values in Alex's table is wrong. Which is it?

A 0 B 2 C 6 D 24

4 A radioactive substance, with initial mass 80 g, disintegrates over time. Which graph represents this?

A

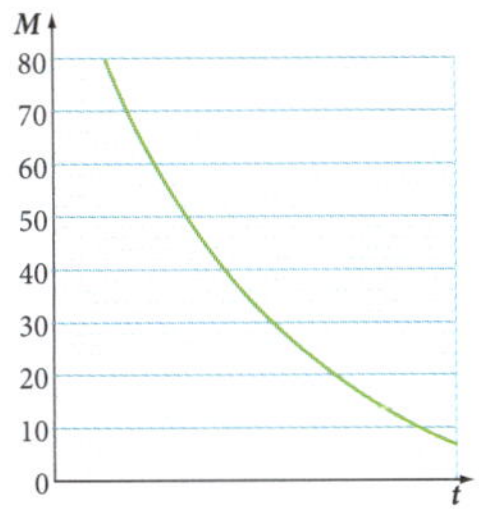

B

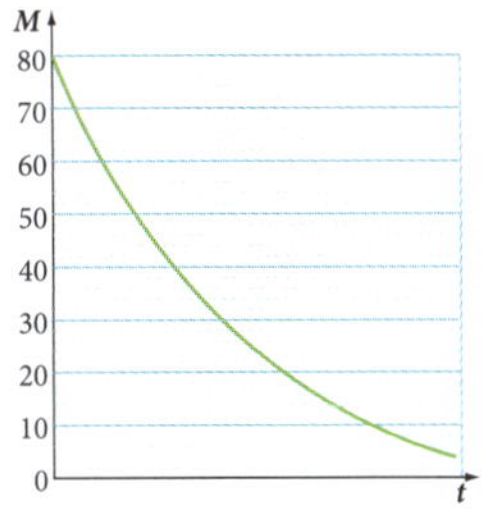

C

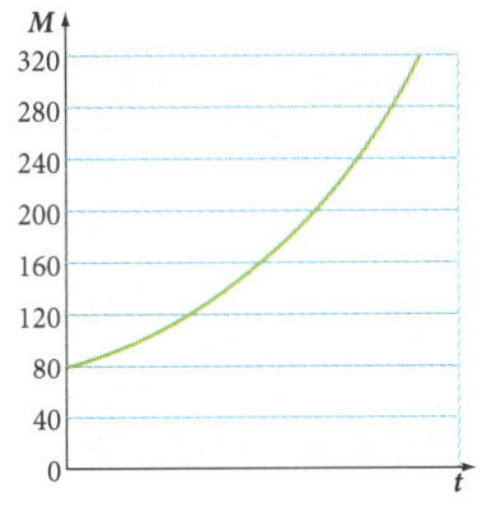

D

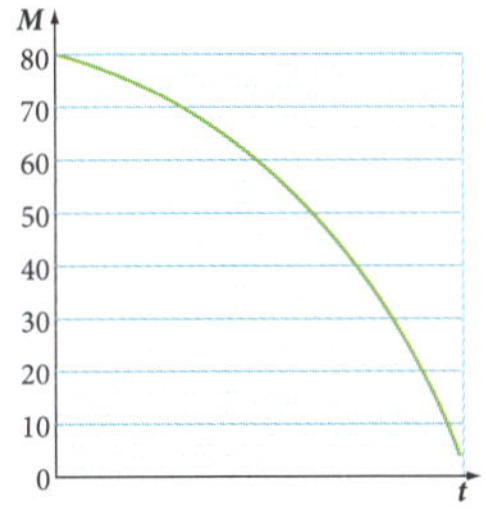

5 Jack and Lewis work in an office answering letters. The graph shows the number of letters from a particular batch that both have answered at different times over a five-hour shift.

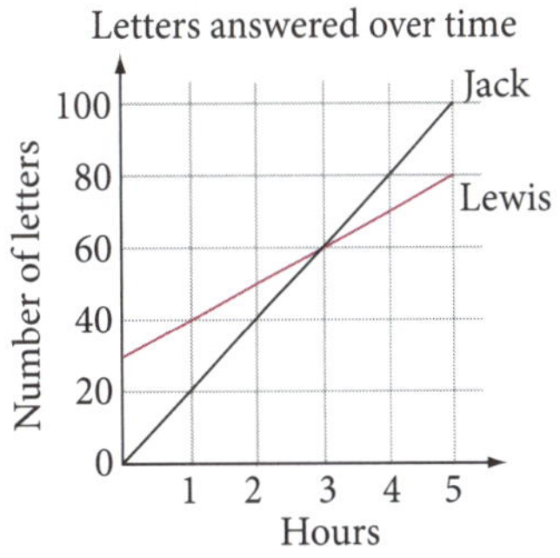

During this shift, how many more letters did Jack answer than Lewis?

A 20

B 30

C 50

D 100

6 Which of these statements could be modelled by this graph?

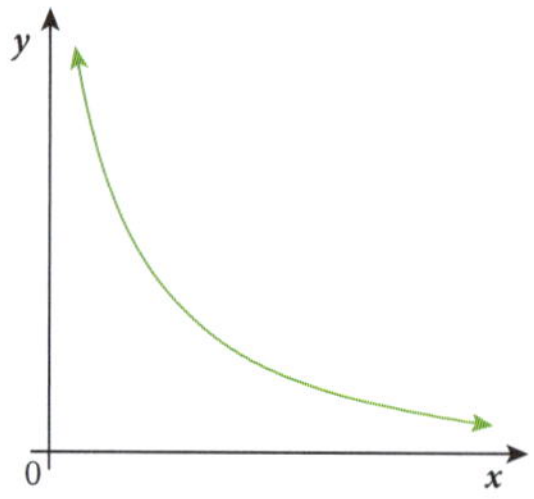

I The time to plant a patch of seedlings varies inversely with the number of people planting.

II The distance travelled at a constant speed varies directly with the time taken.

A I only
B II only
C both I and II
D neither I nor II

7 The graph drawn from the points in this table of values could be an example of what type of relationship?

x	0	1	2
y	7	14	28

A linear
B exponential growth
C exponential decay
D inverse variation

8 Which graph best shows the change in height of liquid in this vat as it is added at a constant rate?

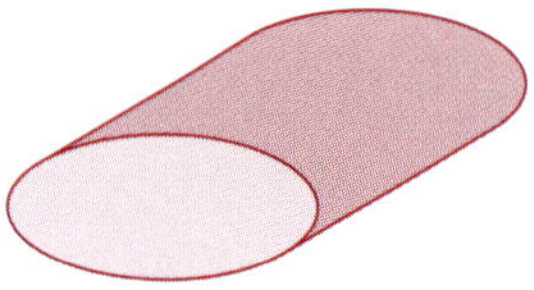

A

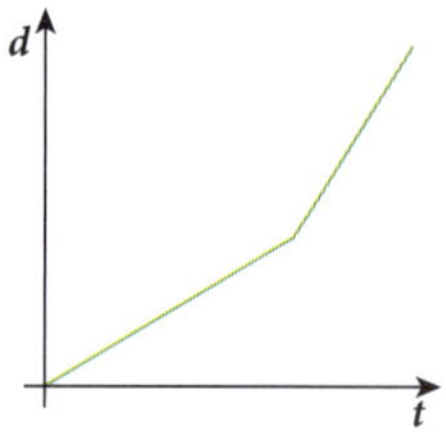

B

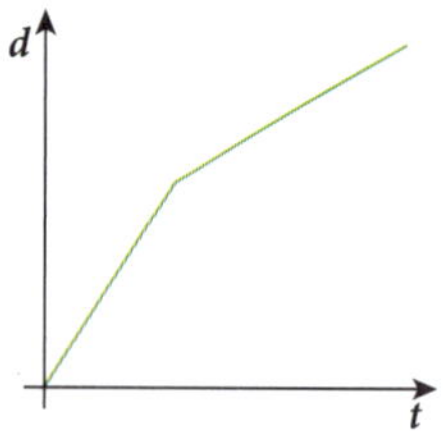

C

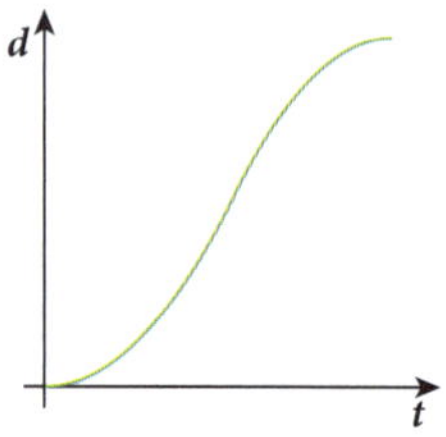

D

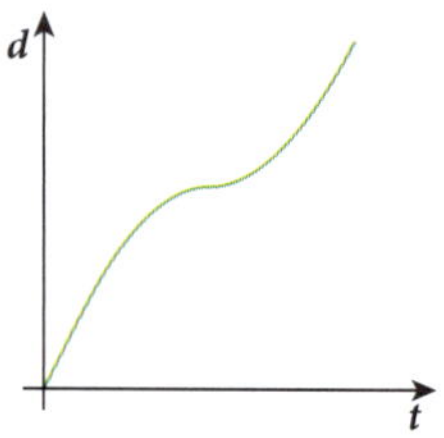

PART B

(Suggested time: 33 minutes)

Show all working.

9 The diagram shows the graphs of costs and income from sales for a particular product.

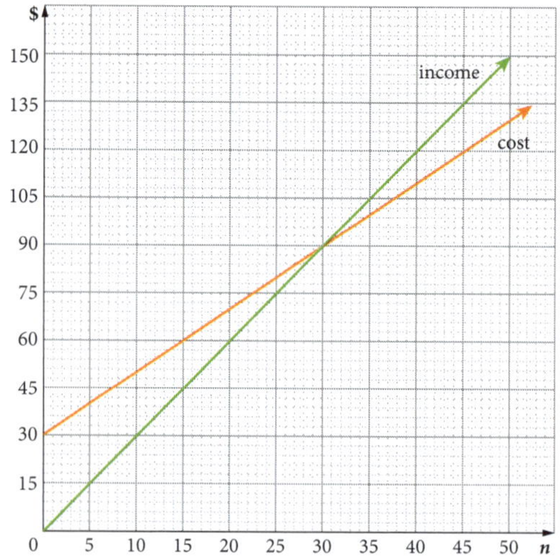

a What is the y-intercept of the cost line and what does it represent? (1 mark)
b What is the gradient of the cost line? (1 mark)
c What does that gradient represent? (1 mark)
d For what value of n does the income equal $72? (1 mark)
e How many items need to be sold to break even? (1 mark)
f How much profit is made if 45 items are sold? (1 mark)
g At what price is the product sold? (1 mark)

10 The diagram shows the graph of $y = \dfrac{600}{x}$ for positive values of x.

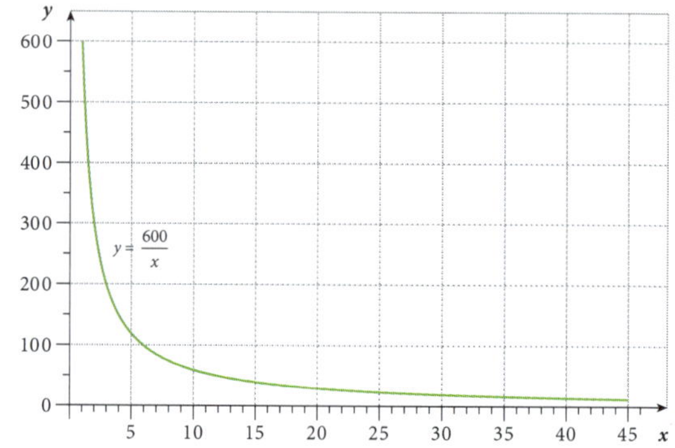

a What is the value of y when $x = 15$? (1 mark)
b What is the value of x when $y = 200$? (1 mark)

The rule $y = \dfrac{600}{x}$ can be used to find the cost per person ($y) if x people share the cost of a present.

c What is the cost per person if 25 people share? (1 mark)
d How many people share the cost if they pay $50 each? (1 mark)
e What restrictions apply to this practical application? (1 mark)

11 The diagram shows the amount of money ($A) in an account after n years.

a What is the value of A when $n = 3$? (1 mark)
b After how many years will the sum of money have doubled? (1 mark)
c Jessie said: 'Based on this graph there will be about $12 000 in the account after 16 years.' Do you agree? Justify your answer. (1 mark)

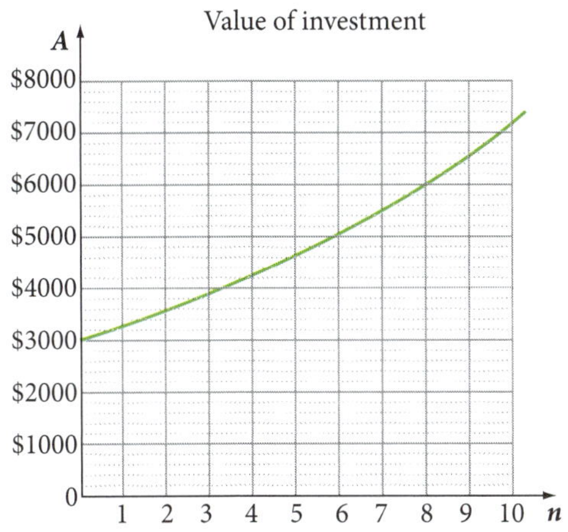

12 The population of a particular species on an island can be approximated by the expression $P = 50(0.84)^t$ where t is the number of years since 2011.

a What was the population in 2011? (1 mark)

b Explain why the expression will work as an approximation for only about 20 years. (1 mark)

13 Patrick leaves point A and runs on a straight road to point B. At the same time Genevieve leaves B and runs to A. The diagram shows their distance from A over time.

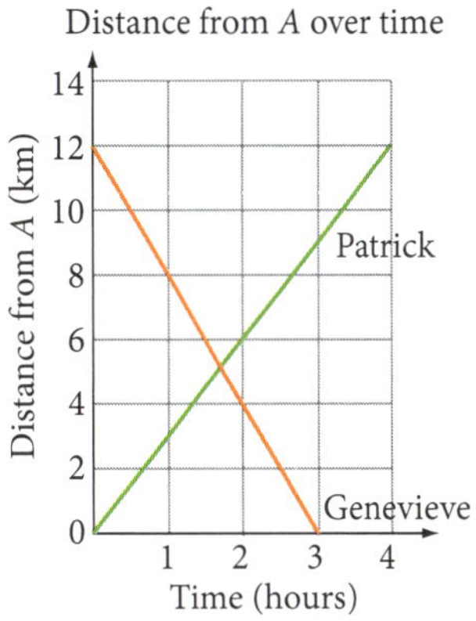

a How far is it from A to B? (1 mark)

b How far from point B are they when they meet? (1 mark)

c Who completes the journey in the shortest time? Justify your answer. (1 mark)

d Write an expression for the distance (d km at t hours) of Genevieve from point A at time t. (2 marks)

A QUICK NOTE! To help you study effectively we have made the **Examples** and **Practice** questions with the same numbers very similar to each other or you will require very similar skills to solve them.

- Please see the sample below: it shows that the **Example 1** question is similar to the **Practice 1** question and the **Example 2** question is similar to the **Practice 2** question.
  - » This is to enable you to **practise** what you have just read in an **Example**.
  - » You can also refer to the **Examples** when you are answering the **Practice** questions as their worked solutions and tips will help you answer the questions.
- The **Practice** questions that have an **asterisk (*)** are different to the other **Practice** questions. They do not match any particular **Example** but are designed to provide you with extra practice.
  - » You can see in the sample below that the **Practice 3** question is a different type to the other questions and is not directly related to any example. It is to give you further practice.

EXAMPLE 1

Some office furniture was purchased for \$4800 and is being depreciated by \$400 per year. What is the value of the furniture after 6 years?

is similar to

EXAMPLE 2

A machine, originally purchased for \$27 500, is being depreciated by the straight-line method. Find the amount of depreciation allowed per year if the salvage value of the machine is \$16 300 after 7 years.

is similar to

This question is different from the other Practice questions.

→ PRACTICE

1 A forklift was purchased for \$26 400 and is being depreciated by \$1850 per year. What is the value of the forklift after 8 years?

2 A machine, originally purchased for \$24 500, is being depreciated by the straight-line method. Find the amount of depreciation allowed per year if the salvage value of the machine is \$15 700 after 8 years.

3* A 6-year-old machine is currently valued at \$4625. If it has been depreciated by \$675 per year, find its purchase price.

PRACTICE

SIMULTANEOUS LINEAR EQUATIONS

1 Graphing lines ➲ p. 5

1 a $y = 4x + 1$

x	0	1	2	3
y	**1**	**5**	**9**	**13**

b

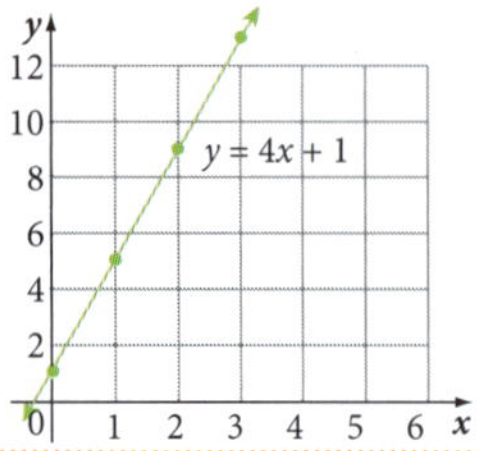

2 $y = -2x + 1$

x	−2	−1	0	1	2
y	**5**	**3**	**1**	**−1**	**−3**

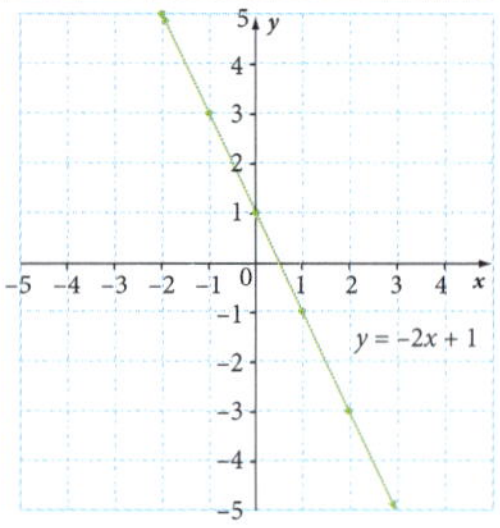

3 $y = 8 - x$

x	0	1	2	3
y	**8**	**7**	**6**	**5**

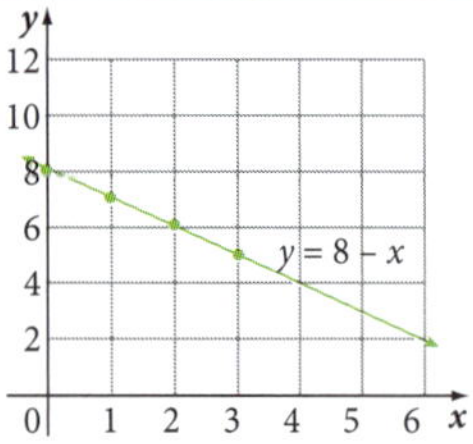

4 $h = 0.8t + 60$

t	0	25	50	75	100
h	**60**	**80**	**100**	**120**	**140**

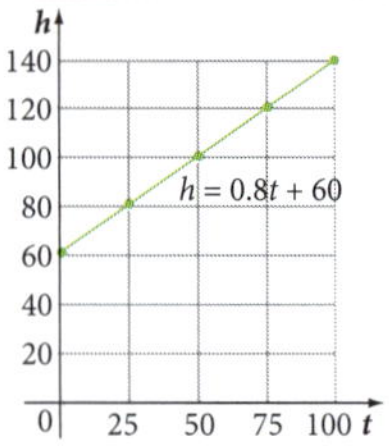

5

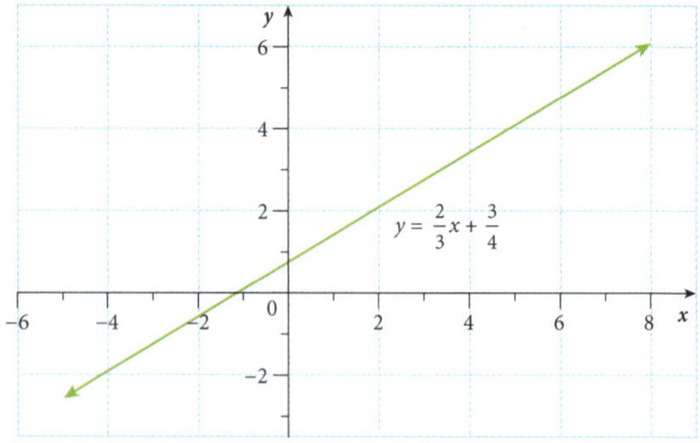

2 Points of intersection ➲ p. 6

1 a When $x = 5$ and $y = 7$

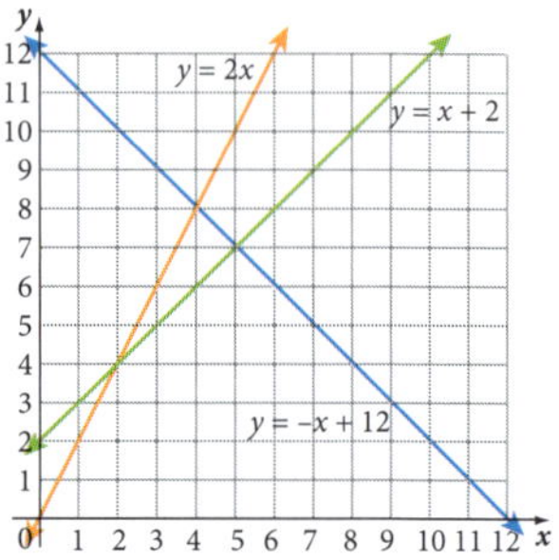

b When $x = 4$ and $y = 8$

c The point of intersection is (2, 4).

2

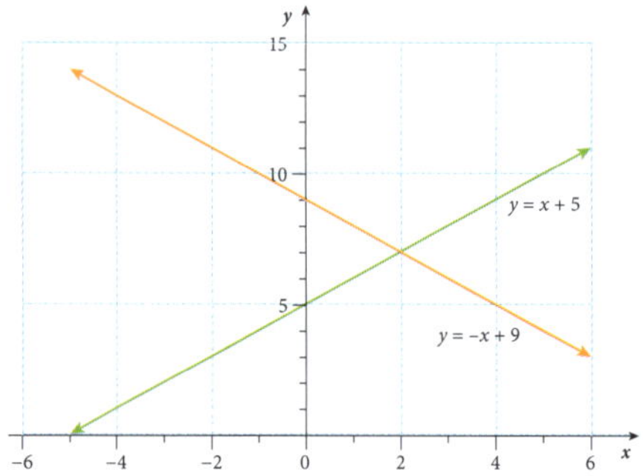

The simultaneous solution of $y = -x + 9$ and $y = x + 5$ is $x = 2$ and $y = 7$.

3 a

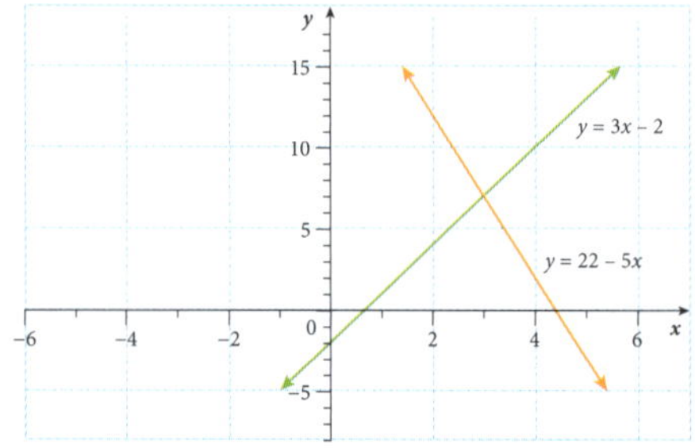

b $x = 3$ and $y = 7$

4

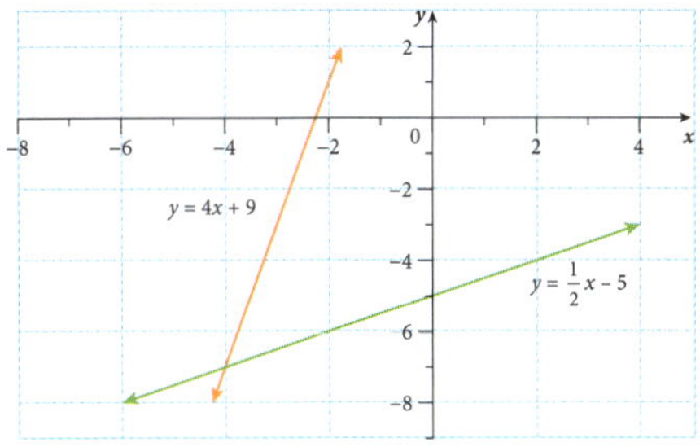

$x = -4$ and $y = -7$

3 Practical interpretations of points of intersection ➲ p. 8

1 a **\$3.60**

b **\$15**

c **Approximately 2.8 km**

d **Candy's Cabs is cheaper but only by about 45 cents.**

2 **After $4\frac{1}{2}$ hours; approximately 280 km from Sydney**

3 a $C = 0.1m + 0.5$

b $C = 0.15m + 0.2$

c

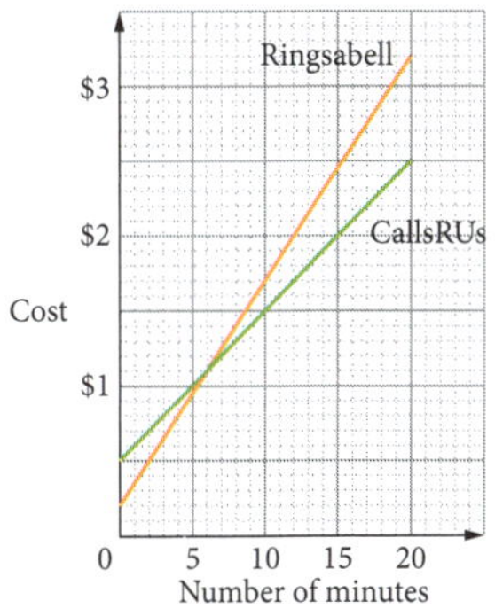

d CallsRUs charges \$1.00.

Ringsabell charges \$0.95.

It is 5 cents cheaper to use Ringsabell.

e **After 6 minutes it is cheaper to use CallsRUs.**

4 a $T = 1.5h - 19$

b Second freezer:

$T = -4h + 25$

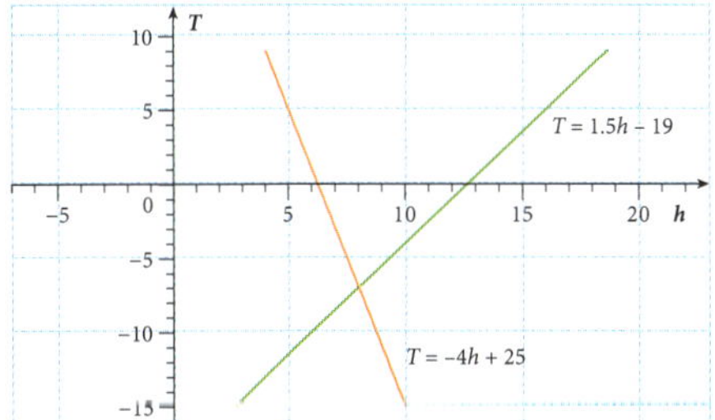

After 8 hours the second freezer is colder than the first so the optimum time to transfer the contents of the first freezer is when $h = 8$.

4 Break-even analysis ➲ p. 10

1 a **A is the 'break-even' point. If 4 people attend, Angus will receive \$50 and so cover his costs.**

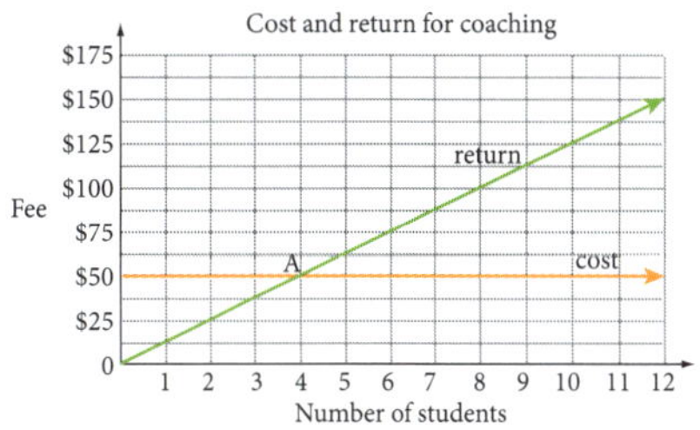

b **Each person is charged \$12.50.**

c If 10 people attend, Angus receives \$125.

Profit = \$125 – \$50

= **\$75**

d If two people attend, Angus receives \$25.

Loss = \$50 – \$25

= **\$25**

2 a $m = \frac{80}{20}$

$= 4$

The gradient is 4 and it represents the extra cost to make each bag of 'sweetbeans'.

b **The y-intercept is 180. It represents the fixed costs.**

c

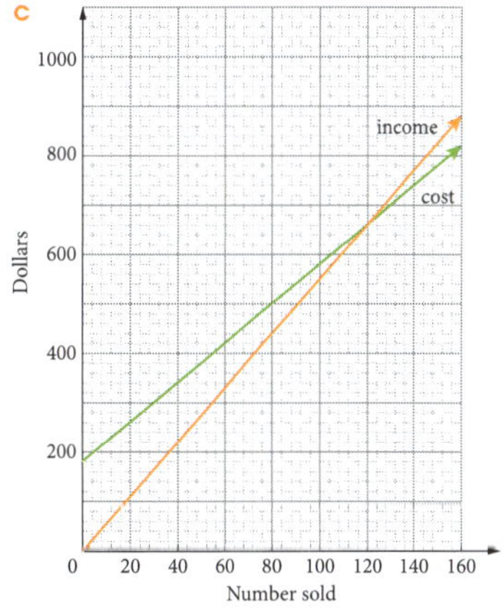

d **The break-even point is when 120 bags of 'sweetbeans' are sold. Both the cost and income are \$660.**

e **The profit when 150 bags are sold is \$45.**

f **If 100 bags are sold there will be a loss of \$30.**

3 a $I = 3.5x$

b $C = 2.75x + 120$

c

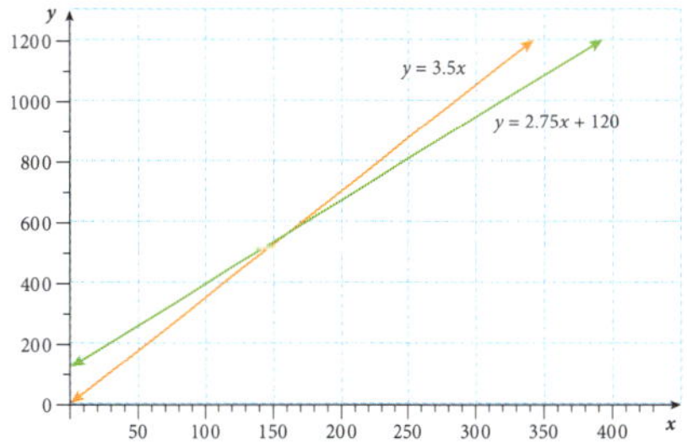

d **Holly will need to sell more than 160 packets to make a profit.**

GRAPHS OF PRACTICAL SITUATIONS

1 Constructing graphs ➲ pp. 11–12

1 a $y = x^2 - 8x + 20$

x	0	1	2	3	4	5	6	7	8
y	**20**	**13**	**8**	**5**	**4**	**5**	**8**	**13**	**20**

b

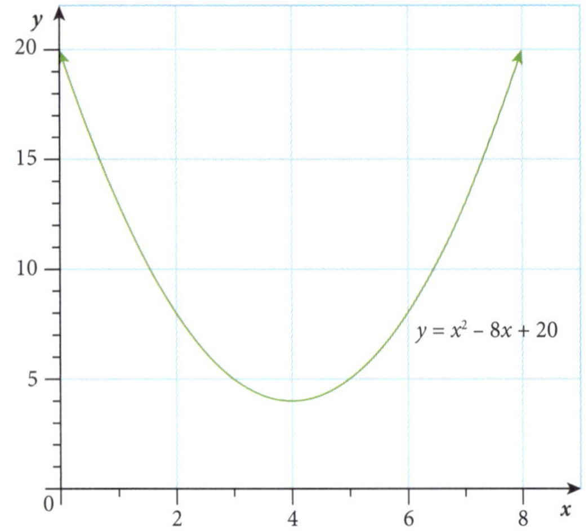

2 a

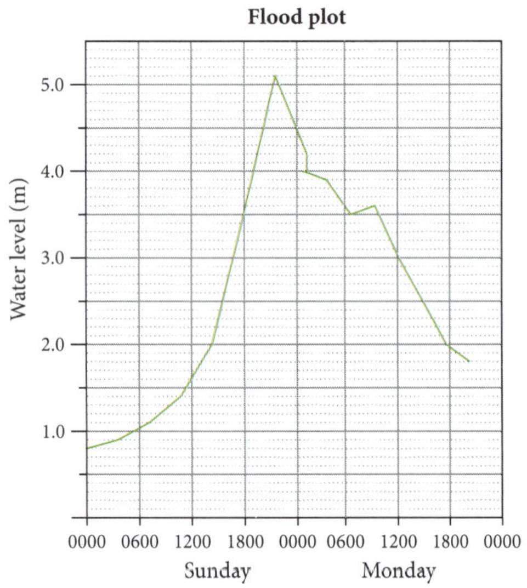

b **Leah will not be able to cross the bridge between 3 pm Sunday and 10 am Monday.**

3 a

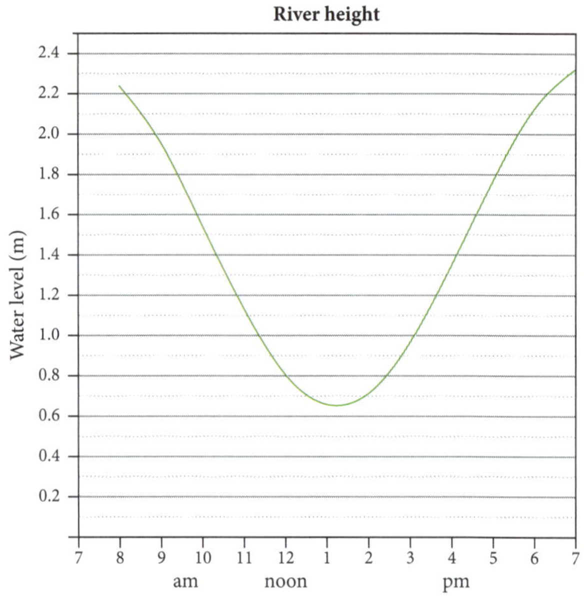

b **From about 10:40 am until 3:50 pm**

2 Using descriptions ➲ pp. 13–14

1 a

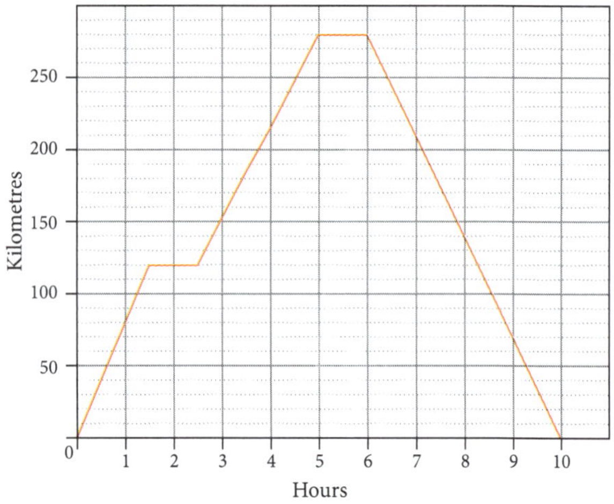

b **About 65 km**

c Speed = (280 ÷ 4) km/h
= **70 km/h**

2 a **linear**

b No. The height will not change at a constant rate. **It will fill more quickly for the next 2 cm, quickest at 8 cm and then more slowly for the next 2 cm.**

c

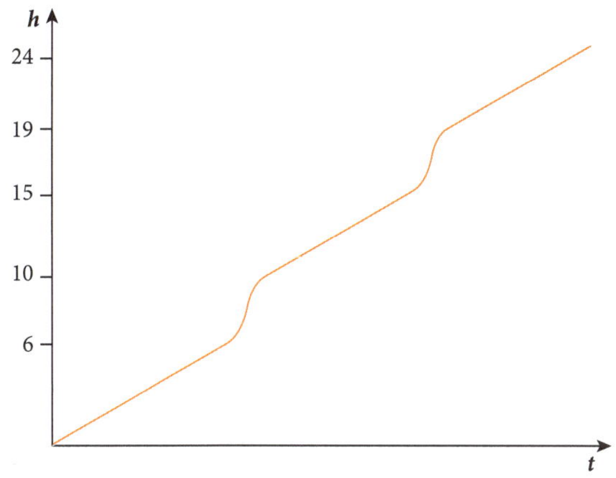

3 a

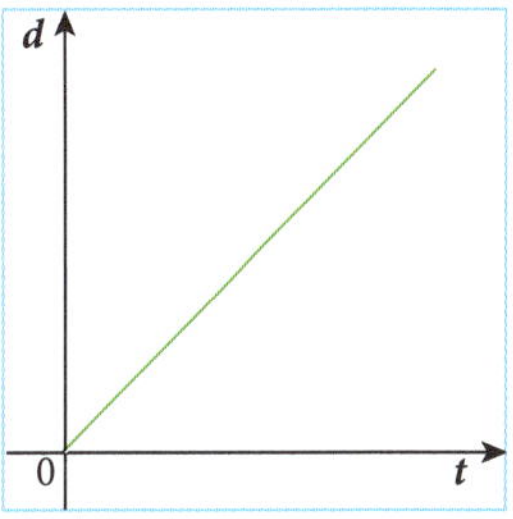

b

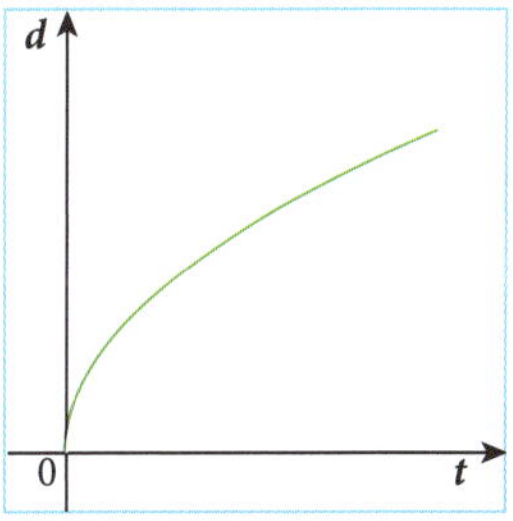

4 a

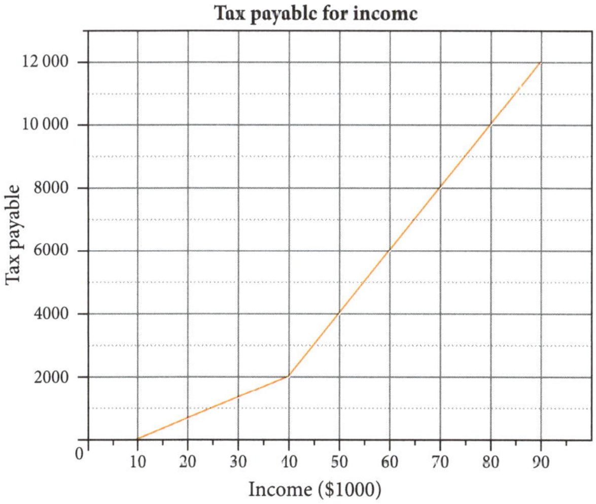

b Gradient $= \frac{2000}{30\,000}$

$= \frac{1}{15}$

c **It gives the amount of tax payable for each extra dollar earned. So for every $15 in income an extra dollar in tax must be paid.**

d **$8000**

e **May would earn about $47 000.**

3 Further linear graphs of practical situations ➲ pp. 15–16

1

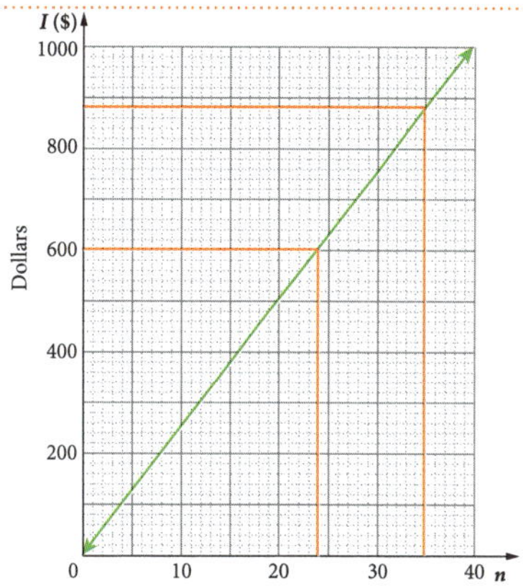

a **When 24 items are sold, the income is $600.**

b **The income will be $875 when 35 items are sold.**

c $m = \frac{\text{vertical change in position}}{\text{horizontal change in position}}$

$= \frac{600}{24}$

$= \mathbf{25}$

d For each item that is sold, the income increases by $25. **This means that the price of each item is $25.**

2 a ***y*-intercept = $26 000**

b The *y*-intercept represents the original value of the car. **It is the price of the new car.**

c $m = \frac{\text{vertical change in position}}{\text{horizontal change in position}}$

$= \frac{-26\,000}{20}$

$= \mathbf{-1300}$

d **The gradient represents the amount by which the value is changing each year. It is decreasing in value by $1300 each year.**

3 a **After 6 years the value is $12 500.**

b **The value is $15 000 after 4 years.**

c **The *y*-intercept is 20 000 and it represents the original value of the car.**

d **The gradient is –1250 and it represents the change in value of the car each year.**

e $V = 20\,000 - 1250t$

4 a **90 °F**

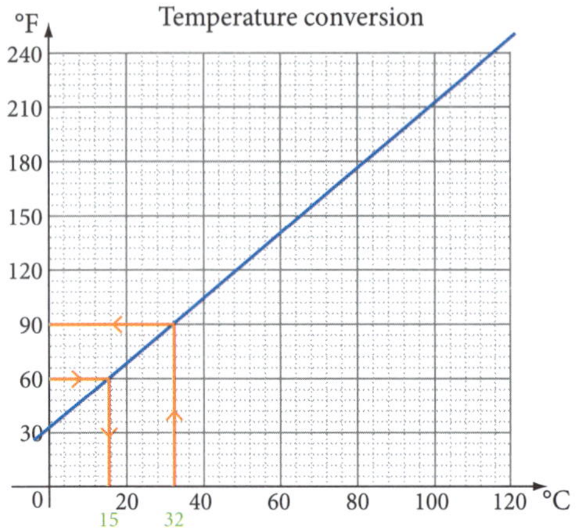

b **15 °C**

c **The vertical intercept is 32.**

It represents the temperature in °F when it is 0 °C. It is the temperature of freezing point in Fahrenheit.

d **The gradient is 1.8. It represents the change in degrees Fahrenheit for each degree change in Celsius.** So an increase of 1° Celsius will mean an increase of 1.8° Fahrenheit.

e **The temperature in Fahrenheit does not double when the temperature in Celsius doubles.** You can see from the graph, for example, that 50 °C is just over 120 °F but 100 °C is about 210 °F, which is not double 120 °F. **If the temperatures were both zero at the same time, Frederick's reasoning would work but 0 °C is 32 °F.**

f $F = 1.8C + 32$

When $C = 180$,

$F = 1.8 \times 180 + 32$

$= 356$

Frederick should have set his oven at about 356 °F.

4 Exponential graphs ➲ pp. 18–19

1 a **After 12 years there is \$4500 in the account.**

b **After 14 years the balance of the account is \$5150.**

c When $n = 8$, $A \approx \$3400$

When $n = 7$, $A \approx \$3200$

Difference = \$3400 − \$3200

= \$200

Approximately \$200 interest is earned during the eighth year.

2 a

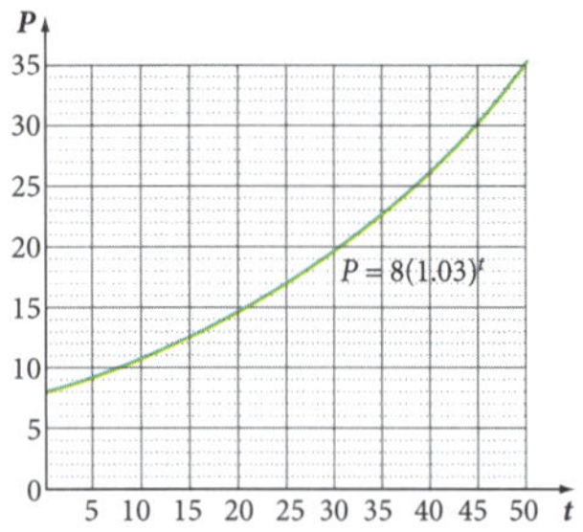

b **The population would reach 25 million in approximately $38\frac{1}{2}$ years.**

c **The y-intercept is 8. This means that the population in 1990 was 8 million.**

3 a **After 6 years the value of the car is about \$12 300.**

b **After about $8\frac{1}{3}$ years the value of the car falls below \$10 000.**

c **The intercept on the vertical axis is \$20 000. It represents the original value of the car.**

d When $n = 1$, $S = \$18\,400$

Depreciation = \$20 000 − \$18 400

= \$1600

The car depreciates by \$1600 in the first year.

4 a $M = 60(0.9)^t$

When $t = 0$,

$M = 60 \times 0.9^0$

$= 60$

The original amount was 60 g.

b

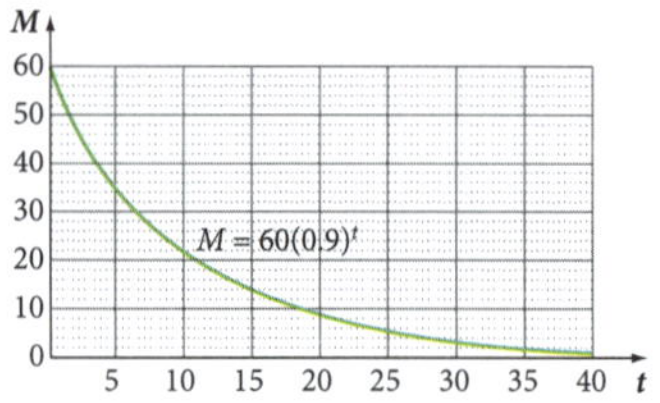

c **After 13 years, approximately 15 g remain.**

d **After 17 years the mass present falls below 10 g.**

e When $M = 30$, $t = 7$

It takes approximately 7 years for half of the mass to disintegrate.

5 a In 2019, $t = 40$

When $t = 40$, $P = 19$

The population was 19 million in 2019.

b $P = 14$ when $t = 20$

The population was 14 million in 1999.

c Nicole might be right. It looks like P will be 22 when $t = 50$ if the graph was continued. As the population is increasing exponentially it might reach 22 million before that. **It is not possible to be sure exactly where the graph will go.**

5 Inverse variation ➲ p. 21

1 a Let n be the number of people and \$$C$ the cost per person.

x	1	2	3	4	5	6	7	8	9
y	**504**	**252**	**168**	**126**	**100.8**	**84**	**72**	**63**	**56**

b

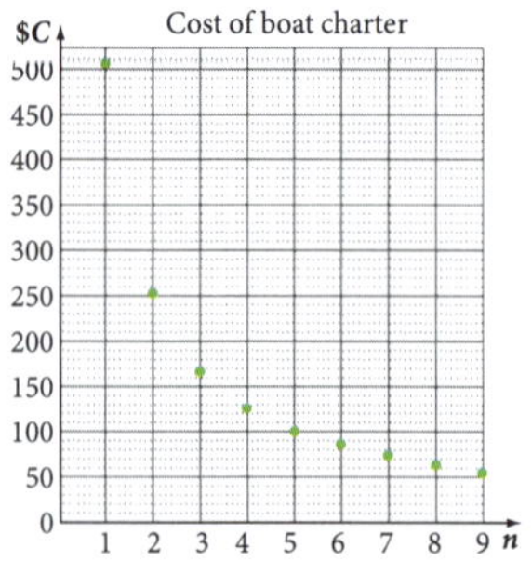

c At least 8 people would need to travel for the cost to be less than \$70 each.

Samuel would need seven other people to take the trip.

2 a $2\frac{1}{2}$ hours = 150 minutes
= 9000 seconds

$T = \frac{a}{s}$

When $s = 8$, $T = 9000$

$9000 = \frac{a}{8}$

$a = 9000 \times 8$

$= 72\,000$

$\therefore \mathbf{T = \frac{72\,000}{s}}$

b $T = \frac{72\,000}{s}$

When $s = 12$,

$T = \frac{72\,000}{12}$

$= 6000$

It takes 6000 seconds to cross the region.

6000 seconds = (6000 ÷ 60) min

= 100 min

= 1 h and 40 min

It takes 1 hour and 40 minutes to cross the region when travelling at 12 m/s.

3 a Let x be the time taken and let y be the average speed and the required formula becomes $y = \frac{12}{x}$. **The graph of $s = \frac{12}{t}$ would be exactly the same, just using different variables.**

b 2 hours and 20 minutes is $2\frac{1}{3}$ hours.

When $x \approx 2.3$, $y \approx 5$

The average speed was about 5 km/h.

c When $y = 1.75$, $x = 7$

So it would take the competitor about 7 hours to complete the race.

4 a

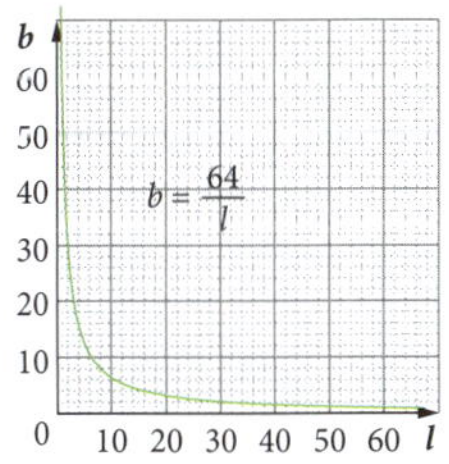

b **When $l = 12$, $b \approx 5.5$.**

c **When $b = 42$, $l \approx 1.5$.**

d **The rectangle is approximately 12.9 cm long and 4.9 cm wide.**

6 Further graphs of practical situations ➲ pp. 22–23

1 a When $x = 4$, $A = 80$.

The area is 80 m².

b When $A = 108$,

$x = 6$ or $x = 18$.

A rectangle of area 108 m² will be 18 m by 6 m.

c **The maximum area is 144 m².**

d When $A = 144$, $x = 12$.

There is only one possible value for the length of the side.

The rectangle is a square of side 12 m.

2 a When $r = 4.5$, $M \approx 550$

The mass of a ball of radius 4.5 cm is approximately 550 g.

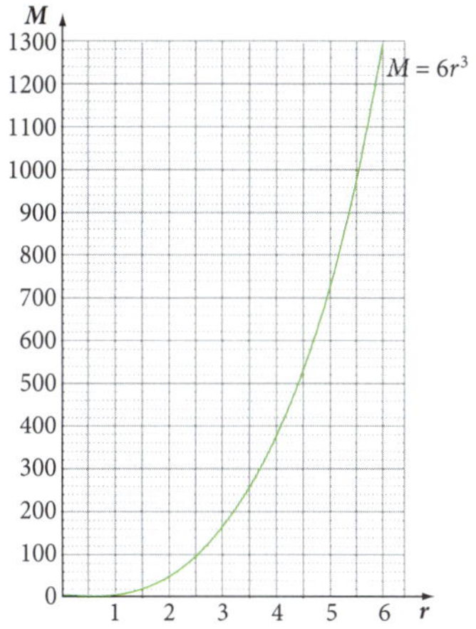

b 1 kg = 1000 g

When $M = 1000$, $r \approx 5.5$

A ball of mass 1 kg has radius approximately 5.5 cm.

c $V = \frac{4}{3}\pi r^3$

When = 2.4,

$V = \frac{4}{3} \times \pi \times (2.4)^3$

$= 57.905\,835\,79\ldots$

$= 58$ (nearest unit)

The volume is 58 cm³ to the nearest cubic centimetre.

d From the graph, when $r = 2.4$, $M \approx 80$

Density $\approx 80 \div 58$

$= 1.379\,310\,345\ldots$

$= 1.4$ (1 d.p.)

The density is approx. 1.4 g/cm³.

3 a

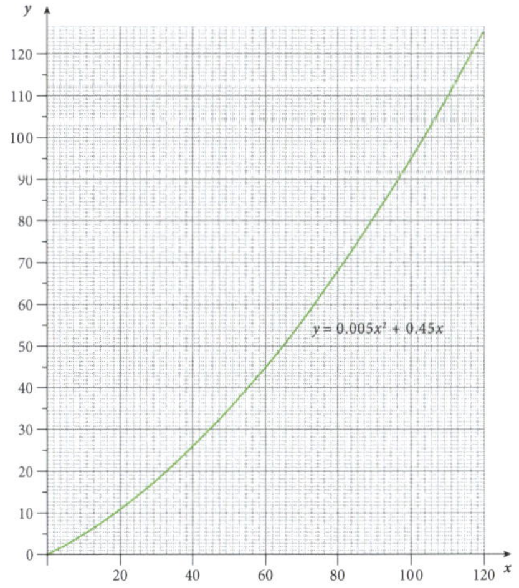

b At 120 km/h the stopping distance is 126 m.

At 100 km/h the stopping distance is about 95 m.

Difference = 126 – 95 = 31

So it takes about 31 m further to stop at 120 km/h than at 100 km/h.

c **To stop in 31 m the maximum speed is 45 km/h.**

7 Limits and restrictions ➲ pp. 25–26

1 a $S = 23\,000 - 1150n$

When $n = 20$,

$S = 23\,000 - 1150 \times 20$

$= 0$

The salvage value is $0 after 20 years.

b The formula must be restricted **to values of n less than or equal to 20**. It makes no sense for the value to be less than zero. **The number of years cannot be negative so n must be greater than 0.**

2 a Vertical change = 24 − 10 = 14

Horizontal change = 3 − 1 = 2

$m = \dfrac{\text{vertical change in position}}{\text{horizontal change in position}}$

$= \dfrac{14}{2}$

$= 7$

$c = 3$

$h = mt + c$

$\mathbf{h = 7t + 3}$

b $h = 7t + 3$

When $t = 6$,

$h = 7 \times 6 + 3$

$= 45$

You would expect the tree to be 45 m high.

c $h = 7t + 3$

When $h = 66$,

$7t + 3 = 66$

$7t = 63$

$t = 9$

You would expect the tree to be 9 years old.

d Trees won't continue to grow at the rate they do to begin with. **They eventually stop growing.**

3 a **When $n = 2$, $S = 6400$**

b **$S \approx 3000$ when $n \approx 5.4$**

c $S = 10\,000(0.8)^n$

When $n = 17$,

$S = 10\,000(0.8)^{17}$

$= 225.179\,981\ldots$

$= \mathbf{225}$ (nearest unit)

d As n gets larger and larger the value of S will get closer and closer to 0. **The minimum value of S is 0, because, although it won't actually reach zero it will get so close we would say the value is zero.**

4 a $h = 96 + 14t - 5t^2$

When $t = 0$,

$h = 96 + 14 \times 0 - 5 \times 0^2$

$= 96$

The height of the ball originally is 96 m. The roof of the building is 96 m high.

b $h = 96 + 14t - 5t^2$

When $t = 6$,

$h = 96 + 14 \times 6 - 5 \times 6^2$

$= \mathbf{0}$

c After 6 seconds the height of the ball is zero. The ball has reached the ground. It cannot continue to fall (and will probably bounce). **The formula gives negative values of h when $t > 6$ and this is impossible.**

5 a Perimeter = 18 m

$2x + 2y = 18$

$x + y = 9$

$\mathbf{y = 9 - x}$

b Area: $A = xy$

$= x(9 - x)$

$= \mathbf{9x - x^2}$

c x m is the length of the yard. **x must be greater than 0 or the yard will not exist.**

d y m is the width of the yard so y must be greater than 0. **This means that x must be less than 9.**

e A is zero when $x = 0$ and $x = 9$.

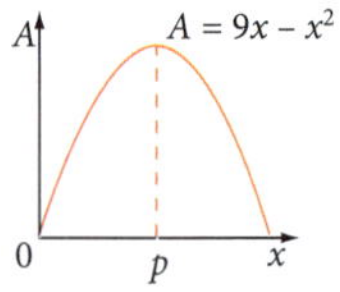

Because the parabola is symmetrical, p is halfway between 0 and 9.

$\mathbf{p = 4.5}$

f The maximum area occurs when $x = 4.5$

$A = 9x - x^2$

When $x = 4.5$,

$A = 9 \times 4.5 - 4.5^2$

$= 20.25$

The maximum area is 20.25 m². Each side will be 4.5 m long.

6 The formula $D = 960 - 80h$ is valid for $0 \le h \le 12$.

It makes no sense for the number of hours to be negative so h must be greater than (or equal to) zero. **It also makes no sense for a train to be a negative distance from Sydney so h must be less than (or equal to) 12.**

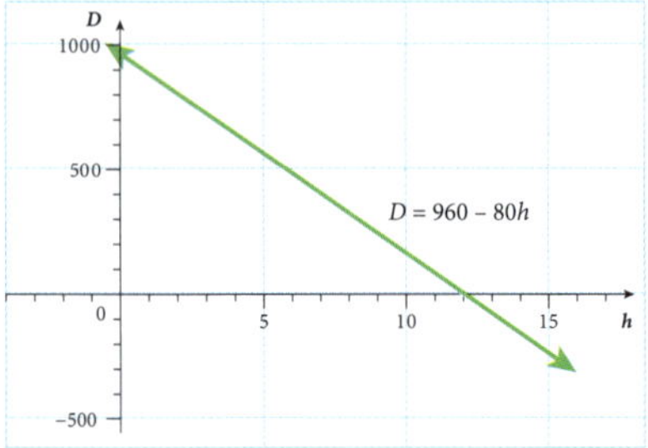

HSC EXAM-TYPE QUESTIONS

Part A

1 **D.** $y = 2x + 1$
When $x = 4$,
$y = 2 \times 4 + 1$
$= 9$
$y = x^2 - 7$
When $x = 4$,
$y = 4^2 - 7$
$= 9$
$y = \dfrac{36}{x}$
When $x = 4$,
$y = \dfrac{36}{4}$
$= 9$
All pass through the point (4, 9).

2 **D.** If 50 packets are sold, the return is \$50 and the cost is \$45.
Ellen makes a profit of \$5.

3 **C.** $y = 3x^2 - x$
When $x = 0$, $y = 3 \times 0^2 - 0 = 0$
When $x = 1$, $y = 3 \times 1^2 - 1 = 2$
When $x = 2$, $y = 3 \times 2^2 - 2 = 10 \neq 6$
When $x = 3$, $y = 3 \times 3^2 - 3 = 24$

4 **B.** The initial mass is 80 g so it is not A. The mass is disintegrating not increasing so it is not C. It is an example of exponential decay and that is graph B.

5 **A.**

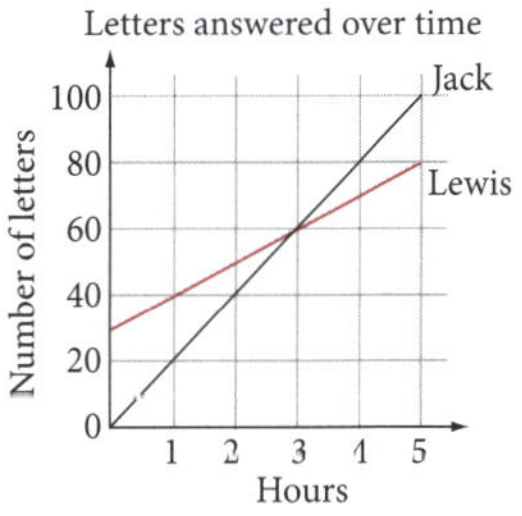

During the shift Jack answers 100 letters and Lewis answers 80.
Difference = 20

6 **A.** The graph is an example of inverse variation.

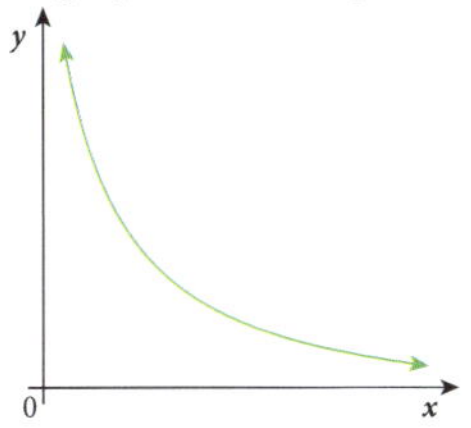

7 **B.** The y-values are increasing as the x-values increase so it is not exponential decay or inverse variation.
There is not a constant rate of increase so it is not linear.
It could be an example of exponential growth.

8 **D.** The depth will not be changing constantly so it is not A or B. The depth will increase quickly at first, then slower at the widest part of the vat and increase in speed after that. So it is D.

Part B

9 a **y-intercept is 30. It represents the fixed costs.** ✓

b $m = \dfrac{\text{vertical change in position}}{\text{horizontal change in position}}$
$= \dfrac{30}{15}$
$= \mathbf{2}$ ✓

c **It represents the extra cost to produce each item.** ✓

d **The income is \$72 when 24 items are sold.** ✓

e **Thirty items need to be sold to break even.** ✓

f **If 45 items are sold, the profit is \$15.** ✓

g **The items are sold for \$3 each.** [The gradient of the income line is 3.] ✓

10 a **When $x = 15$, $y = 40$.** ✓

b **When $y = 200$, $x = 3$.** ✓

c **If 25 people share, the cost is \$24.00 per person.** ✓

d **Twelve people share the cost if they pay \$50 each.** ✓

e **For this practical application, x must be a positive whole number.** ✓

11 a **When $n = 3$, $A \approx 3850$.** ✓

b When $A = 6000$, $n \approx 8$.
The sum of money will have doubled in a little over 8 years. ✓

c Yes, Jessie is right.
The sum of money doubles roughly every 8 years. **There is approximately \$6000 in the account after 8 years so there will be about twice that amount in another 8 years.** ✓

12 $P = 50(0.84)^t$

a In 2011, $t = 0$
$P = 50(0.84)^{20}$
$= 50$
In 2011 the population was 50. ✓

b When $t = 20$,
$P = 50(0.84)^{20}$
$= 1.529 \ldots$
The population has fallen below 2. **There cannot be any change except for the demise of the remaining animal.** ✓

13 a **It is 12 km from A to B.** ✓

b When they meet they are approx. 5 km from A. **They are 7 km from point B.** ✓

c Genevieve completes the journey in the shortest time.
She begins from B whereas Patrick starts from A.
Genevieve reaches A in 3 hours but Patrick takes 4 hours to reach B. ✓

d $m = \dfrac{\text{vertical change in position}}{\text{horizontal change in position}}$
$= \dfrac{-12}{3}$
$= -4$ ✓
Vertical intercept = 12
$\therefore \mathbf{d = -4t + 12}$ ✓

CHAPTER 2 RIGHT-ANGLED TRIANGLES

TOPIC: MEASUREMENT

1 Pythagoras' theorem

- **A right-angled triangle is any triangle** that has one angle measuring 90°.
- **The hypotenuse is the longest side of a right-angled triangle**. It is the side opposite the right angle.

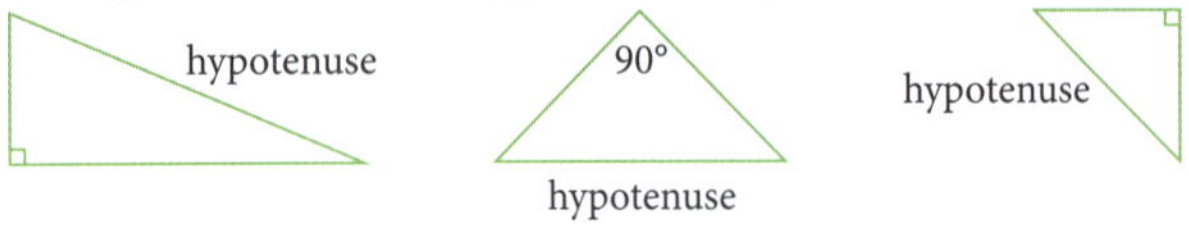

- **Upper-case letters (capital letters) are used to name the angles and lower-case letters are used to name the sides**. Opposite sides and angles use the same letter.

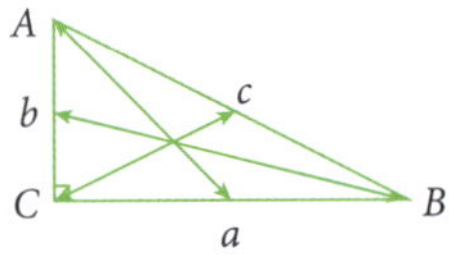

- **Pythagoras was an ancient Greek mathematician. His theorem states**: In any right-angled triangle, the square on the hypotenuse is equal to the sum of the squares on the other two sides. So $c^2 = a^2 + b^2$.

EXAMPLE 1

What is the length of the hypotenuse of this triangle?

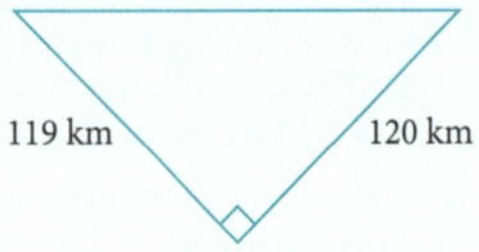

Let the length of the hypotenuse be x km.

$x^2 = 119^2 + 120^2$
$= 28\,561$
$x = \sqrt{28\,561} \quad (x > 0)$
$= 169$

The length of the hypotenuse is 169 km.

A length cannot be negative so we don't need to consider the negative square root.

EXAMPLE 2

Find the width of this dam to the nearest metre.

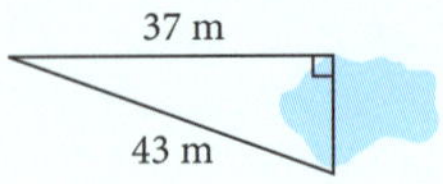

Let x m be the width of the dam.

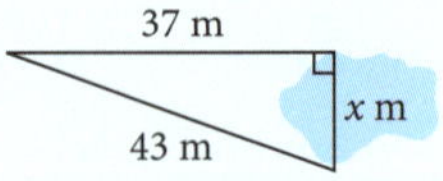

$x^2 = 43^2 - 37^2$
$= 480$
$x = \sqrt{480} \quad (x > 0)$
$= 21.908\,902\ldots$

So the width of the dam is 22 m to the nearest metre.

By Pythagoras' theorem $x^2 + 37^2 = 43^2$ so we subtract 37^2 from both sides.

EXAMPLE 3

A rectangular gate made from pipe is 4.7 m long and 1.2 m high. It has a central pipe and is to be braced with two further diagonal pieces of pipe, of equal length, as shown in the diagram.

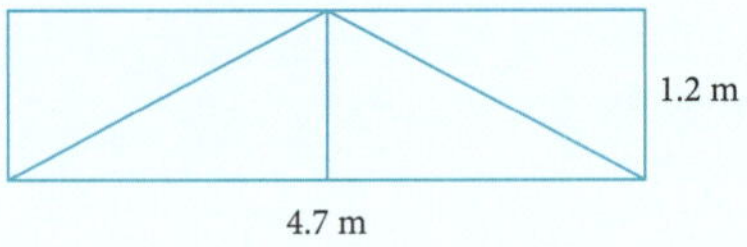

How much pipe is needed to make the gate?

Let x m be the length of each diagonal pipe.
Now 4.7 m ÷ 2 = 2.35 m
So $x^2 = 2.35^2 + 1.2^2$
$= 6.9625$
$x = \sqrt{6.9625}$
$= 2.638\,6549\ldots$
$= 2.64 \quad$ (2 d.p.)

The total length of pipe
$= (2 \times 4.7 + 3 \times 1.2 + 2 \times 2.64)$ m
$= 18.28$ m

So around 18.3 m of pipe will be needed to make the gate.

The gate is rectangular so the angles will be right angles.

→ PRACTICE

1 **What is the length of the hypotenuse of this triangle?**

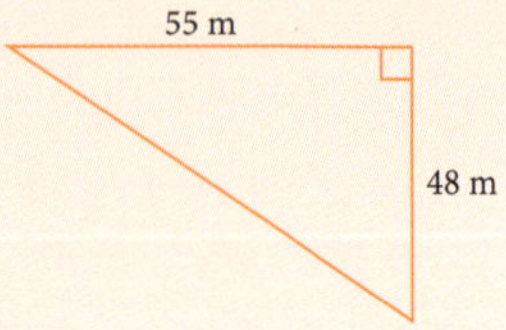

2 **Find the length of this swamp.**

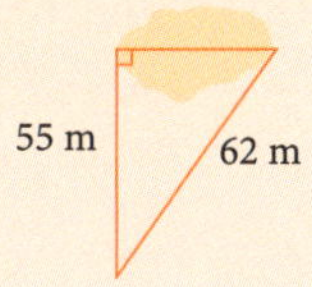

3 **A rectangular gate made from pipe is 5.4 m long and 1.3 m high. It is to be braced with three further pieces of pipe, of equal length, as shown in the diagram.**

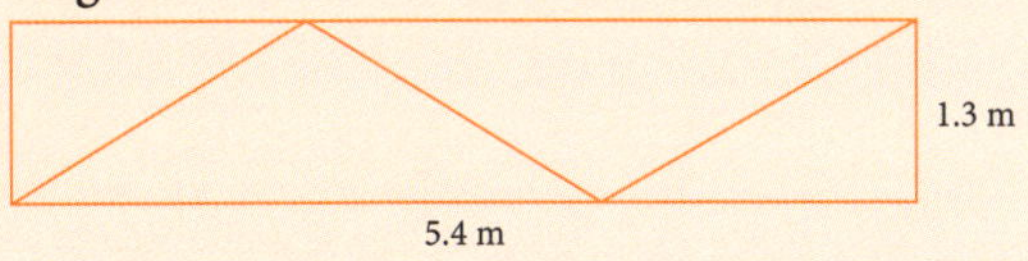

How much pipe is needed to make the gate?

4* **A ladder is 1.85 m long. How far up a wall will it reach if its foot is placed 60 cm from the foot of the wall?**

Answers ➲ p. 48

2 Using trigonometry to find sides of right triangles

➔ **Sin (sine), cos (cosine) and tan (tangent)** are trigonometric ratios or trig ratios.

➔ **The hypotenuse is always the longest side of a right-angled triangle**, the side opposite the right angle.

➔ **The location of the opposite and adjacent sides depends upon the angle being considered.** (Adjacent means 'next to' so the adjacent side is the side, apart from the hypotenuse, that is next to the angle.)

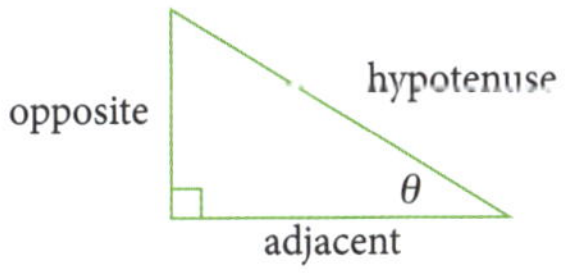

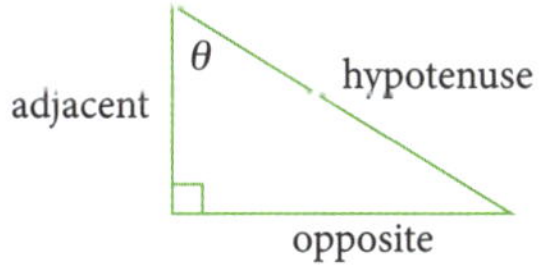

➔ $\sin\theta = \dfrac{\text{opposite}}{\text{hypotenuse}}$ $\cos\theta = \dfrac{\text{adjacent}}{\text{hypotenuse}}$ $\tan\theta = \dfrac{\text{opposite}}{\text{adjacent}}$.

EXAMPLE 1

Find the value of x to one decimal place.

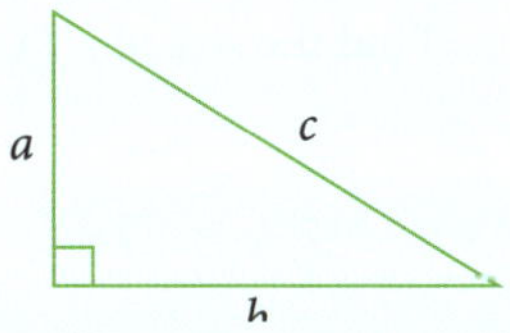

$$\tan 32° = \frac{x}{9}$$

$$x = 9 \times \tan 32°$$

$$= 5.623\,824\,16\ldots$$

$$= 5.6 \quad (1 \text{ d.p.})$$

The side marked x is opposite the given angle and we know the side that is adjacent to that angle, so the ratio is the one linking opposite and adjacent: tan.

EXAMPLE 2

Find the length of side PR. (Give the answer correct to two decimal places.)

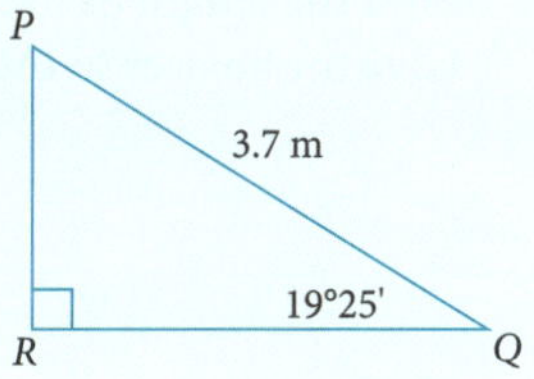

$$\sin 19°25' = \frac{PR}{3.7}$$

$$PR = 3.7 \times \sin 19°25'$$

$$= 1.230\,011\,31\ldots$$

$$= 1.23 \quad (2 \text{ d.p.})$$

The length of PR is 1.23 m, correct to two decimal places.

The unknown side is opposite the angle and we have the hypotenuse, so the ratio to use links opposite and hypotenuse: sin.

EXAMPLE 3

Find the length of the hypotenuse of the triangle. Give the answer to the nearest metre.

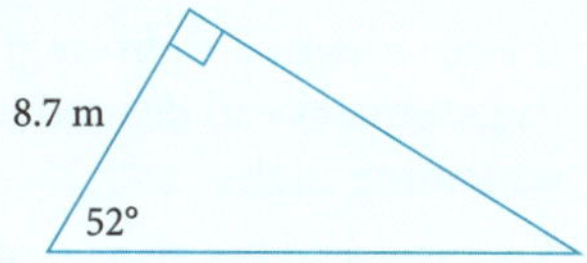

Let the length of the hypotenuse be x m.

$$\cos 52° = \frac{8.7}{x}$$

$$x \times \cos 52° = 8.7$$

$$x = \frac{8.7}{\cos 52°}$$

$$= 14.131\,1424\ldots$$

$$= 14 \quad (\text{nearest unit})$$

The length of the hypotenuse is 14 m, to the nearest metre.

We multiply both sides by x and then divide both sides by cos 52°. This could all be done in one step.

➔ PRACTICE

1 **Find the value of x to one decimal place.**

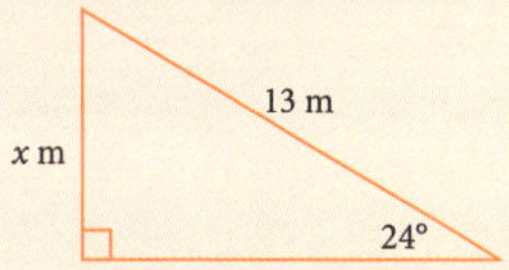

2 **Find the length of side BC. (Give the answer correct to two decimal places.)**

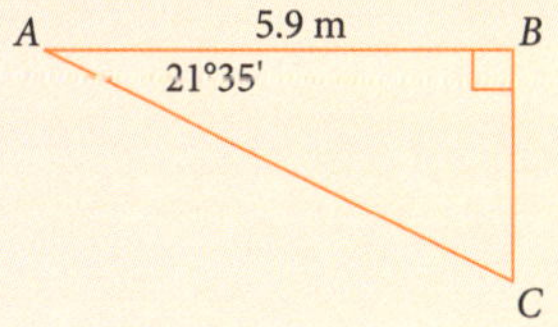

3 Find the length of the hypotenuse of the triangle. Give the answer to the nearest metre.

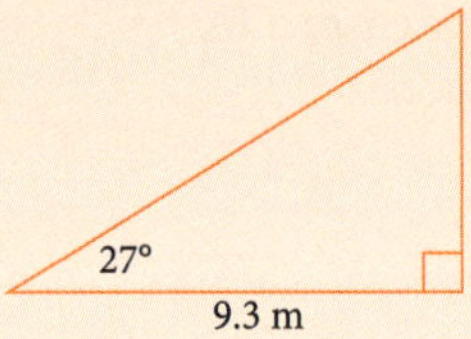

4* Find the length of side PQ. Give the answer in centimetres correct to the nearest centimetre.

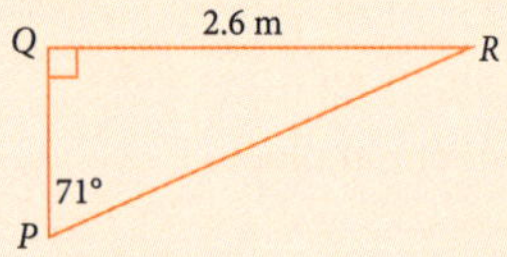

Answers ➲ p. 48

3 Finding angles

- **If two sides of a right-angled triangle are known** then trigonometry can be used to find the size of either of the remaining angles.
- **An answer to the size of an angle might be required** in whole degrees or in degrees and minutes.
- **There are 60 minutes in a degree and 60 seconds in a minute.**

EXAMPLE 1

Find the size of angle QPR, to the nearest degree.

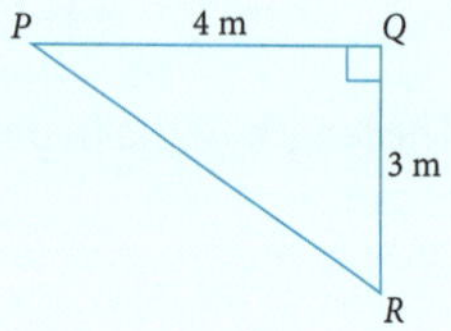

$\tan P = \dfrac{3}{4}$

$P = \tan^{-1}\left(\dfrac{3}{4}\right)$

$= 36.869\,8976\ldots°$

$= 37°$ (nearest degree)

$\therefore \angle QPR = 37°$ to the nearest degree.

The inverse (or opposite) operation to tan is $\tan^{-1}$.

EXAMPLE 2

Find the size of θ to the nearest degree.

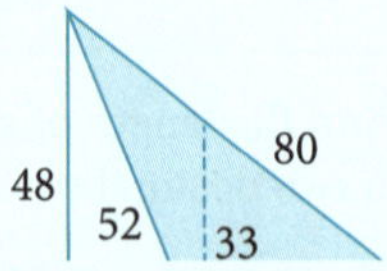

$\sin\theta = \dfrac{2}{5}$

$\theta = \sin^{-1}\left(\dfrac{2}{5}\right)$

$= 23.578\,4784\ldots°$

$= 24°$ (nearest degree)

θ (theta) is a Greek letter. It is often used for an unknown angle.

EXAMPLE 3

Find the size of $\angle ABC$, giving the answer to the nearest minute.

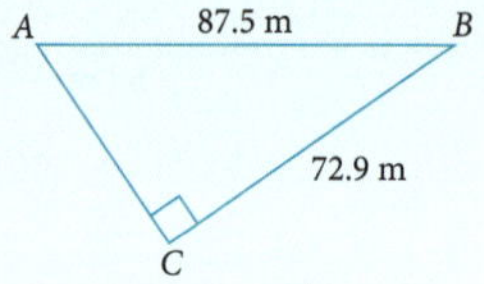

$\cos B = \dfrac{72.9}{87.5}$

$B = \cos^{-1}\left(\dfrac{72.9}{87.5}\right)$

$= 33°34'37.37''$

$= 33°35'$ (nearest minute)

$\therefore \angle ABC = 33°35'$ to the nearest minute.

There are 60 minutes in a degree and 60 seconds in a minute. So if there are 30 or more seconds you must round up to the next minute.

➔ PRACTICE

1 Find the size of angle BCA, to the nearest degree.

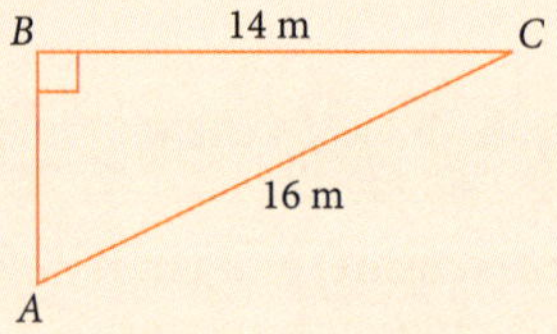

2 Find the size of θ to the nearest degree.

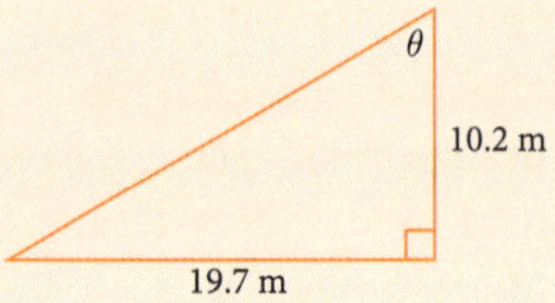

3 Find the size of $\angle LNM$, giving the answer correct to the nearest minute.

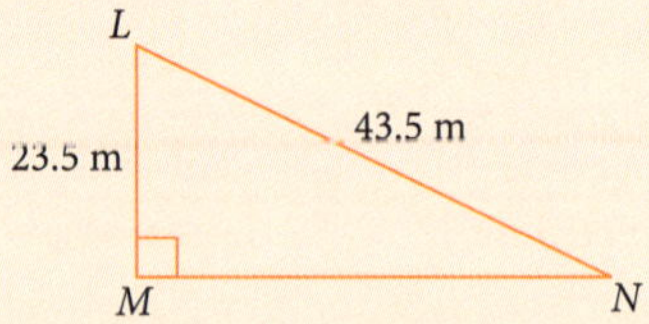

4* Find the size of $\angle BCA$ to the nearest minute.

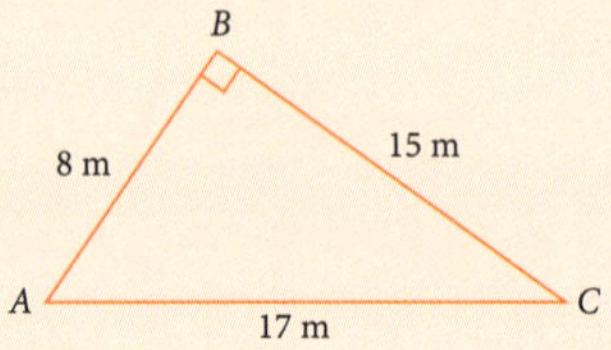

Answers ➲ p. 48

4 Solving problems involving trigonometry

- **When solving problems, draw a large clear diagram** and mark on it all the relevant information.
- **Angles of elevation and depression** are always measured from the horizontal.

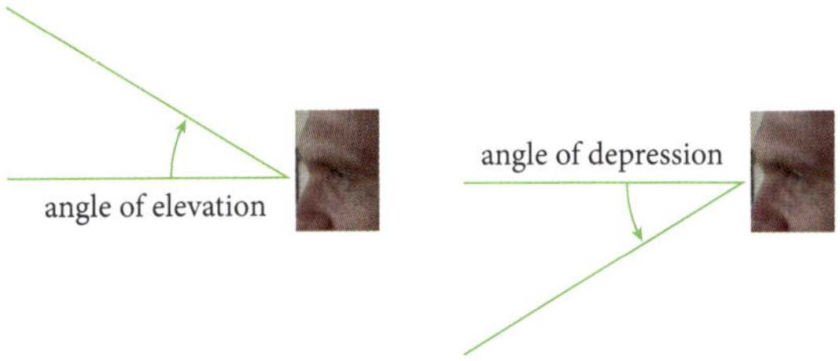

EXAMPLE 1

A 3 m ladder leans against a wall, reaching 2.6 m up the wall. Find the angle, to the nearest degree, between the ladder and the wall.

Let θ be the angle between the ladder and the wall.

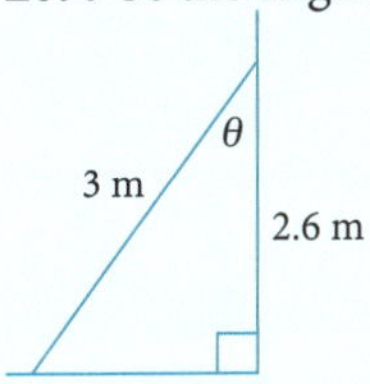

$$\cos\theta = \frac{2.6}{3}$$

$\theta = 29.926\,4348\ldots°$

$= 30°$ (nearest degree)

The angle between the wall and the ladder is 30°, to the nearest degree.

> The ladder forms the hypotenuse of the triangle and the wall is adjacent to the required angle.

EXAMPLE 2

A ramp is to be built. One end of the ramp will be on the ground and will make an angle of 20° with the horizontal. The other end of the ramp will be at a height of 1.2 m. Find the length of the ramp (up the incline), to the nearest centimetre.

Let x m be the length of the ramp.

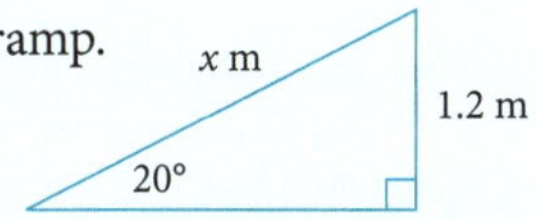

$$\sin 20° = \frac{1.2}{x}$$

$$x = \frac{1.2}{\sin 20°}$$

$= 3.508\,5652\ldots$

$= 3.51$ (2 d.p.)

The length of the ramp is 3.51 m, to the nearest centimetre.

> There are 100 cm in a metre, so a measurement in metres rounded off to the nearest centimetre is rounded off to the nearest hundredth or two decimal places.

EXAMPLE 3

From a lookout at the top of a 128 m high tower, the angle of depression of a campsite is 13°.

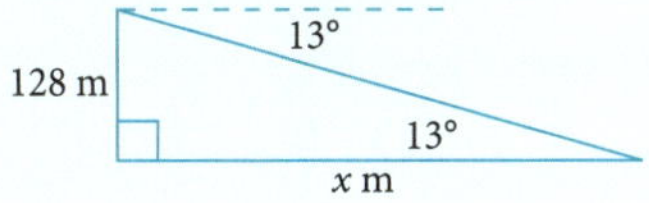

How far is the campsite from the base of the tower (assuming that the ground is level)?

Let x m be the distance from the base of the tower to the campsite.

$$\tan 13° = \frac{128}{x}$$

$$x = \frac{128}{\tan 13°}$$

$= 554.428\,911\ldots$

$= 554$ (nearest unit)

The campsite is 554 m from the base of the tower, to the nearest metre.

> When working with angles of depression, we often make use of the fact that alternate angles, formed when a transversal cuts a pair of parallel lines, are equal.

EXAMPLE 4

A building that is 12 m tall casts a shadow on level ground that is 18 m long. Find the angle of elevation of the sun to the nearest degree.

Let θ be the angle of elevation.

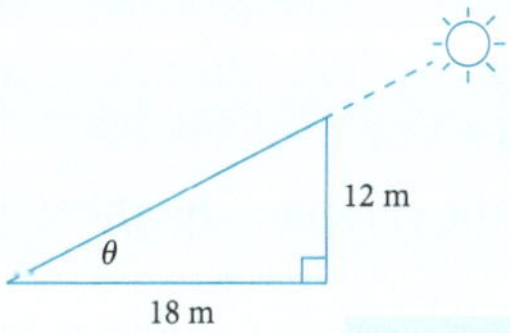

$$\tan\theta = \frac{12}{18}$$

$\theta = 33.690\,0675\ldots°$

$= 34°$ (nearest degree)

The angle of elevation of the sun is 34°, to the nearest degree.

> A shadow is found on the ground so the length of the shadow is the horizontal distance.

PRACTICE

1. A 2.4 m long ladder leans against a wall. The foot of the ladder is 1.1 m from the base of the wall. Find the angle, to the nearest degree, between the ladder and the wall.
2. A ramp is to be built. It will be inclined at an angle of 25° from the ground and will be 3 m long up the incline. How high will the ramp be at the highest point? Give the answer to the nearest 10 cm.

3 **From a lookout at the top of a cliff, the angle of depression of a boat is 20°. If the boat is 650 m from the base of the cliff, find the height of the lookout.**

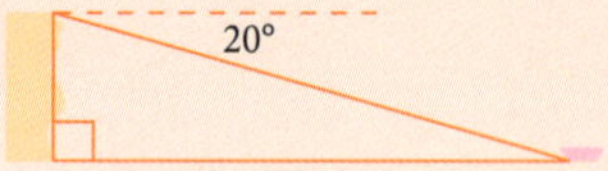

4 **A 9 m cable, attached to the top of a pole, reaches the ground at a point 5.2 m from the base of the pole. What is the angle of elevation of the top of the pole from the point where the cable reaches the ground?**

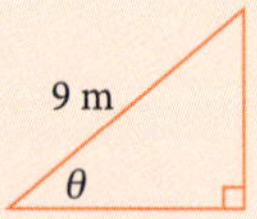

5* **A building casts a 9.6 m long shadow on level ground. If the angle of elevation of the sun is 52°, find the height of the building to the nearest metre.**

Answers ➲ p. 49

5 Navigational methods

➔ **Navigational methods are ways of determining position in the world and directions for movement.** The Global Positioning System (GPS) uses satellites to give exact positions. Stars have been used to navigate for a long time. First Nations Australians used storylines that linked landscape with the night sky.

➔ **The points of the compass should all be known.**

➔ **Compass bearings are measured from either north or south** in either an easterly or westerly direction,

e.g. N50°E S20°W S70°E N48°W

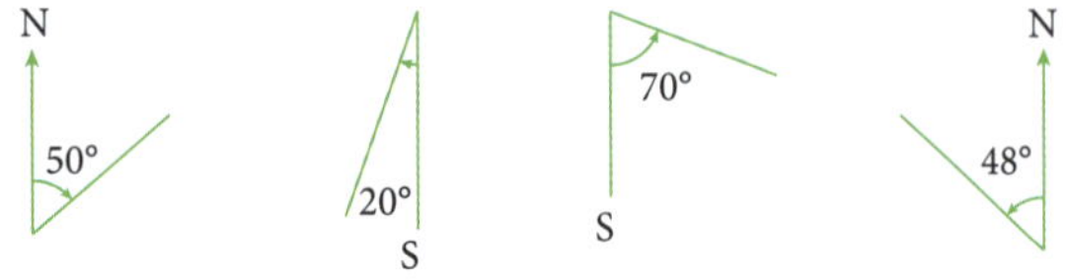

➔ **True bearings are measured in a clockwise direction from north**. They are always given using three digits, e.g. the bearing of Q from P is:

050° 157° 270°

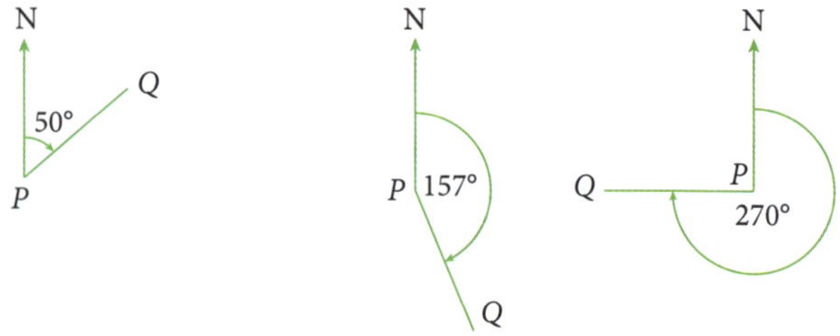

EXAMPLE 1

P is 75 km S35°E of Q.

a **What is the true bearing of P from Q?**

b **What is the compass bearing of Q from P?**

a True bearing of P from Q = 180° – 35°
= 145°

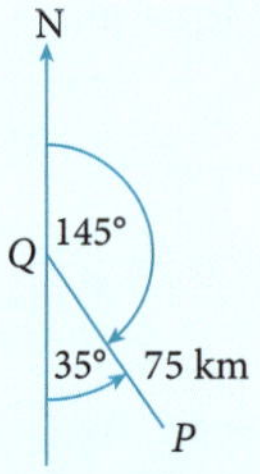

Use a diagram to help see what is happening.

b The compass bearing of Q from P is N35°W.

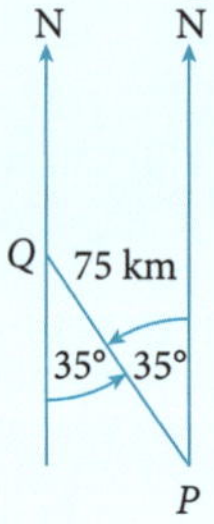

We often use the properties of angles formed by parallel lines when working with bearings. In this case the alternate angles are equal.

EXAMPLE 2

The bearing of Ashwood from Brenton is 310°T. What is this as a compass bearing?

Let $x°$ be the angle between AB and the direction of due north.

Now $x = 360 - 310$
$= 50$

So the compass bearing is N50°W.

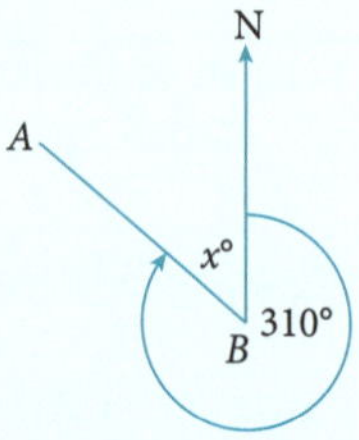

The T after the degrees shows that it is a true bearing. The T is only sometimes used. If a bearing is given with three figures (and no other directions), it is a true bearing.

EXAMPLE 3

A plane flies 50 km from Jindalee (J) due South to Kurraville (K). From Kurraville it flies 50 km south-west to Lincoln (L) and then 120 km due east to Mackston (M) and then straight back to Jindalee.

a **Draw a diagram showing the path of the plane.**

b **In what general direction will the plane fly on the last leg?**

a

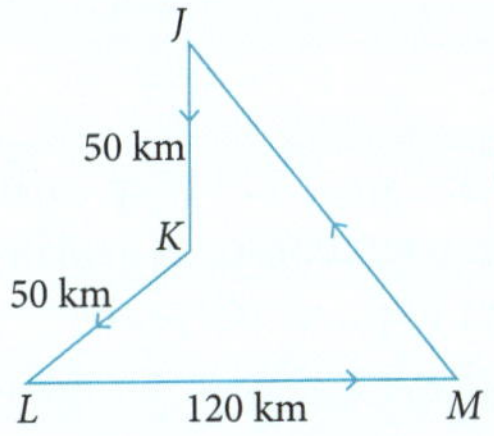

Travelling straight between points means in a straight line. It doesn't mean that the line has to be horizontal.

b The plane will be flying north-west on the last leg.

EXAMPLE 4

The bearing of R from P is 140°. Q is due east of P and due north of R. Find the true bearing of P from R.

Let X be a point due north of P.

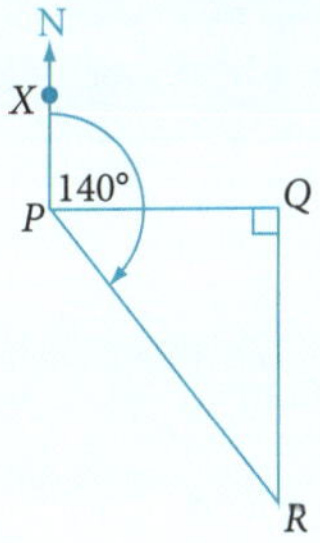

$\angle XPR = 140°$

$\angle QRP + \angle XPR = 180°$ (co-interior angles, parallel lines)

$\therefore \angle QRP = 180° - 140°$

$= 40°$

Required bearing $= 360° - 40°$

$= 320°$

The bearing of P from R is 320°.

The required bearing is the reflex angle QRP.

EXAMPLE 5

A boat sails from port P on a bearing of 150° to an island Q. From Q it sails on a bearing of 050° to point R, which is due east of P.

a **Draw a diagram showing the information.**

b **Find the size of $\angle RPQ$.**

c **Find the size of $\angle PQR$.**

d **Find the true bearing of Q from R.**

a

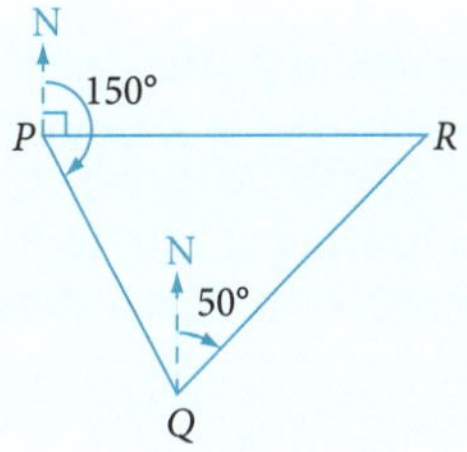

b Let X be a point due north of P.

$\angle XPQ = 150°$

$\angle XPR = 90°$ (R is due east of P)

$\therefore \angle RPQ = 150° - 90°$

$= 60°$

c Let Y be a point due north of Q.

$\angle YQR = 50°$

$\angle PQY + \angle XPQ = 180°$ (co-interior angles, parallel lines)

$\angle PQY = 180° - 150°$

$= 30°$

$\angle PQR = 30° + 50°$

$= 80°$

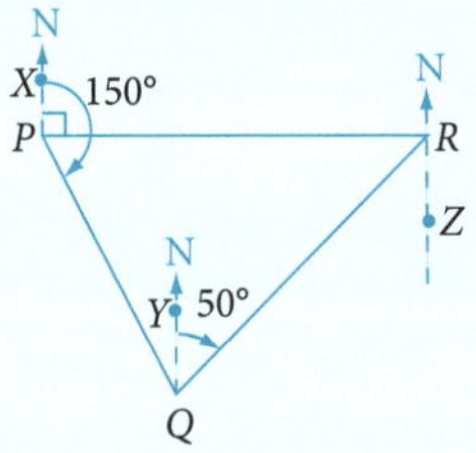

d Let Z be a point due south of R.

$\angle ZRQ = \angle YQR$ (alternate angles, parallel lines)

$\angle ZRQ = 50°$

The bearing of Q from $R = 180° + 50°$

$= 230°$

A new diagram can always be drawn if it helps you see what is required and how to get there.

→ PRACTICE

1 **P is 75 km N58°E of Q.**

a **What is the true bearing of P from Q?**

b **What is the compass bearing of Q from P?**

2 **The bearing of Crestwood from Denton is 265°T. What is this as a compass bearing?**

3 **A helicopter leaves P and flies 32 km due west to Q. From Q it flies 15 km due north to R and then 17 km due west to S. It then flies 24 km north-east to T, then straight back to P.**

a **Draw a diagram showing the path of the helicopter.**

b **In what general direction did the helicopter fly on the last leg?**

4 X is due west of Y and due north of Z. The bearing of Y from Z is 074°. Find the bearing of Z from Y.

5 A plane left A and flew on a bearing of 220° to B. From B it flew on a bearing of 290° to C, which is due west of A.

a Draw a diagram showing the information.

b Find the size of $\angle CAB$.

c Find the size of $\angle CBA$.

d Find the compass bearing of B from C.

6* Johnson (J), Kellyville (K) and Lowndes (L) are three towns. Johnson is 87 km from Kellyville on a bearing of 165°. Lowndes is 102 km from Kellyville. The bearing of Kellyville from Lowndes is 065°.

a Draw a diagram to show the position of the three towns.

b What is the size of $\angle LKJ$?

Answers ➲ pp. 49–50

6 Problems involving bearings

➔ **Trigonometry can be used to solve problems** involving bearings.

➔ **Remember that it might not always be necessary to use trigonometry** to answer every part of a question. Sometimes simple geometry is all that is required.

➔ **Always draw a diagram**, even if one is supplied, and mark on it all the given information. As much as is possible use the correct directions and make sure measurements are roughly to scale.

EXAMPLE 1

Town A is 58 km due south of town B. Town C is on a bearing of 290° from A and is due west of B.

a **Find the size of $\angle BAC$.**

b **Find the distance between towns A and C, to the nearest kilometre.**

a

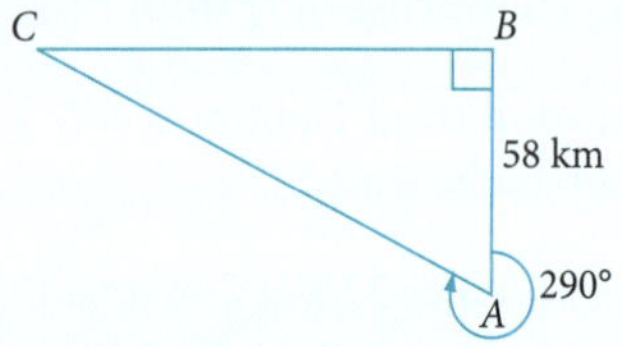

$\angle BAC = 360° - 290°$
$= 70°$

b $\cos 70° = \dfrac{58}{AC}$

$AC = \dfrac{58}{\cos 70°}$

$= 169.580\,655\ldots$

$= 170$ (nearest unit)

The distance between A and C is 170 km, to the nearest kilometre.

Read the question carefully and make sure you have all the directions correct. It is very easy to make mistakes.

EXAMPLE 2

A ship sails from port 95 km due west and then turns and sails 67 km due south. On what compass bearing should it sail to return straight to port? (Give the answer to the nearest minute.)

$\tan \theta = \dfrac{95}{67}$

$\theta = \tan^{-1}\left(\dfrac{95}{67}\right)$

$= 54°48'21.93''$

$= 54°48'$ (nearest minute)

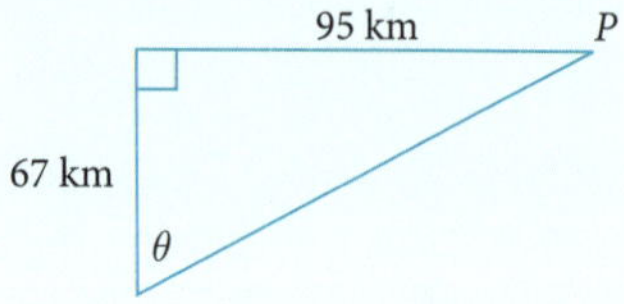

The ship should sail on a bearing of N54°48'E.

Always stop to consider whether the answer makes sense. From the diagram we have a rough idea of what the answer should be.

EXAMPLE 3

The bearing of P from O is 056°. Q is 108 km due south of P and on a bearing of 146° from O.

a **Draw a diagram showing the above information.**

b **Find the size of $\angle POQ$.**

c **Find the distance between O and Q, to the nearest kilometre.**

a

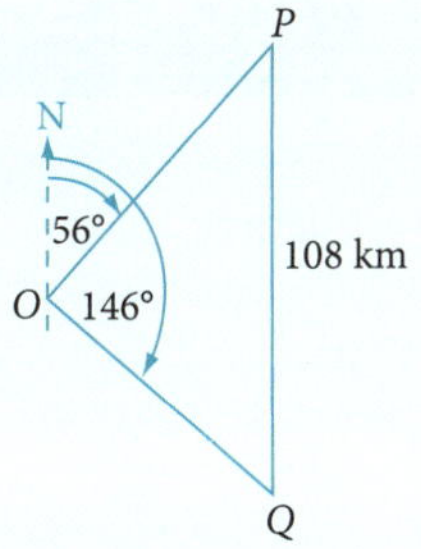

b $\angle POQ = 146° - 56°$

$= 90°$

c $\angle OPQ = 56°$ (alternate angles, parallel lines)

$\sin 56° = \dfrac{OQ}{108}$

$OQ = 108 \times \sin 56°$

$= 89.5360578\ldots$

$= 90$ (nearest unit)

The distance between O and Q is 90 km, to the nearest kilometre.

> Alternatively, we could use co-interior angles to find $\angle OQP$ and use cos instead of sin.

→ PRACTICE

1 Town P is 72 km due east of town Q. Town R is due south of Q and on a bearing of 240° from P.

- **a** Find the size of $\angle QPR$.
- **b** Find the distance between towns Q and R, to the nearest kilometre.

2 A helicopter flies from its base 63 km due south and then it flies 84 km due east. On what compass bearing should it fly to return directly to base? (Give the answer to the nearest minute.)

3 The bearing of X from P is 023°. X is 95 km due north of Y, which is on a bearing of 113° from P.

- **a** Draw a diagram showing the above information.
- **b** Find the size of $\angle XPY$.
- **c** Find the distance between X and P, to the nearest kilometre.

4* A is southwest of B. C is S65°E of A and B is N25°E of C. Find the distance of C from A given that the distance of B from A is 135 km.

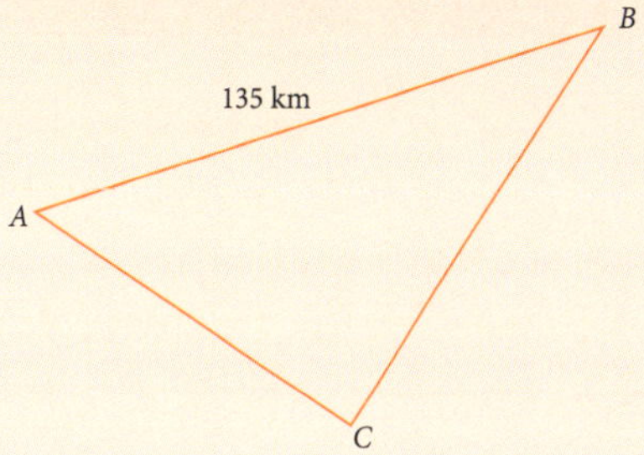

Answers ➲ p. 50

HSC EXAM-TYPE QUESTIONS

Total marks: 30

PART A

(Suggested time: 12 minutes)

(1 mark each)

1 What is the length of LM to the nearest metre?

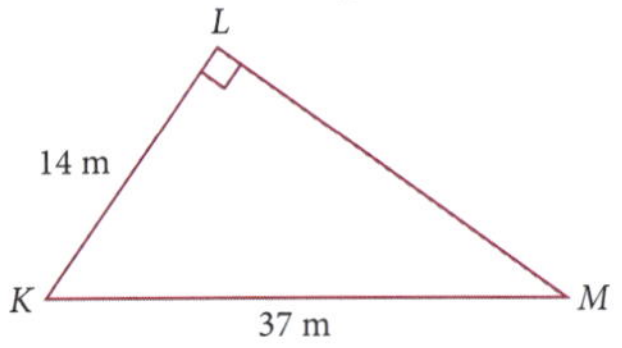

A 28 m **B** 31 m
C 34 m **D** 40 m

2 A compass bearing of S40°E is equivalent to a true bearing of:

A 220° **B** 240°
C 130° **D** 140°

3 If $\sin\theta = 0.704$, what is the size of θ to the nearest minute?

A 44°44' **B** 44°45'
C 44°55' **D** 44°56'

4 Town P is 32 km from Q on a bearing of 025°. Town R is 19 km from Q on a bearing of 143°.

Which diagram best represents this information?

A

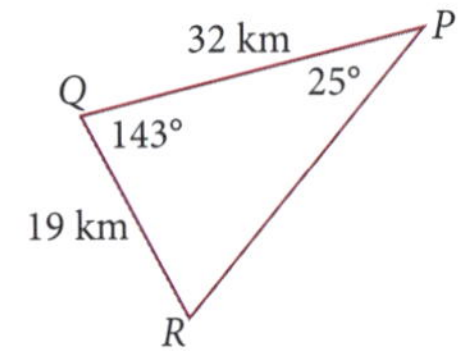

B

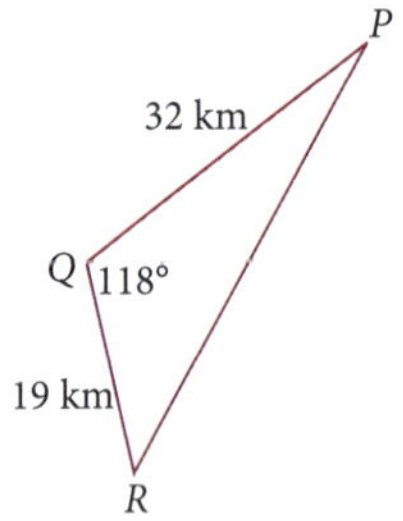

C

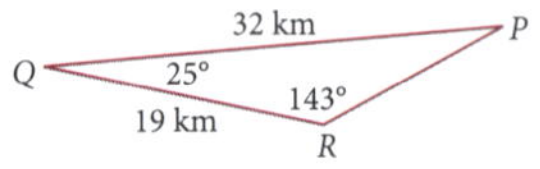

D

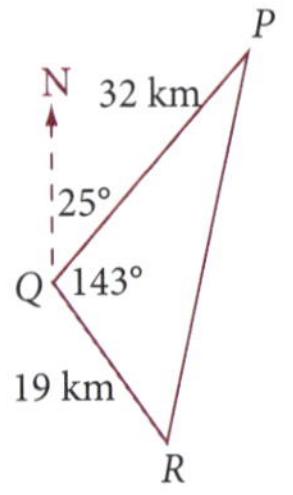

5 What is the size of θ, to the nearest degree?

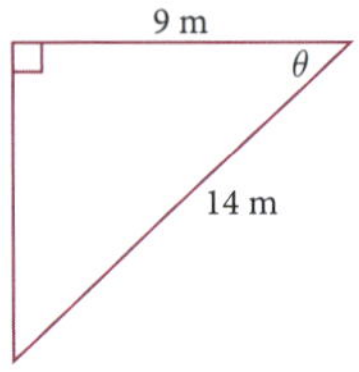

A 40° **B** 41°
C 49° **D** 50°

6 A plane flies 28 km due north and then 195 km due east. How far must it fly to return straight home?

A 197 km **B** 203 km
C 213 km **D** 217 km

7 What is the length of QR, to the nearest metre?

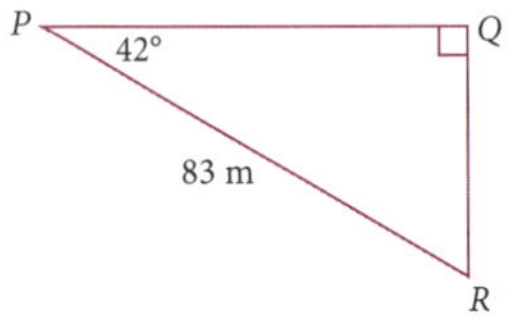

A 55 m **B** 56 m
C 61 m **D** 62 m

8 From the top of a building, the angle of depression of an object on the ground, 300 m from the base of the building on level ground, is 40°. Which expression will find the height (h m) of the building?

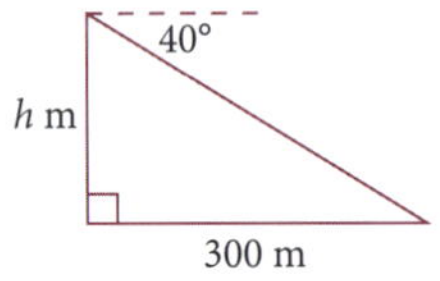

A 300 tan 40° **B** 300 tan 50°
C $\dfrac{300}{\tan 40°}$ **D** $\dfrac{\tan 50°}{300}$

PART B

(Suggested time: 33 minutes)

Show all working.

9 What is the length of the hypotenuse to the nearest centimetre? (2 marks)

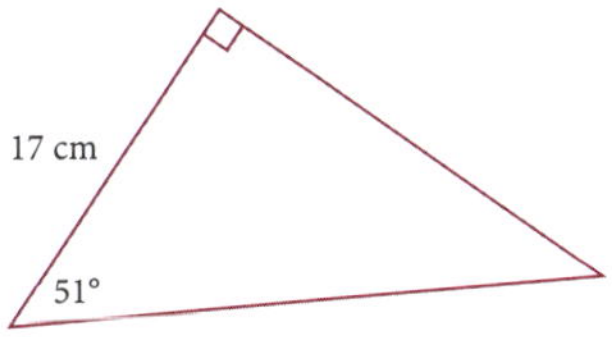

10 What is the size of $\angle BAC$ to the nearest minute? (2 marks)

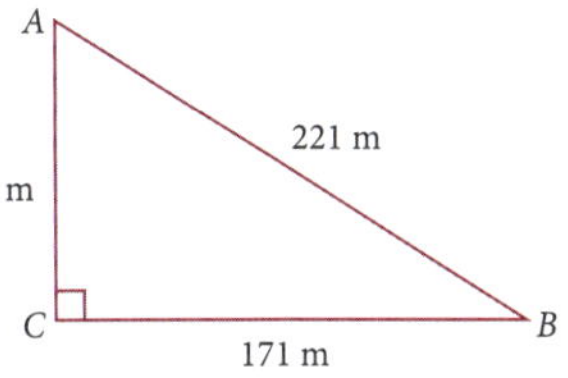

11 From a point 50 m from the base of a building, on level ground, the angle of elevation of the top of the building is 35°. How high is the building to the nearest metre? (2 marks)

12 C is 90 km due south of A. The bearing of A from B is 324° and of C from B is 234°.

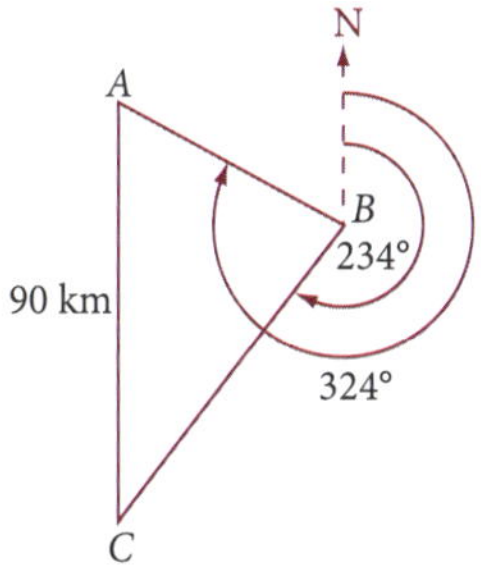

- **a** Find the size of $\angle ABC$. (1 mark)
- **b** Find the size of $\angle BAC$. (1 mark)
- **c** Find the distance from B to C, to the nearest kilometre. (1 mark)

13 Find the area of this triangle. (2 marks)

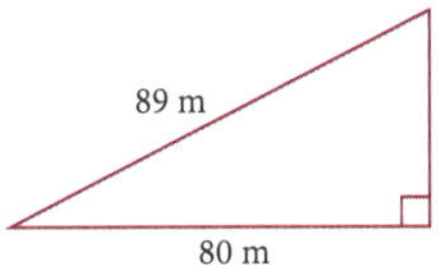

14 a Find the length of AB, correct to one decimal place. (1 mark)

b Find the length of CD. Give the answer to the nearest metre. (2 marks)

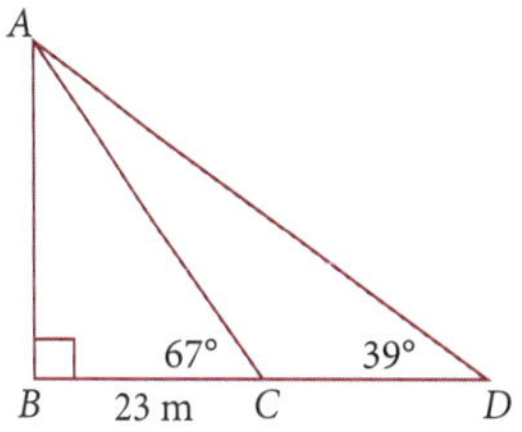

15 A helicopter flies 125 km due north and then 200 km due west. On what compass bearing should it fly to return straight to its starting point? (2 marks)

16

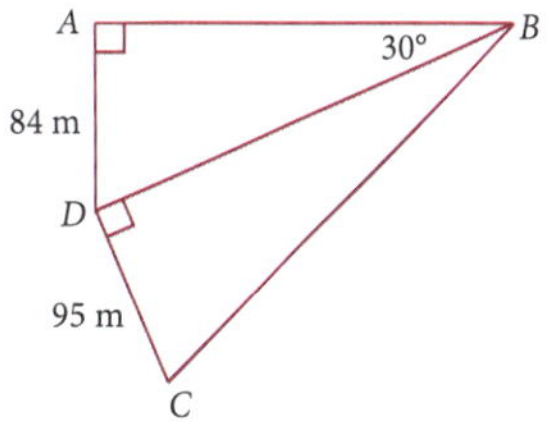

- **a** Find the length BC. (3 marks)
- **b** Is $\angle DBC$ larger or smaller than $\angle ABD$? Justify your answer with appropriate calculations. (3 marks)

ANSWERS

PRACTICE

1 Pythagoras' theorem ➲ pp. 38–39

1 Let the length of the hypotenuse be x m.

$x^2 = 55^2 + 48^2$
$= 5329$
$x = \sqrt{5329} \quad (x > 0)$
$= 73$

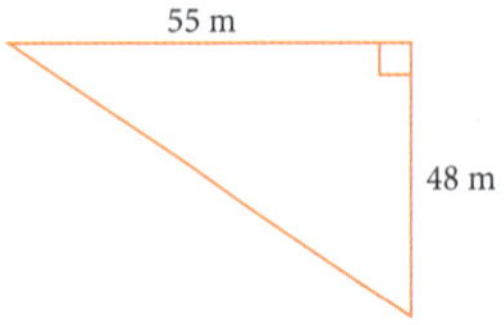

The length of the hypotenuse is 73 m.

2 Let x m be the length of the swamp.

$x^2 + 55^2 = 62^2$
$x^2 + 3025 = 3844$
$x^2 = 3844 - 3025$
$= 819$
$x = \sqrt{819} \quad (x > 0)$
$= 28.618\,1760\ldots$
$= 28.6 \quad (1 \text{ d.p.})$

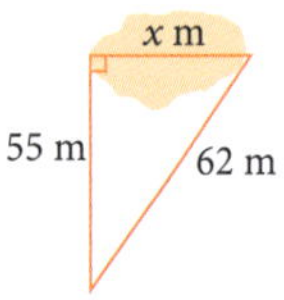

The length of the swamp is 28.6 m, correct to one decimal place.

3 Let x m be the length of each diagonal pipe.

Now 5.4 m ÷ 3 = 1.8 m

So $x^2 = 1.8^2 + 1.3^2$
$= 4.93$
$x = \sqrt{4.93}$
$= 2.220\,360\,33\ldots$
$= 2.22 \quad (2 \text{ d.p.})$

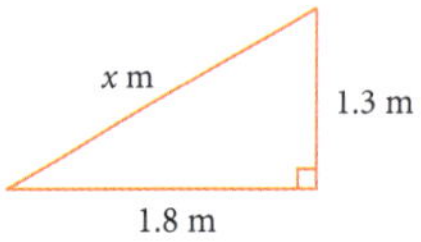

The total length of pipe
$= (2 \times 5.4 + 2 \times 1.3 + 3 \times 2.22)$ m
$= 20.06$ m

So around 20 m of pipe will be needed to make the gate.

4 60 cm = 0.6 m

Let x m be the length the ladder reaches up the wall.

$x^2 = 1.85^2 - 0.6^2$
$= 3.0625$
$x = \sqrt{3.0625}$
$= 1.75$

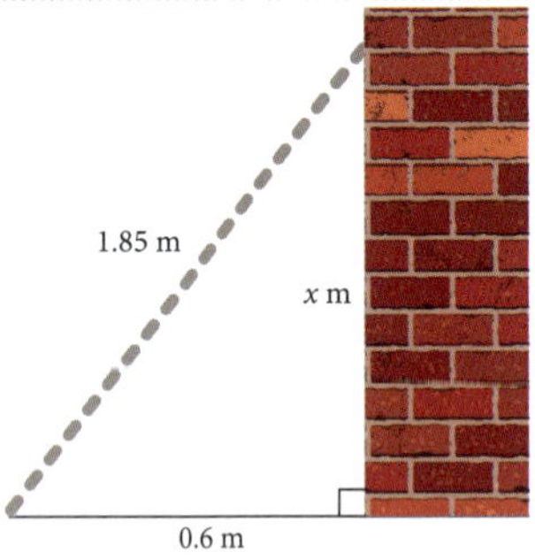

The ladder will reach 1.75 m up the wall.

2 Using trigonometry to find sides of right triangles ➲ pp. 39–40

1 $\sin 24° = \dfrac{x}{13}$

$x = 13 \times \sin 24°$
$= 5.287\,5763\ldots$
$= \mathbf{5.3} \quad (1 \text{ d.p.})$

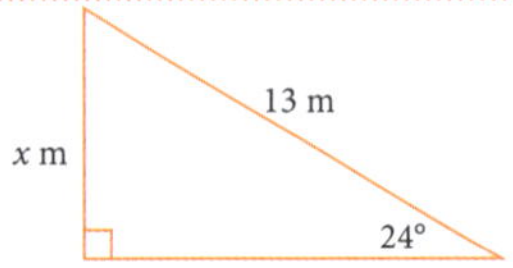

2 $\tan 21°35' = \dfrac{BC}{5.9}$

$BC = 5.9 \times \tan 21°35'$
$= 2.333\,990\,20\ldots$
$= 2.33 \quad (2 \text{ d.p.})$

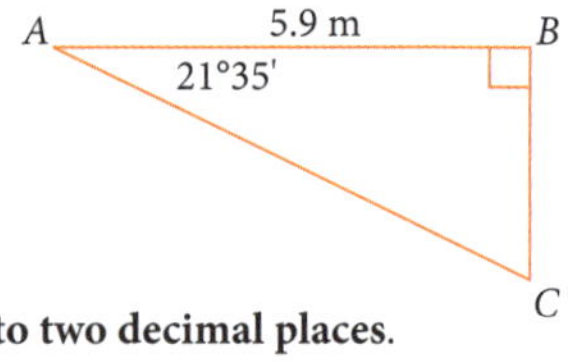

The length of BC is 2.33 m, correct to two decimal places.

3 Let the length of the hypotenuse be x m.

$\cos 27° = \dfrac{9.3}{x}$

$x = \dfrac{9.3}{\cos 27°}$
$= 10.437\,6340\ldots$
$= 10 \quad (\text{nearest unit})$

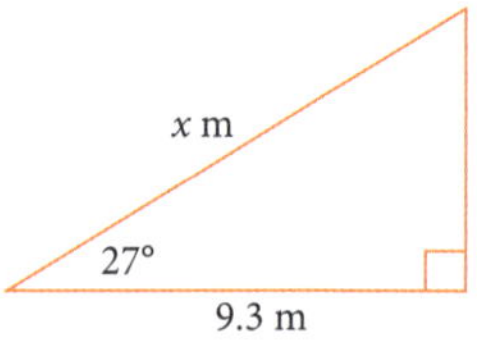

The length of the hypotenuse is 10 m, to the nearest metre.

4 $\tan 71° = \dfrac{2.6}{PQ}$

$PQ = \dfrac{2.6}{\tan 71°}$
$= 0.895\,251\,794\ldots$
$= 0.90 \quad (2 \text{ d.p.})$

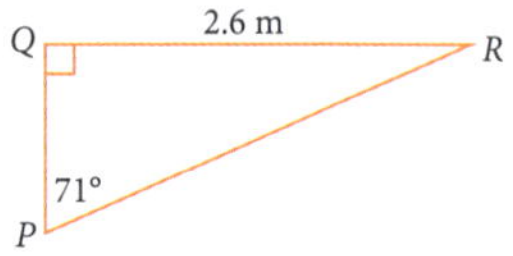

Now 0.90 m = 90 cm.

The length of side PQ is 90 cm, to the nearest centimetre.

3 Finding angles ➲ p. 40

1 $\cos C = \dfrac{14}{16}$

$C = \cos^{-1}\left(\dfrac{14}{16}\right)$
$= 28.955\,0243\ldots°$
$= 29° \quad (\text{nearest degree})$

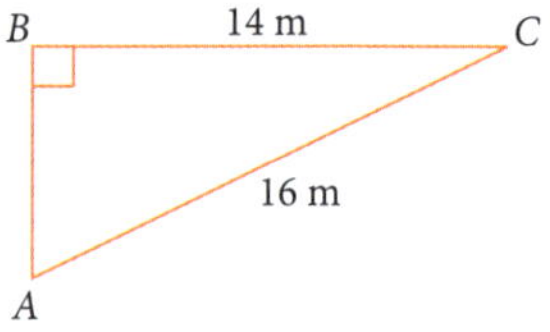

∴ ∠BCA = 29° to the nearest degree

2 $\tan \theta = \dfrac{19.7}{10.2}$

$\theta = \tan^{-1}\left(\dfrac{19.7}{10.2}\right)$
$= 62.626\,3927\ldots°$
$= \mathbf{63°} \quad (\text{nearest degree})$

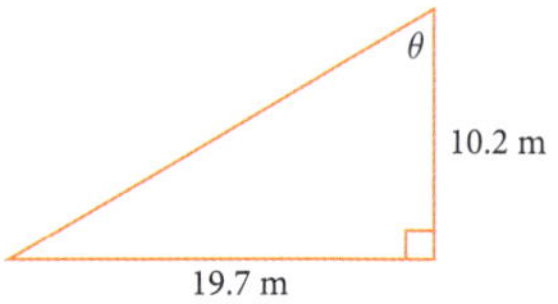

3 $\sin N = \dfrac{23.5}{43.5}$

$N = \sin^{-1}\left(\dfrac{23.5}{43.5}\right)$
$= 32°41'57.44''$
$= 32°42' \quad (\text{nearest minute})$

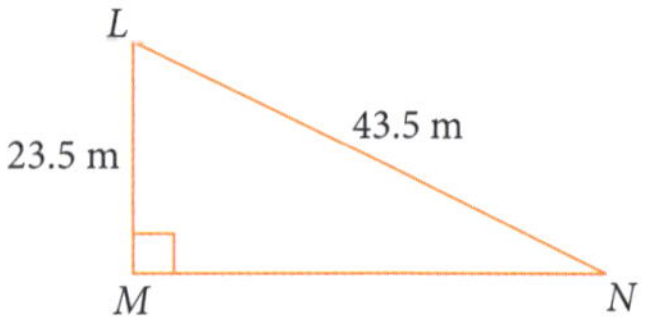

∴ ∠LNM = 32°42' to the nearest minute

4 $\sin C = \dfrac{8}{17}$

$C = \sin^{-1}\left(\dfrac{8}{17}\right)$
$= 28°4'20.95''$
$= 28°4' \quad (\text{nearest minute})$

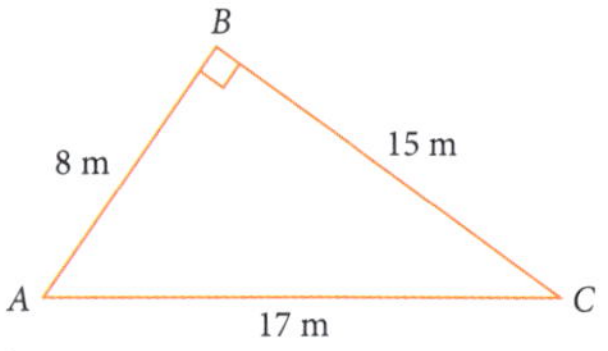

So ∠BCA = 28°4' to the nearest minute.

(Cos or tan could have been used instead of sin.)

4 Solving problems involving trigonometry ➲ pp. 41–42

1 Let θ be the angle between the ladder and the wall.

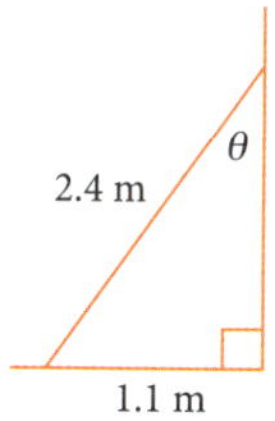

$\sin\theta = \dfrac{1.1}{2.4}$

$\theta = \sin^{-1}\left(\dfrac{1.1}{2.4}\right)$

$= 27.2796127\ldots°$

$= 27°$ (nearest degree)

The angle between the wall and the ladder is 27°, to the nearest degree.

2 Let x m be the height of the ramp.

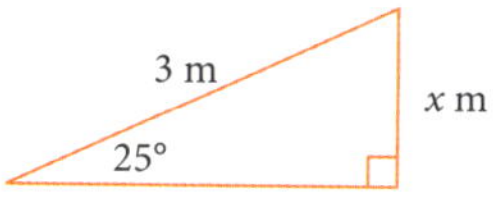

$\sin 25° = \dfrac{x}{3}$

$x = 3 \times \sin 25°$

$= 1.26785478\ldots$

$= 1.3$ (1 d.p.)

The ramp will be 1.3 m high, to the nearest 10 cm.

3 Let x m be the height of the lookout.

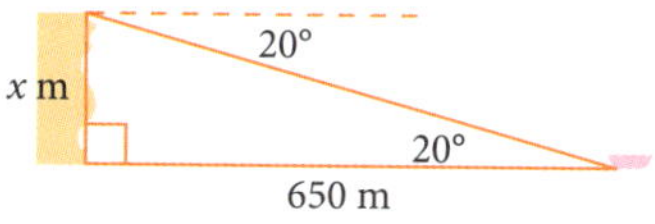

$\tan 20° = \dfrac{x}{650}$

$x = 650 \times \tan 20°$

$= 236.580652\ldots$

$= 237$ (nearest unit)

The height of the lookout is 237 m, to the nearest metre.

4 Let θ be the angle of elevation.

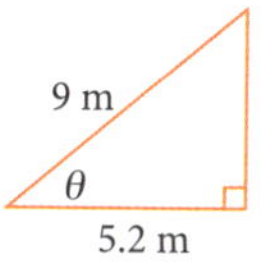

$\cos\theta = \dfrac{5.2}{9}$

$\theta = \cos^{-1}\left(\dfrac{5.2}{9}\right)$

$= 54.7056053\ldots°$

$= 55°$ (nearest degree)

The angle of elevation of the top of the pole is 55°, to the nearest degree.

5 Let h m be the height of the building.

$\tan 52° = \dfrac{h}{9.6}$

$h = 9.6 \times \tan 52°$

$= 12.2874396\ldots$

$= 12.3$ (1 d.p.)

The height of the building is 12.3 m, correct to one decimal place.

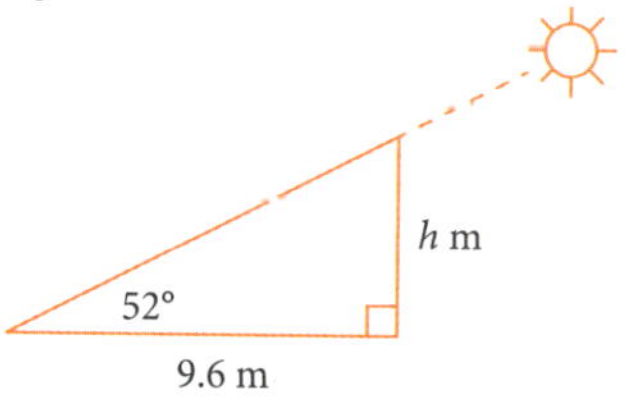

5 Navigational methods ➲ pp. 43–44

1 a **True bearing of P from Q is 058°.**

b **The compass bearing of Q from P is S58°W.**

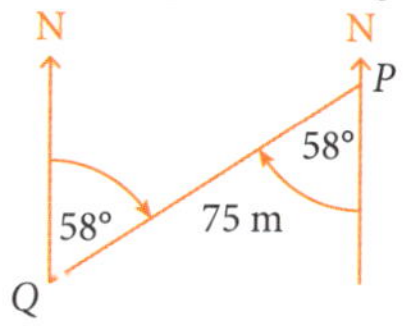

2 Let $x°$ be the size of the angle between due south and the line joining C to D.

Now $x = 265 - 180$

$= 85$

So a true bearing of 265° is equivalent to a compass bearing of S85°W.

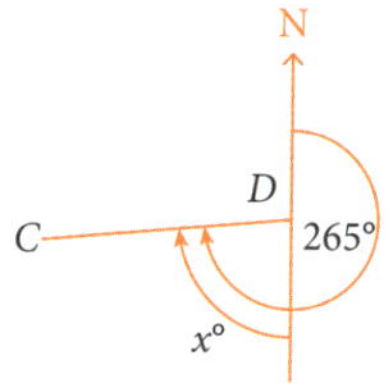

3 a

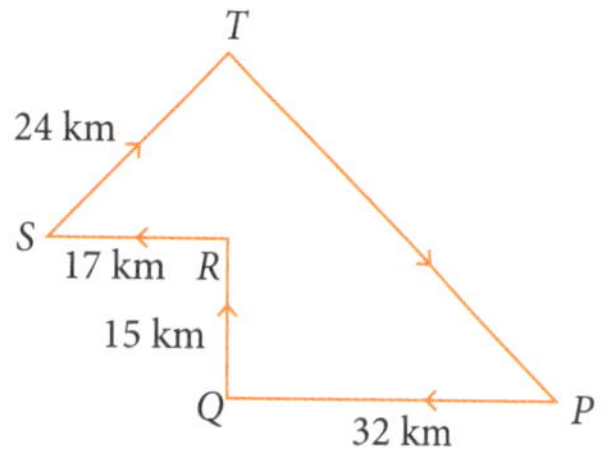

b **The helicopter flew south-east on its last leg.**

4 Let P be a point due south of Y.

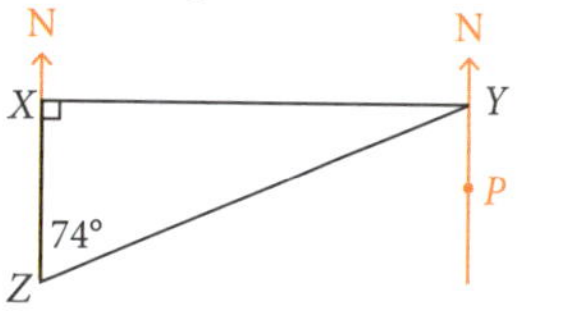

$\angle PYZ = \angle XZY$ (alternate, parallel lines)

$\therefore \angle PYZ = 74°$

The bearing of Z from Y is 180° + 74° or 254°.

5 a

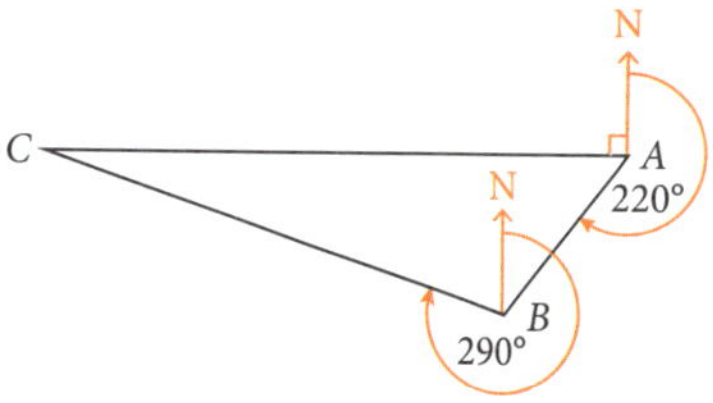

b C is due west of A

$\therefore$ the bearing of C from A is 270°.

$\angle CAB = 270° - 220°$

$= \mathbf{50°}$

c Let X be a point due north of A, and let Y be a point due north of B.

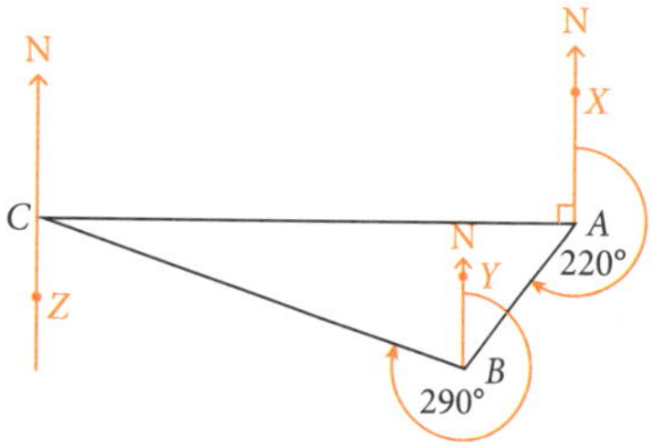

$\angle XAB = 360° - 220°$

$= 140°$

$\angle XAB + \angle YBA = 180°$ (co-interior angles, parallel lines)

$\angle YBA = 180° - 140°$

$= 40°$

$\angle CBY = 360° - 290°$

$= 70°$

$\angle CBA = 40° + 70°$

$= \mathbf{110°}$

d Let Z be a point due south of C.

$\angle ZCB = \angle CBY$ (alternate angles, parallel lines)

$\angle ZCB = 70°$

The compass bearing of B from C is S70°E.

6 a

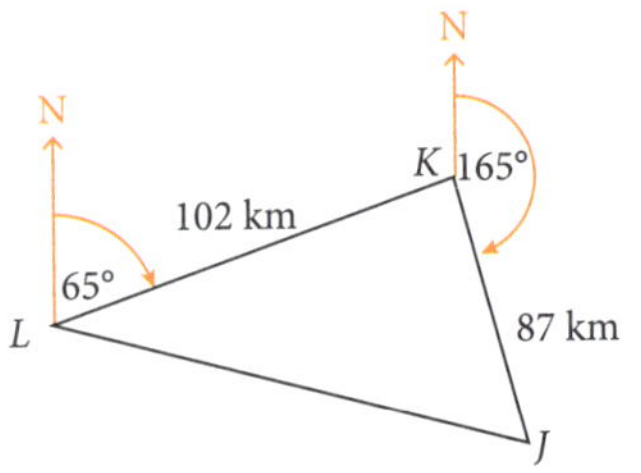

b Let X be a point due north of K and let Y be a point due north of L.

$\angle XKL + \angle KLY = 180°$ (co-interior angles, parallel lines)

$\angle XKL = 180° - 65°$

$= 115°$

$\angle LKJ + 115° + 165° = 360°$ (angles at a point)

$\angle LKJ = 360° - 280°$

$= \mathbf{80°}$

6 Problems involving bearings ➲ p. 45

1 a $\angle QPR = 270° - 240°$

$= \mathbf{30°}$

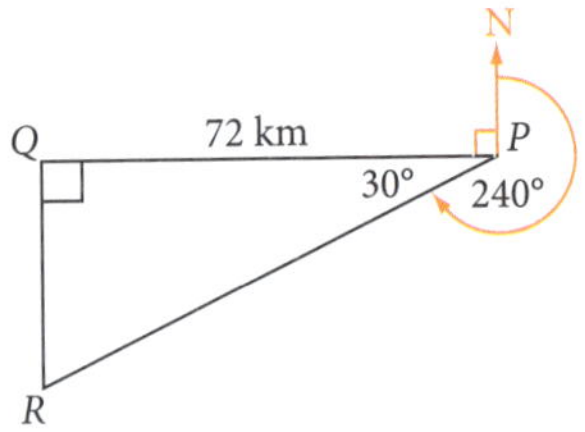

b $\tan 30° = \dfrac{QR}{72}$

$QR = 72 \times \tan 30°$

$= 41.569\,2193\ldots$

$= 42$ (nearest unit)

The distance between Q and R is 42 km, to the nearest kilometre.

2 Let θ be the angle between BQ and due north at Q.

$\therefore \angle PBQ = \theta$ (alternate angles, parallel lines)

$\tan \theta = \dfrac{84}{63}$

$\theta = \tan^{-1}\left(\dfrac{84}{63}\right)$

$= 53°7'48.37''$

$= 53°8'$ (nearest minute)

The helicopter should fly N53°8'W.

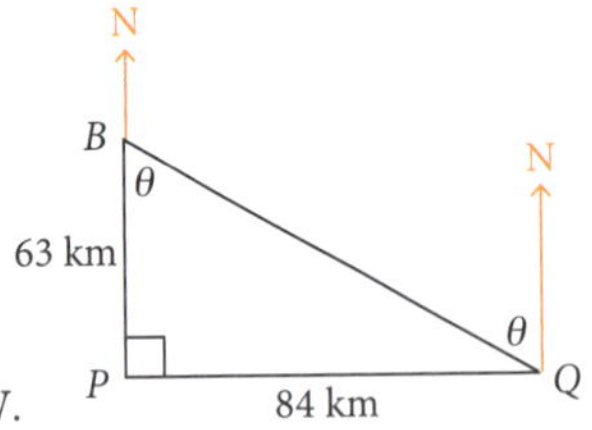

3 a

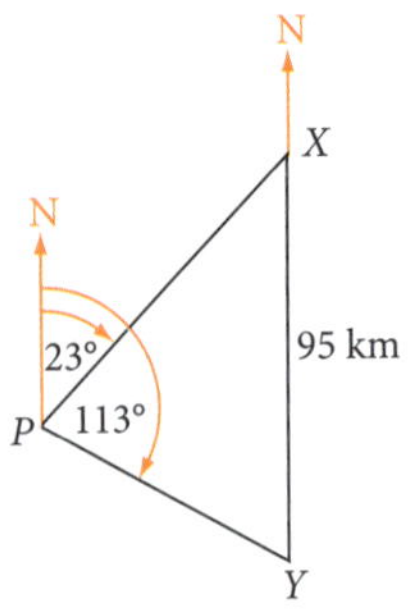

b $\angle XPY = 113° - 23°$

$= \mathbf{90°}$

c $\angle PXY = 23°$ (alternate angles, parallel lines)

$\cos 23° = \dfrac{PX}{95}$

$PX = 95 \times \cos 23°$

$= 87.447\,9610\ldots$

$= 87$ (nearest unit)

The distance between P and X is 87 km, to the nearest kilometre.

4

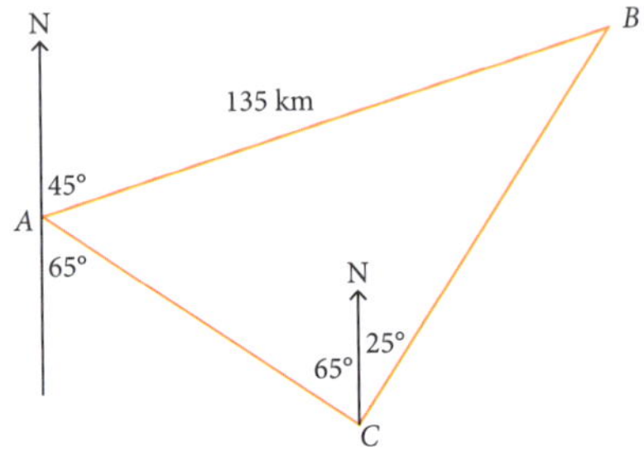

$\angle ACB = 65° + 25°$

$= 90°$

$\angle BAC = 180° - (45 + 65)°$

$= 70°$

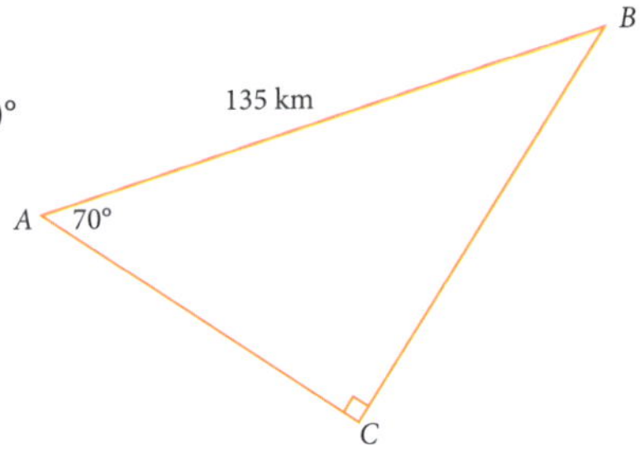

$\cos 70° = \dfrac{AC}{135}$

$AC = 135 \times \cos 70°$

$= 46.172\,7193\ldots$

$= 46.2$ (1 d.p.)

So C is 46.2 km from A to one decimal place.

HSC EXAM-TYPE QUESTIONS

Part A

1 C. $k^2 = 37^2 - 14^2$

$= 1173$

$k = \sqrt{1173}$

$= 34.249\,0875\ldots$

$= 34$ m (nearest metre)

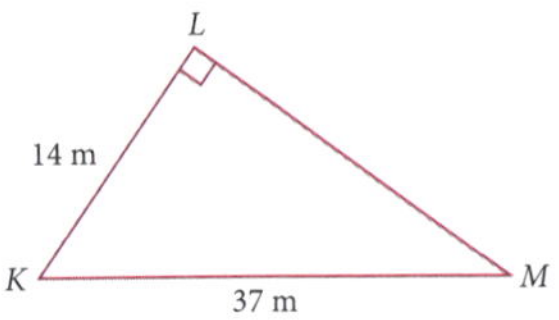

2 D. S40°E

Now $180° - 40° = 140°$

So S40°E is equivalent to a true bearing of 140°.

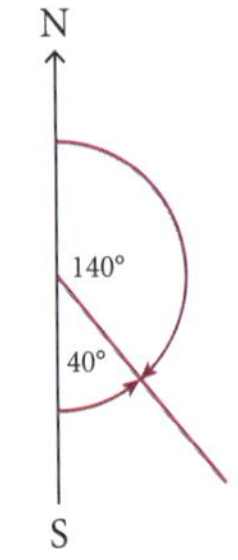

3 B. $\sin\theta = 0.704$

$\theta = \sin^{-1}(0.704)$

$= 44°44'55.72''$

$= 44°45'$ (nearest minute)

4 B. $\angle PQR = 143° - 25°$

$= 118°$

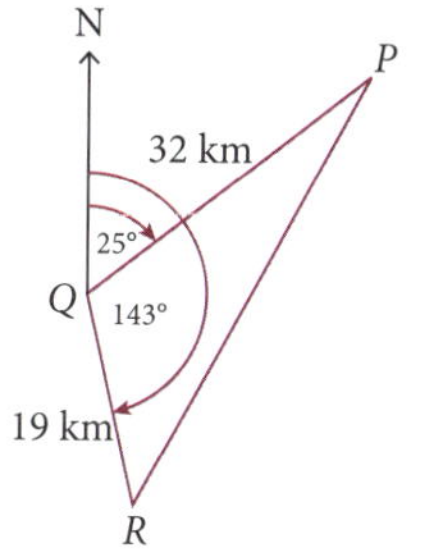

5 D. $\cos\theta = \dfrac{9}{14}$

$\theta = \cos^{-1}\left(\dfrac{9}{14}\right)$

$= 49.994\,7991\ldots°$

$= 50°$ (nearest degree)

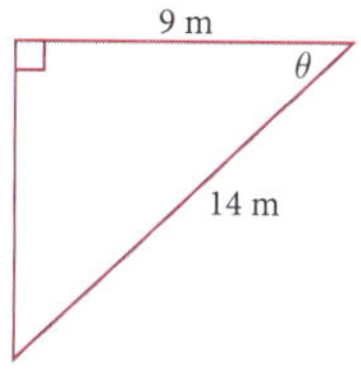

6 A. Let x km be the distance to fly home.

$x^2 = 28^2 + 195^2$

$= 38\,809$

$x = \sqrt{38\,809}$

$= 197$

The plane must fly 197 km home.

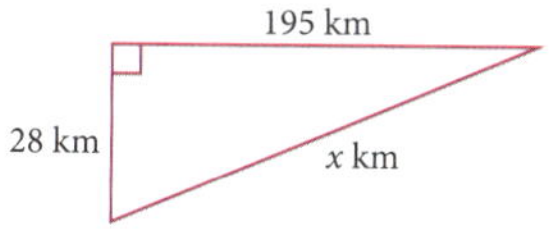

7 B. $\sin 42° = \dfrac{QR}{83}$

$QR = 83 \times \sin 42°$

$= 55.537\,8403\ldots$

$= 56$ (nearest unit)

The length of QR is 56 m to the nearest metre.

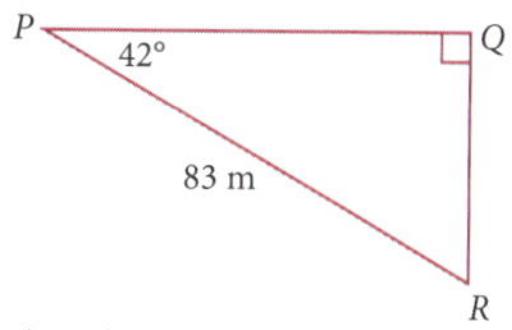

8 A.

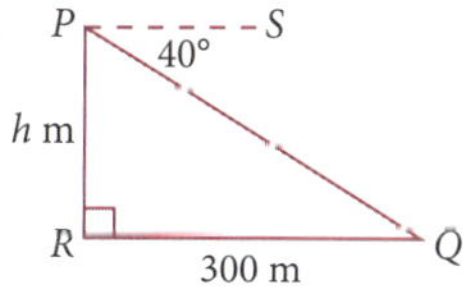

$\angle PQR = \angle SPQ$ (alternate angles, parallel lines)

$\angle PQR = 40°$

$\tan 40° = \dfrac{h}{300}$

$\therefore h = 300 \tan 40°$

Part B

9 Let x cm be the length of the hypotenuse.

$\cos 51° = \dfrac{17}{x}$

$x = \dfrac{17}{\cos 51°}$ ✓

$= 27.013\,2673\ldots$

$= 27$ (nearest unit)

The length of the hypotenuse is 27 cm, to the nearest centimetre. ✓

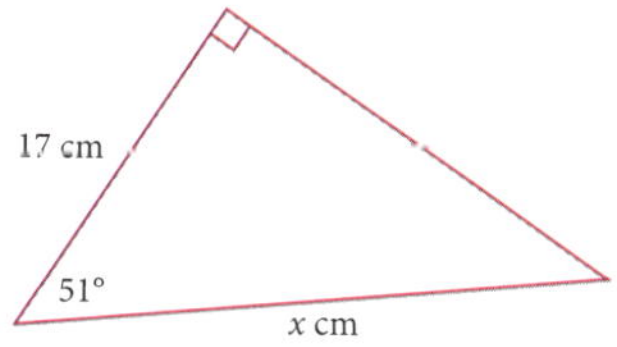

10 $\sin A = \dfrac{171}{221}$

$A = \sin^{-1}\left(\dfrac{171}{221}\right)$

$= 50°41'32.47''$ ✓

$= 50°42'$ (nearest minute)

So $\angle BAC = 50°42'$ to the nearest minute. ✓

(Either cos or tan could have been used instead of sin.)

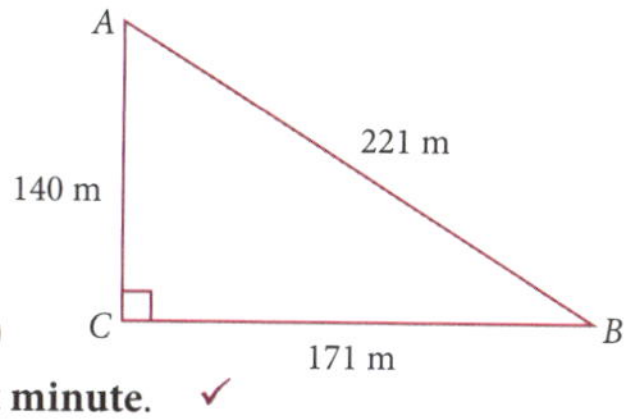

11 Let the height of the building be x m.

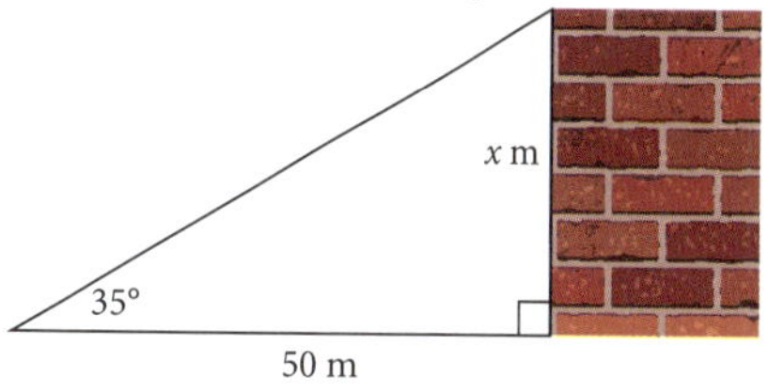

$\tan 35° = \dfrac{x}{50}$

$x = 50 \times \tan 35°$ ✓

$= 35.010\,3769\ldots$

$= 35$ (nearest unit)

The building is 35 m high, to the nearest metre. ✓

12 a $\angle ABC = 324° - 234°$

$= \mathbf{90°}$ ✓

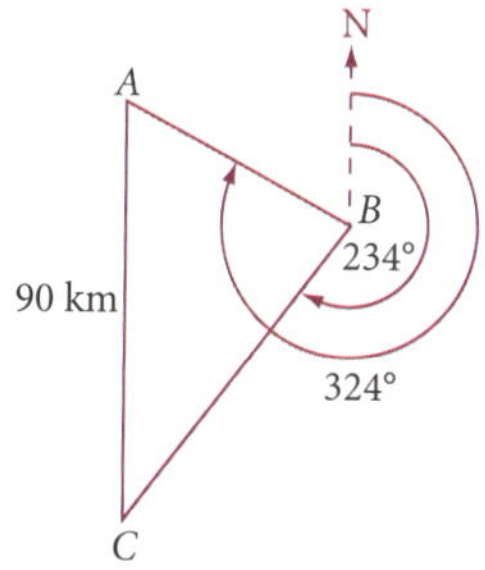

b Let X be a point due north of B.

$\angle XBA = 360° - 324°$

$= 36°$

$\angle BAC = \angle XBA$ (alternate angles, parallel lines)

$\angle \mathbf{BAC = 36°}$ ✓

c $\sin 36° = \dfrac{BC}{90}$

$BC = 90 \times \sin 36°$

$= 52.900\,6727\ldots$

$= 53$ km (nearest km)

The distance from B to C is 53 km, to the nearest kilometre. ✓

13 Let the height of the triangle be h m.

$h^2 = 89^2 - 80^2$

$= 1521$

$h = \sqrt{1521}$

$= 39$

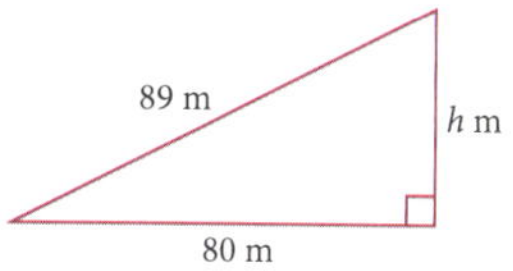

So the height of the triangle is 39 m. ✓

$A = \dfrac{1}{2}bh$

$= \dfrac{1}{2} \times 80 \times 39$

$= 1560$

The area of the triangle is 1560 m^2. ✓

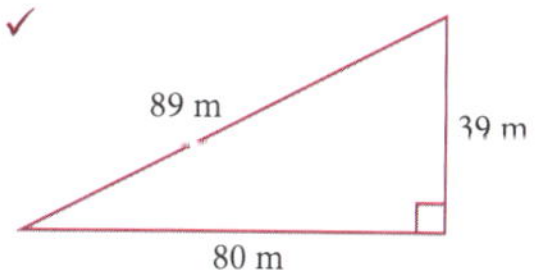

14 a In ΔABC,

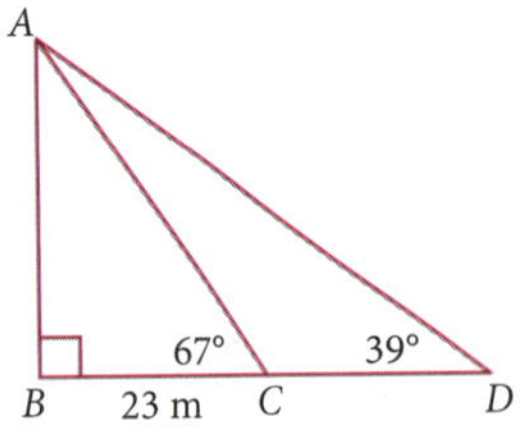

$$\tan 67° = \frac{AB}{23}$$

$$AB = 23 \times \tan 67°$$
$$= 54.1846044\ldots$$
$$= 54.2 \quad (1\text{ d.p.})$$

The length of *AB* is 54.2 m to one decimal place. ✓

b In ΔABD,

$$\tan 39° = \frac{AB}{BD}$$

$$BD = \frac{AB}{\tan 39°}$$
$$= 66.9124139\ldots\text{ m} \quad ✓$$

$$CD = BD - BC$$
$$= 66.9124139\ldots - 23$$
$$= 43.9124139\ldots$$
$$= 44\text{ m} \quad (\text{nearest metre})$$

The length of *CD* is 44 m, to the nearest metre. ✓

15

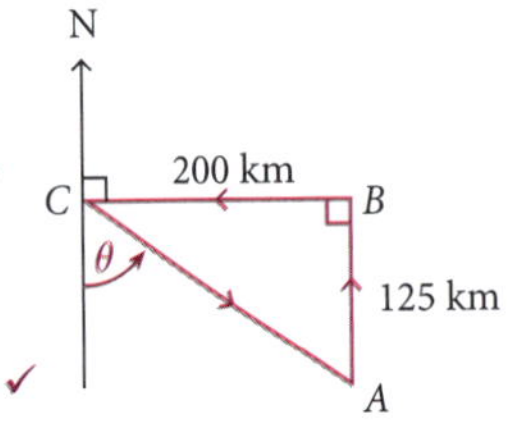

$$\tan C = \frac{125}{200}$$

$$C = 32.0053832\ldots°$$

$\therefore \angle BCA = 32°$ to the nearest degree. ✓

$$\theta = 90° - 32°$$
$$= 58°$$

The helicopter will need to fly S58°E. ✓

16 a In $\angle ABD$,

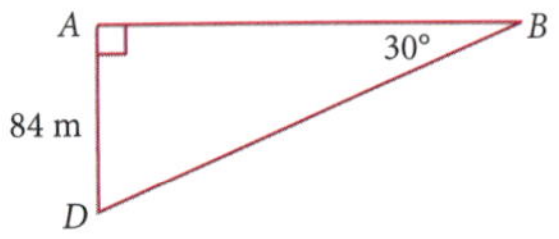

$$\sin 30° = \frac{84}{BD} \quad ✓$$

$$BD = \frac{84}{\sin 30°}$$
$$= 168\text{ m} \quad ✓$$

In $\angle BDC$,

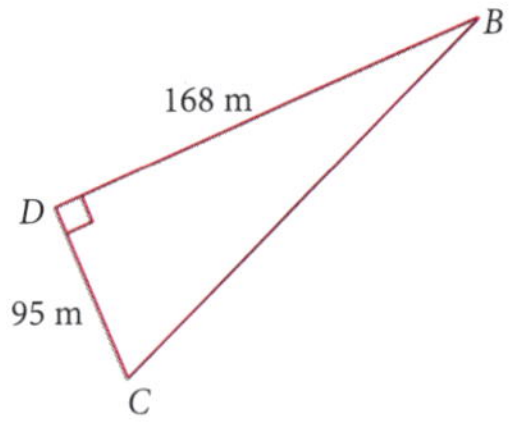

$$BC^2 = 168^2 + 95^2$$
$$= 37\,249$$
$$BC = \sqrt{37\,249}$$
$$= 193$$

So the length of *BC* is 193 m. ✓

b In $\angle BDC$,

$$\tan B = \frac{95}{168}$$

$$B = \tan^{-1}\left(\frac{95}{168}\right) \quad ✓$$
$$= 29.4871256\ldots°$$
$$= 29° \quad (\text{nearest degree})$$

$\angle DBC = 29°$ to the nearest degree. ✓

But $\angle ABD = 30°$.

So $\angle DBC$ is slightly smaller than $\angle ABD$. ✓

CHAPTER 3 RATES

TOPIC: MEASUREMENT

1 Using rates

- **Rates compare quantities of different kinds.**
- **A rate is usually expressed as the amount of one item** for each unit of the other.
- **There are many examples of different types of rates,** most with practical applications.

EXAMPLE 1

It takes half an hour to fill with water a tank that holds 600 L. At what rate, in litres per minute, is the water flowing?

Rate = 600 L ÷ 30 min
= 20 L/min

Many rates are expressed per unit of time. Take care converting between units of time.

EXAMPLE 2

A patient must receive 1.5 L of fluid at a flow rate of 125 mL/h. How long will this take?

1.5 L = 1500 mL
Time taken = (1500 ÷ 125) h
= 12 h

It is important to understand what is required. A flow rate of 125 mL/h means that every hour 125 mL flows to the patient. We need to know how many hours it will take for 1500 mL to flow to the patient.

EXAMPLE 3

The population of a country is 24.64 million. The country has an area of 7.7 million square kilometres. What is the population density in people per square kilometre?

Population density = (24.64 ÷ 7.7) people/km^2
= 3.2 people/km^2

Read the question carefully. A rate is not the same if expressed the other way around. The square kilometres per person is different to the people per square kilometre.

EXAMPLE 4

A swimming pool that was previously empty was filled in 5 hours 25 minutes. If the water was flowing at 72 L/min, how much water did the pool hold?

5 h 25 min = (5 × 60 + 25) min
= 325 min
Amount of water = 325 × 72 L
= 23 400 L

Every minute 72 L flowed into the pool. So you need to multiply the number of minutes by 72.

→ PRACTICE

1. **It takes a quarter of an hour to empty a drum filled with 360 L of water. At what rate, in litres per minute, is the water flowing?**
2. **A patient must receive 2.4 L of fluid at a flow rate of 150 mL/h. How long will this take?**
3. **The population of a country is 127.4 million. The country has an area of 378 000 square kilometres. What is the population density in people per square kilometre?**
4. **A swimming pool that was previously full was drained in 7 hours 35 minutes. If the water was flowing at 75 L/min, how much water did the pool hold?**
5. * **A patient must receive 0.2 g of a drug. The available medication has 50 mg in 2 mL. How many millilitres of the medication should be given?**

Answers ➲ p. 64

2 Best buys

- **The price of an item is often expressed as a rate.**
- **Finding the best buy is an application of rates.**
- **Best buys are determined by comparing unit prices.**

EXAMPLE 1

Kiwi fruit are selling for $6.80/kg. What will you pay for 2.35 kg?

Price paid = 2.35 × $6.80
= $15.98

The rate is the price per kilogram.

EXAMPLE 2

Find the price per litre if you buy 825 mL for $3.96.

825 mL = 0.825 L
Price per litre = $3.96 ÷ 0.825 L
= $4.80/L

Knowing the price per litre, comparisons could be made with other buys.

EXAMPLE 3

Which is the better buy: 3.5 kg for $5.95 or 4.5 kg for $8.10?

3.5 kg for $5.95 = 1 kg for $(5.95 ÷ 3.5)
= $1.70/kg
4.5 kg for $8.10 = 1 kg for $(8.10 ÷ 4.5)
= $1.80/kg
So 3.5 kg for $5.95 is the better buy.

The better buy is the cheaper of the two.

EXAMPLE 4

Which is the best buy?
- **600 mL for $1.95**
- **1.25 L for $4.55**
- **2.4 L for $8.15**
- **3.5 L for $11.55**

600 mL for $1.95 = $3.25/L
1.25 L for $4.55 = $3.64/L
2.4 L for $8.15 = $3.395 8333…/L
3.5 L for $11.55 = $3.30/L
The best price per litre is $3.25.
The best buy is 600 mL for $1.95.

600 mL = 0.6 L;
$1.95 ÷ 0.6 = $3.25

➔ PRACTICE

1 **Bananas are selling for $7.50/kg. What will you pay for 2.1 kg?**

2 **Find the price per litre if you buy 600 mL for $2.70.**

3 **Which is the better buy: 2.8 kg for $9.66 or 3.6 kg for $12.24?**

4 **Which is the best buy?**
- **400 mL for $1.56**
- **1.25 L for $4.75**
- **2.75 L for $10.12**
- **6 L for $22.50**

5* **What is the cheapest way of buying 10 rolls if they cost $23.75 each or can be bought in packs of four for $91.00 or $139.95 for half a dozen?**

Answers ➲ p. 64

3 Costs involving rates

➔ **The costs associated with different types of work** are often quoted as rates.

➔ **A cost for a tradesperson might involve a fixed amount**, such as a call-out fee, plus a rate per a period of time. There might be a travel component if the person has to travel some distance to reach the job.

EXAMPLE 1

A company advertised timber flooring at $112 per square metre.

a **How much would you need to pay for this type of flooring for a room 3.2 m long and 3 m wide?**

b **For a slightly different type of flooring the company charged $1771 for a 4.4 m by 3.5 m room. What is the cost per square metre?**

a Area = 3.2 × 3
= 9.6 m^2
Cost = $112 × 9.6
= $1075.20

b Area = 4.4 × 3.5
= 15.4 m^2
Rate = $1771 ÷ 15.4
= $115 per m^2

In practice, costs quoted per square metre might need to be rounded up to avoid, for example in flooring, cuts and joins.

EXAMPLE 2

A plumber charges a call-out fee of $75 plus a rate of $120 per hour or part thereof. How much would Amy pay for a plumber to come and fix an issue that takes 3 hours and 20 minutes?

Cost = \$75 + 4 × \$120
= \$555

Amy would need to pay \$555 in total.

> Although the time is closer to 3 hours than 4 hours, part of the 4th hour has been taken up so 4 hours are charged for.

EXAMPLE 3

A taxi company charges a \$2.50 booking fee, \$3.60 flag fall plus \$2.20 per kilometre. There is also a wait time cost of 95 cents per minute. Tilly books a taxi to take her from her home to her bank, a distance of 6.7 km, and return home. If the taxi had to wait 7 minutes at the bank, what would be the total cost to Tilly?

Booking fee = \$2.50
Travel cost = \$3.60 + 2 × 6.7 × \$2.20
= \$33.08
Wait cost = 7 × \$0.95
= \$6.65
Total cost = \$2.50 + \$33.08 + \$6.65
= \$42.23

> Flag fall is the fixed part of the travel cost charged by taxis and some other businesses.

→ PRACTICE

1 **A company advertised carpet at \$75 per square metre.**
   a **How much would you need to pay for this type of flooring for a room 4.25 m long and 3.8 m wide?**
   b **For a slightly different type of carpet the company charged \$1768 for a 5.2 m by 4 m room. What is the cost per square metre?**

2 **An electrician charges a call-out fee of \$60 plus a rate of \$125 per hour or part thereof. How much would Andy pay for an electrician to come and fix an issue that takes an hour and a quarter?**

3 **A taxi company charges a \$2.50 booking fee, \$3.00 flag fall and \$2.25 per kilometre. There is also a wait time cost of 90 cents per minute. Hugo booked a taxi to travel 8 km to collect a friend (where there was a wait of 9 minutes) and then take both to the airport a further 17 km. What would be the total cost to Hugo?**

4* **A vet charges \$17.50 per quarter-hour or part thereof, plus \$3 per kilometre to travel. To treat a horse she travelled 56 km and charged a total of \$509, including an amount of \$148.50 for medication. For how long did she charge for treating the horse?**

Answers ➲ p. 64

4 Speed

→ **Average speed, S, is the rate at which distance, D, is covered over time T.**

→ $S = \dfrac{D}{T}$

→ **Average speed differs from instantaneous speed, which is the actual speed at any given time.**

EXAMPLE 1

A car left at 8:38 am and travelled 126 km arriving at 10:23 am. What was the average speed in kilometres per hour?

Time taken = 1 h 45 min
$= 1\frac{3}{4}$ h
Average speed = 126 km ÷ $1\frac{3}{4}$ h
= 72 km/h

> Remember: 1 h 45 min is not 1.45 h.
> 1 h 45 min = $1\frac{3}{4}$ h = 1.75 h

EXAMPLE 2

A train leaves town A at 5:25 pm and travels at an average speed of 116 km/h to town B, a distance of 261 km. At what time will it arrive?

Time taken = (261 ÷ 116) h
= 2.25 h
= 2 h 15 min

2 h 15 min after 5:25 pm is 7:40 pm.
So the train will arrive at 7:40 pm.

> Rearranging the formula, time equals distance divided by speed.

EXAMPLE 3

Joe drew this travel graph of his recent trip for work.

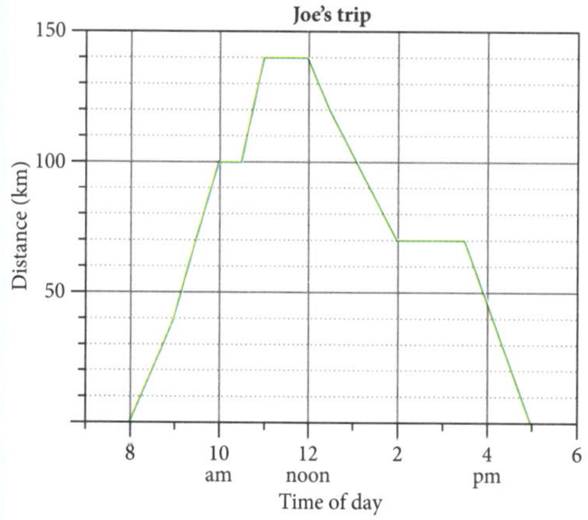

a **Joe was stopped for 3 hours altogether. How is this shown on the graph?**

b **What was Joe's average speed for the total time he was travelling?**

c Joe said that his speed at 9 am was 40 km/h. Is that correct? Justify your answer.

d Between what times was Joe travelling at the fastest average time?

a The times that Joe is stopped are shown by horizontal lines on the graph.

The distances stay the same over small periods of time.

b From 8 am until 5 pm is 9 hours.

So Joe was travelling for 6 hours altogether.

Total distance = 2 × 140 km
= 280 km

Average speed = 280 ÷ 6
= $46.\dot{6}$ km/h

c Joe averaged 40 km/h between 8 am and 9 am but the graph does not give any information about the exact speed at 9 am. Joe might or might not be correct.

d Between 10:30 am and 11:00 am

The steepest part of the graph is where the speed is greatest, so we just consider where the gradient is greatest (either positive or negative).

EXAMPLE 4

Mary ran 100 m in 14.4 seconds. What was her average speed in kilometres per hour?

Speed = (100 ÷ 14.4) m/s
= 6.944 444… m/s
= 416.6666… m/min
= 25 000 m/h
= 25 km/h

To convert m/s to km/h we multiply by 3.6. We multiply by 60 to change seconds to minutes, multiply by 60 again to change minutes to hours and then divide by 1000 to change metres to kilometres. 60 × 60 ÷ 1000 = 3.6. It is important to understand what you are doing rather than just remembering to multiply by 3.6.

EXAMPLE 5

A car is travelling at 110 km/h. If the driver is distracted for 5 seconds, how many metres will the car travel in that time?

To change km/h to m/s multiply by 1000 to change kilometres to metres and then divide by 3600 (60 × 60).

Speed = 110 km/h
= 110 000 m/h
= 30.5555… m/s

Distance travelled = 5 × 30.5555… m
= 152.7777… m

The car will travel about 153 m.

→ PRACTICE

1 **A car left at 2:15 pm and travelled 143 km, arriving at 4:05 pm. What was the average speed in kilometres per hour?**

2 **A train leaves town A at 7:35 am and travels at an average speed of 96 km/h to town B, a distance of 312 km. At what time will it arrive?**

3 **Jo drew this travel graph of her recent trip for work.**

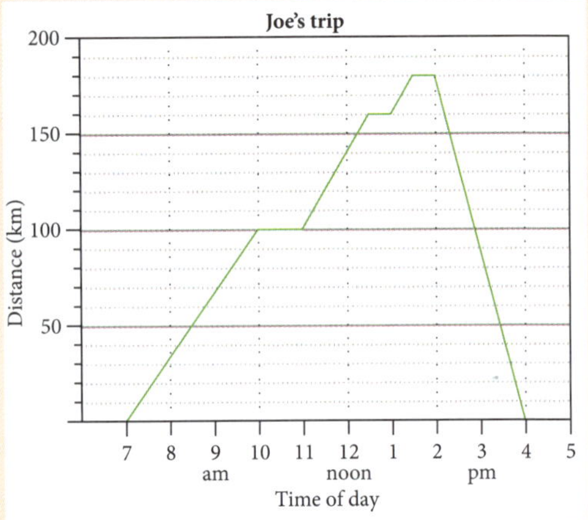

a **Jo was stopped for 2 hours altogether. How is this shown on the graph?**

b **What was Jo's average speed for the total time she was travelling?**

c **Jo said that her speed at 12 noon was 140 km/h. Is that correct? Justify your answer.**

d **Between what times was Jo travelling at the fastest average time?**

4 **Davy ran 200 m in 30 seconds. What was his average speed in kilometres per hour?**

5 **A car is travelling at 100 km/h. If the driver is distracted for 6 seconds, how many metres will the car travel in that time?**

6* **Jilly goes on a car trip. She averages 75 km/h for the first 180 km and 67.5 km/h for the remaining 135 km. What was the average speed for the trip?**

Answers ➲ pp. 64–65

5 Fuel consumption

→ **Fuel consumption is usually given in litres used per 100 km travelled.**

→ **The lower the amount of fuel used, the better the fuel consumption.**

→ **Fuel consumption is often quoted differently for highway travel and city travel.**

EXAMPLE 1

A car has a fuel consumption rate of 7.4 L/100 km. The capacity of the fuel tank is 58 L. How far, to the nearest 10 km, will the car travel on a full tank of fuel?

Distance travelled = (58 ÷ 7.4) × 100 km
= 783.783 783… km
= 780 km (nearest 10 km)

We know that 7 × 8 = 56 so we should have a rough idea that the distance travelled will be about 800 km.

EXAMPLE 2

A motorist fills her car and sets the trip meter to record how far she travels. When she next fills the car she sees that it has travelled 673 km. She pays $59.40 for petrol at $1.285 per litre. What was her car's fuel consumption?

673 km = 6.73 × 100 km
Amount of fuel bought = (59.40 ÷ 1.285) L
= 46.225 68… L
Fuel consumption = (46.225 68… ÷ 6.73) L/100 km
= 6.868 60… L/100 km
= 6.9 L/100 km (1 d.p.)

So for each 100 km travelled she used about 6.9 L of petrol.

EXAMPLE 3

Two similarly sized and priced cars have claimed fuel consumption of 5.8 L/100 km and 7.4 L/100 km. How much more fuel would be used by the second car over a year if both travelled 18 000 km in that time?

Difference = (7.4 – 5.8) L/100 km
= 1.6 L/100 km
18 000 km = 180 × 100 km
Extra fuel used = 180 × 1.6 L
= 288 L

The amount of fuel used by each car could be found and the difference calculated but it is simpler to find the difference in the rate of consumption first.

→ PRACTICE

1 A car has a fuel consumption rate of 7.2 L/100 km. The capacity of the fuel tank is 56 L. How far, to the nearest 10 km, will the car travel on a full tank of fuel?

2 A motorist fills his car and sets the trip meter to record how far he travels. When he next fills the car he sees that it has travelled 735 km. He pays $65.60 for petrol at $1.332 per litre. What was his car's fuel consumption?

3 Two similarly sized and priced cars have claimed fuel consumption of 7.6 L/100 km and 6.2 L/100 km. How much less fuel would be used by the second car over a year if both travelled 32 000 km in that time?

4* The claimed petrol consumption of a new car was 6.9 L/100 km for highway and 8.2 L/100 km for city travel. How much further should the car travel on a full tank of 78 L of petrol if it travelled on highways rather than in the city?

Answers ➲ p. 65

6 Heart rates

- **A person's heart rate is also known as their pulse.**
- **Heart rates are measured in beats per minute (bpm).**
- **The resting heart rate is the person's usual heart rate when they are at rest.** The normal range for adults is between 60 and 100 beats per minute, but can be as low as 40 bpm.
- **Heart rates can be affected by fitness levels, physical activity, age and emotions.** The temperature and medication can also have impact.
- **A person's maximum heart rate is said to be 220 – age of the person.**
- **A target heart rate range is the range of heart rate values aimed to be achieved during exercise.**

EXAMPLE 1

Toby has a resting heart rate of 72 bpm. At this rate how many times will Toby's heart beat in a day?

Beats per day = 72 × 60 × 24
= 103 680
Toby's heart would beat 103 680 times in a day.

The heart won't always be beating at the resting rate, so this is only an estimate and probably at the lower end of the scale.

EXAMPLE 2

Susan is 40 years old and has a resting heart rate of 80 bpm. Her target heart rate during a certain exercise is between 75% and 85% of her maximum heart rate. What is Susan's target heart rate range?

Maximum heart rate = 220 − 40
= 180

Now 75% of 180 = 0.75 × 180 = 135

and 85% of 180 = 0.85 × 180 = 153

So Susan's target heart rate range is between 135 bpm and 153 bpm.

Achieving a target heart rate range maximises the benefits of exercise and not exceeding the range minimises the risk of overdoing it.

EXAMPLE 3

The maximum heart rate is said to be found by subtracting a person's age from 220. The heart rate reserve is found by subtracting the resting heart rate from the maximum heart rate. The targeted heart rate range is found by multiplying the heart rate reserve by the desired training intensity and then adding the resting heart rate. Bree has a resting heart rate of 65 beats per minute. If Bree is 28 years old and her desired training intensity is 40% to 60%, find her targeted heart rate range.

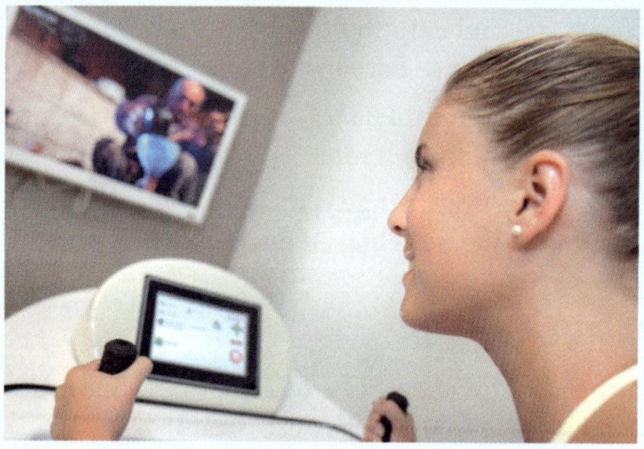

Maximum heart rate = 220 − 28
= 192 bpm

Heart rate reserve = 192 − 65
= 127 bpm

Minimum targeted heart rate = 127 × 0.4 + 65
= 115.8

Maximum targeted heart rate = 127 × 0.6 + 65
= 141.2

So Bree's targeted heart rate range is between about 116 bpm and 141 bpm.

Target heart rates are used as a measure of exercise intensity. Research shows that exercise is most effective when the heart rate is kept within the target range.

➔ PRACTICE

1 **Ruby has a resting heart rate of 81 bpm. At this rate how many times will Ruby's heart beat in a week?**

2 **Shaun is 30 years old and has a resting heart rate of 70 bpm. His target heart rate during a certain exercise is between 50% and 70% of his maximum heart rate. What is Shaun's target heart rate range?**

3 **The maximum heart rate is said to be found by subtracting a person's age from 220. The heart rate reserve is found by subtracting the resting heart rate from the maximum heart rate. The targeted heart rate range is found by multiplying the heart rate reserve by the desired training intensity and then adding the resting heart rate. Barney has a resting heart rate of 55 beats per minute. If Barney is 34 years old and his desired training intensity is 35% to 45%, find his targeted heart rate range.**

4* **Sam is 50 years old and has a resting heart rate of 68 bpm. During exercise he aims to have his heart rate between 102 and 136 bpm. What percentage range of his maximum heart rate is Sam aiming for?**

Answers ➲ p. 65

Note: Always rely on advice from medical professionals if you have any concerns about heart rates or blood pressure.

7 Blood pressure

➔ **Blood pressure measures the pressure of blood in the arteries.**

➔ **The unit of pressure is mmHg**; a millimetre of mercury.

➔ **A blood pressure reading is made up of two parts.** The first figure is the **systolic** blood pressure, which is a measure of the pressure while the heart pumps blood as it beats. The second figure is the **diastolic** blood pressure when the heart is between beats.

➔ **Blood pressure readings vary from day to day and fluctuate during the day.** A reading of 140 over 90 is considered high (hypertension) and a reading of 90 over 60 is considered low (hypotension). A normal reading would lie between these two.

EXAMPLE 1

Anna used a blood pressure monitor to take her blood pressure and pulse at a certain time every day for a fortnight. She drew this graph.

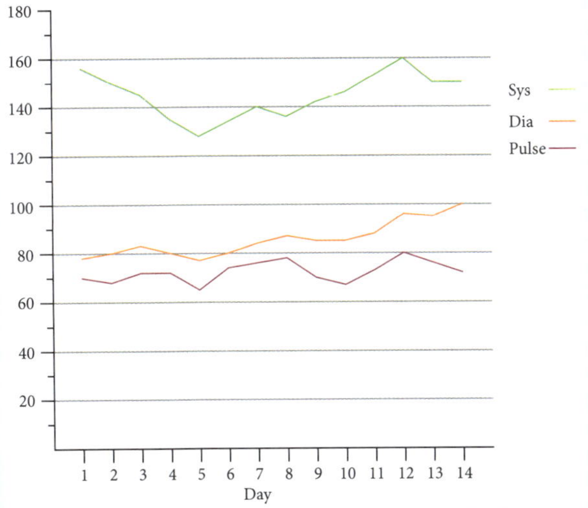

a **What was Anna's blood pressure on the last day?**

b **On what day was the biggest difference between the systolic and diastolic readings?**

c **What trends with Anna's blood pressure can you see from the graph?**

a 150 over 100

b The biggest difference was on the first day.

c The lowest blood pressure was on day 5. After that day her blood pressure on the whole steadily increased.

If blood pressure readings are being taken it should be under the same conditions every day.

EXAMPLE 2

A health company published this chart.

BLOOD PRESSURE TABLE			
Category	Systolic mm Hg		Diastolic mm Hg
Hypertensive emergency	> 180	or	> 110
Moderate hypertension	160–180	or	100–110
Mild hypertension	140–159	or	90–99
Prehypertension	121–139	or	81–89
Normal	80–120	and	60–80
Hypotension (Low)	< 80	or	< 60

a **According to the chart in what category is someone with blood pressure of 115 over 75?**

b **Jessica said: 'My blood pressure is 155 over 105 so I have mild hypertension'. Is Jessica correct? Justify your answer.**

c **Leon's blood pressure was 127 over 47. Briefly explain why the table is not appropriate to categorise Leon's blood pressure.**

a Normal

The systolic reading of 115 is between 80 and 120 and the diastolic reading of 75 is between 60 and 80.

b No, Jessica is not correct. The systolic reading would fall into the mild hypertension category but the diastolic reading means that Jessica would be categorised as having moderate hypertension.

c The table would categorise Leon's systolic reading as prehypertension but his diastolic reading as hypotension. As it is impossible to have both high and low blood pressure at the same time, trying to categorise Leon's blood pressure using the table is impossible.

→ PRACTICE

1 **Arnold used a blood pressure monitor to take his blood pressure and pulse at a certain time every day for a fortnight. He drew this graph.**

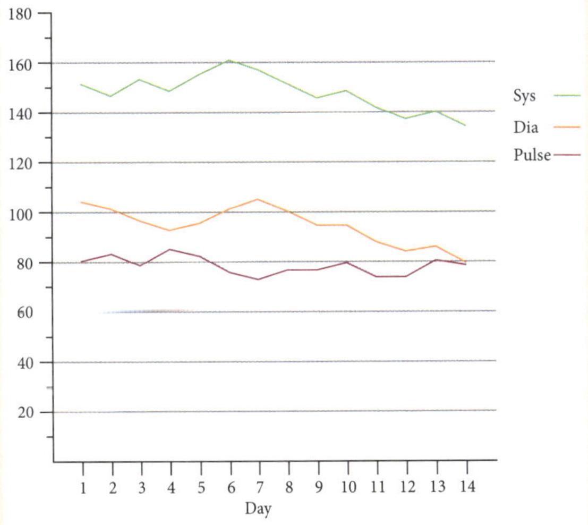

a **What was Arnold's blood pressure on the first day?**

b **On what day was the biggest difference between the systolic and diastolic readings?**

c **What trends with Arnold's blood pressure can you see from the graph?**

2 A health association published this table.

BLOOD PRESSURE READINGS			
Systolic (mm Hg)		Diastolic (mm Hg)	Recommended action
< 85	or	< 55	Too low See a doctor
85–129	and	55–85	Optimal No action required
130–139	or	85–89	Make lifestyle changes to develop heart-healthy habits
140–179	or	90–109	Too high See a doctor
over 180	or	over 110	Emergency medical attention required

a According to the chart what action is required for someone with blood pressure of 135 over 87?

b Jesse said: 'My blood pressure is 122 over 86 so that is optimal blood pressure. I don't need to take any action'. Is Jesse correct? Justify your answer.

c Leanne's blood pressure was 128 over 92. Briefly explain what action you would recommend that Leanne should take.

3* The graph shows the normal systolic and diastolic measures by age.

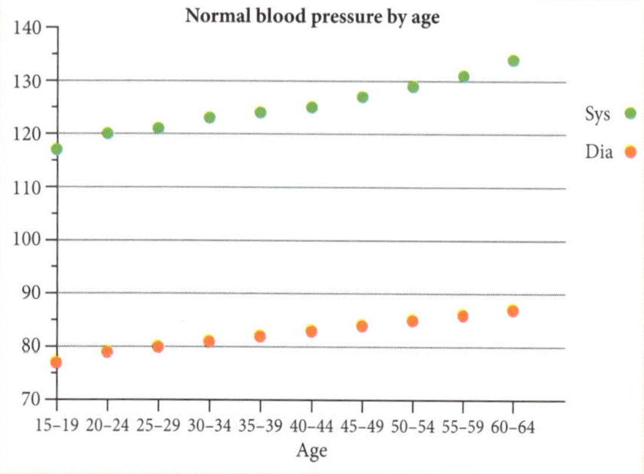

a What is the normal blood pressure of a 23 year old?

b For what age range is the normal blood pressure 129 over 85?

c Briefly comment on any observations that can be made.

Answers ➲ p. 65

8 Further applications of rates

- **There are many applications of rates in various fields;** for example, agriculture, health and energy efficiency.
- **In practical applications limits and restrictions need to be taken into account** and the reasonableness of answers should always be considered.

EXAMPLE 1

Lime is applied to a paddock at the rate of 2.4 tonnes per hectare. What would be the cost of lime needed to cover a 15-hectare paddock at $89 per tonne?

Amount of lime required = 15 × 2.4 t
= 36 t

Cost = 36 × $89
= $3204

Alternatively, the cost per hectare is 2.4 × $89 = $213.60. Multiply this by 15 to find the total cost.

EXAMPLE 2

A 300 mL bottle of vodka with lime and soda contains 4.5% alcohol per volume.

a How many millilitres of alcohol are in each bottle?

b If a standard drink contains 12 mL of alcohol, how many standard drinks are in each bottle?

a Amount of alcohol = 4.5% of 300 mL
= 13.5 mL

b Standard drinks per bottle = 13.5 ÷ 12
= 1.125

Many alcoholic drinks contain more than one 'standard' drink.

EXAMPLE 3

A patient receives an infusion over two hours through an IV set giving 60 drops/mL. The drip rate is 25 drops/min.

a How many drops will the patient receive?

b How much fluid will be given?

c If there is 250 mg of medication in 100 mL, how much medication will the patient receive?

a 2 hours = 120 minutes

Total drops = 25 × 120
= 3000

The patient will receive 3000 drops.

b Volume of fluid = (3000 ÷ 60) mL
= 50 mL

c 250 mg in 100 mL = 125 mg in 50 mL

So 125 mg of medication will be received by the patient.

The flow rate of an intravenous infusion is usually measured in millilitres per hour. The drip rate is the flow rate expressed in drops per minute.

EXAMPLE 4

Jana drew this table to show how much water she estimates that she uses each day.

Where used	Number of litres
Kitchen	25
Bathroom	110
Laundry	50
Outside	55
Other	10

a If Jana's water supply is a 10 000 L water tank, for how many days can Jana use water without the tank being refilled?

b Jana has a 4-minute shower every day. She estimates that her shower uses 15 L of water per minute. If she replaces the showerhead so that it uses 9 L per minute how many kilolitres of water would be saved in a year?

a Total litres per day = 25 + 110 + 50 + 55 + 10
= 250

Number of days usage = 10 000 ÷ 250
= 40

Without the tank being refilled, Jana can use water for 40 days.

b Saving per minute = (15 – 9) L
= 6 L

Saving per shower = 4 × 6 L
= 24 L

Yearly saving = 365 × 24 L
= 8760 L
= 8.76 kL

Remember: 1 kL = 1000 L.

Jana would save 8.76 kL in a year.

→ PRACTICE

1 A mineral mix is applied to a garden at the rate of 1.2 kilograms per square metre. What would be the cost of the mix needed to cover a 25 m^2 garden at $2.40/kg?

2 A 750 mL bottle of red wine contains 13% alcohol per volume.

a How many millilitres of alcohol are in each bottle?

b If a standard drink contains 12 mL of alcohol, how many standard drinks are in each bottle?

3 A patient receives an infusion over 10 hours through an IV set giving 20 drops/mL. The drip rate is 30 drops/min.

a How many drops will the patient receive?

b How much fluid will be given?

c If there is 125 mg of medication in 100 mL, how many grams of the medication will the patient receive?

4 Jono drew this table to show how much water he estimates that he uses each day.

Where used	Number of litres
Kitchen	20
Bathroom	125
Laundry	40
Outside	60
Other	30

a If Jono's water supply is a 33 000 L water tank, for how many days can Jono use water without the tank being refilled?

b Jono uses a hose for 3 minutes to rinse his car every day. He estimates that his hose uses 8 L of water per minute. If he replaces the nozzle on the hose so that it uses 5 L per minute, how many kilolitres of water would he save in a year?

5* A solution has 100 mg of active ingredient in 5 mL.

a How much of the active ingredient is in 1 L of solution?

b How many millilitres of the solution are needed to deliver 750 mg of active ingredient?

Answers ➲ p. 66

HSC EXAM-TYPE QUESTIONS

Total marks: 30

PART A

(Suggested time: 12 minutes)

(1 mark each)

1 A car uses 42.5 L of petrol to travel 625 km. What is the fuel consumption in litres/100 km?

A 6.8 B 7.3
C 8.4 D 14.7

2 A family has 25 000 L in their rainwater tank. If they use an average of 320 L per day, how many weeks will the water last if they receive no further rain?

A 8 B 11
C 13 D 18

3 James leaves home at 7.05 am and travels 234 km. When he arrives at his destination, James calculates that his average speed was 72 km/h. What time did he arrive?

A 11:15 am B 10:30 am
C 10:20 am D 10:15 am

4 A drug is to be given to a child weighing 46 kg. The drug is available in a solution with 30 mg per 5 mL. 1.5 mg per kg is ordered for the child. How many millilitres should be given to the child?

A 5.1 B 11.5
C 184 D 414

5 Mick is 56 years old and has a resting heart rate of 70 bpm. During exercise he aims to increase his resting heart rate by between 20 and 45 bpm. What percentage of his maximum heart rate is Mick's target heart rate range? (Maximum heart rate is 220 – age.)

A 50 to 65 B 55 to 65
C 50 to 70 D 55 to 70

6 Which is the best buy?

A 4 L for $14.80 B 600 mL for $2.25
C 1.5 L for $5.70 D 2.5 L for $9.00

7 Joe charges a call-out fee of $90 and an hourly rate of $80 per hour. Frank charges a call-out fee of $40 and $90 per hour. For how many hours will the cost be the same?

A 5 B 6
C 7 D 8

8 Referring to the chart, which statement is correct?

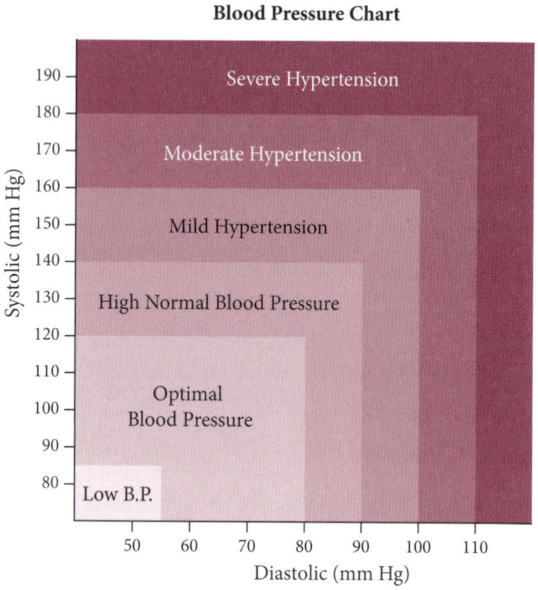

A Blood pressure of 147 over 92 is an example of high normal blood pressure.
B Blood pressure of 121 over 88 is an example of optimal blood pressure.
C Blood pressure of 156 over 99 is an example of mild hypertension.
D Blood pressure of 175 over 102 is an example of severe hypertension.

PART B

(Suggested time: 33 minutes)

Show all working.

9 Weeds on a paddock are to be sprayed using a chemical spray at the rate of 0.8 L per hectare. The cost per litre of the spray is $82.50.

- a What is the cost per hectare of the spray? (1 mark)
- b How many litres of spray will be needed to spray an area of 15 ha? (1 mark)
- c It takes 12 minutes per hectare to apply the spray and the total cost of machinery and labour has been calculated to be $93 per hour. Find the total cost to spray an area of 15 ha. (2 marks)

10 A car travels 465 km in 6 hours and 40 minutes.

- a What was the average speed? (1 mark)
- b What is the cost of petrol for the journey if the car's fuel consumption was 7.0 L/100 km and the petrol cost $1.40 per litre? (2 marks)

11 Angus charges a $30 call-out fee, $5 per kilometre to travel and $70 per hour or part thereof.
What will Angus charge for a job that takes $6\frac{1}{4}$ hours and involves travelling 48 km? (2 marks)

12 A patient is to receive 400 mg of a drug in 240 mL of solution. The drug needs to be delivered to the patient at the rate of 25 mg/h.

- a How many hours will the infusion take? (1 mark)
- b What is the flow rate in mL/h? (1 mark)
- c If the IV delivers at 20 drops/mL, what is the required drip rate in drops/min? (2 marks)

13
- a Convert 99 km/h to m/s. (1 mark)
- b At 99 km/h Kim's car will have a braking distance of 66 m. How many seconds will it take the car to stop from when the brakes are first applied? (1 mark)
- c The car is travelling at 99 km/h when Kim sees an accident ahead. If she takes 3 seconds to react, how far altogether will the car travel before stopping? (1 mark)

14 Andrea used a blood pressure monitor and it showed these results.

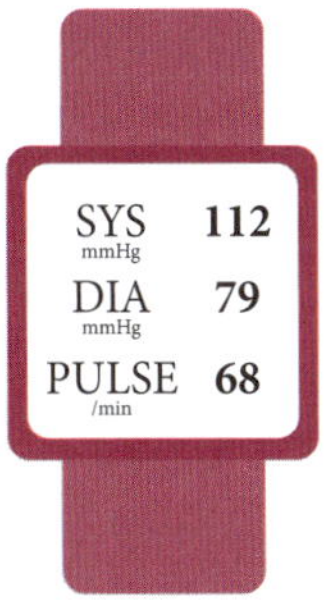

- a Would you describe Andrea's blood pressure as low, high or normal? (1 mark)
- b What is Andrea's heart rate? (1 mark)
- c During exercise Andrea's heart rate doubled. Is that a good thing? Comment. (1 mark)

15 At a cricket match each team batted for 20 overs. The first team scored 175 runs in their 20 overs.

- a What was the run rate in runs per over for the first team? (1 mark)
- b At the end of the eighth over, the second team were scoring at a rate of 6.625 runs per over. How many runs had they scored then? (1 mark)
- c At what rate did the second team need to score for the remaining overs to win the game? (1 mark)

ANSWERS

PRACTICE

1 Using rates ➲ p. 53

1 Rate = 360 L ÷ 15 min
= **24 L/min**

2 2.4 L = 2400 mL
Time = (2400 ÷ 150) h
= **16 h**

3 Population density = 127 400 000 ÷ 378 000
= 337.037 03…
= 337.0 people/km² (1 d.p.)
The population density is 337 people per square kilometre.

4 7 h 35 min = 455 min
Amount of water = 455 × 75 L
= **34 125 L**

5 0.2 g = 200 mg
= 4 × 50 mg
So required medication = 4 × 2 mL
= 8 mL
The patient should be given 8 mL of the medication.

2 Best buys ➲ p. 54

1 Price paid = 2.1 × $7.50
= **$15.75**

2 600 mL = 0.6 L
Price per litre = $2.70 ÷ 0.6 L
= **$4.50/L**

3 2.8 kg for $9.66 = 1 kg for $(9.66 ÷ 2.8)
= $3.45/kg
3.6 kg for $12.24 = 1 kg for $(12.24 ÷ 3.6)
= $3.40/kg
So 3.6 kg for $12.24 is the better buy.

4 400 mL for $1.56 = $3.90/L
1.25 L for $4.75 = $3.80/L
2.75 L for $10.12 = $3.68/L
6 L for $22.50 = $3.75/L
The best price per litre is $3.68.
The best buy is 2.75 L for $10.12.

5 4 for $91.00 = $22.75 each
6 for $139.95 = $23.325 each
So the cheapest cost will involve packs of 4.
Cost of 2 packs of 4 and 2 single rolls
= 2 × $91.00 + 2 × $23.75
= $229.50
Cost of 1 pack of 6 and 1 pack of 4
= $139.95 + $91.00
= $230.95
The cheapest way to buy 10 rolls is to buy 2 packs of 4 and 2 single rolls.

3 Costs involving rates ➲ p. 55

1 a Area = 4.25 × 3.8
= 16.15 m²
Cost = $75 × 16.15
= **$1211.25**

b Area = 5.2 × 4
= 20.8 m²
Rate = $1768 ÷ 20.8
= **$85 per m²**

2 Cost = $60 + 2 × $125
= $310
Andy would need to pay $310 in total.

3 Booking fee = $2.50
Travel cost = $3.00 + (8 + 17) × $2.25
= $59.25
Wait cost = 9 × $0.90
= $8.10
Total cost = $2.50 + $59.25 + $8.10
= **$69.85**

4 Travel cost = 56 × $3.00
= $168.00
Time charge = $509 − $168.00 − $148.50
= $192.50
Time = ($192.50 ÷ $17.50) quarter-hours
= 11 quarter-hours
= **2 h 45 min**

4 Speed ➲ p. 56

1 Time taken = 1 h 50 min
$= 1\frac{5}{6}$ h
Average speed = 143 km ÷ $1\frac{5}{6}$ h
= **78 km/h**

2 Time taken = (312 ÷ 96) h
= 3.25 h
= 3 h 15 min
3 h 15 min after 7:35 am is 10:50 am.
So the train will arrive at 10:50 am.

3 a **The times that Jo is stopped are shown by horizontal lines on the graph.**

b From 7 am until 4 pm is 9 h.
So Jo was travelling for 7 h altogether.
Total distance = 2 × 180 km
= 360 km
Average speed = 360 ÷ 7
= 51.428 571…
= **51.4 km/h** (1.d.p.)

c No, it is probably not correct. Jo was 140 km from home at 12 noon. She averaged 40 km/h between 11 am and 12:30 pm. **The graph does not give any information about the exact speed at 12 noon but as 140 km/h is well above the speed limit Jo is not likely to be correct.**

d **Between 2 pm and 4 pm**

4 Speed = (200 ÷ 30) m/s
= 6.666 666… m/s
= 400 m/min
= 24 000 m/h
= **24 km/h**

5 Speed = 100 km/h
= 100 000 m/h
= 27.777 77… m/s

Distance travelled = 6 × 27.777 77… m
= 166.666 66… m

The car will travel about 167 m.

6 Time for first part = (180 ÷ 75) h
= 2.4 h

Time for second part = (135 ÷ 67.5) h
= 2 h

Total time = 4.4 h

Total distance = (180 + 135) km
= 315 km

Average speed = (315 ÷ 4.4) km/h
= 71.590 909… km/h

The average speed for the trip was about 71.6 km/h.

5 Fuel consumption ➲ p. 57

1 Distance travelled = (56 ÷ 7.2) × 100 km
= 777.777 77… km
= **780 km** (nearest 10 km)

2 735 km = 7.35 × 100 km

Amount of fuel bought = (65.60 ÷ 1.332) L
= 49.249 2492… L

Fuel consumption = (49.249 2492… ÷ 7.35) L/100 km
= 6.700 5781… L/100 km
= **6.7 L/100 km** (1 d.p.)

3 Difference = (7.6 − 6.2) L/100 km
= 1.4 L/100 km

32 000 km = 320 × 100 km

Difference in fuel used = 320 × 1.4 L
= 448 L

So the second car would use about 448 L less fuel than the first car.

4 Distance travelled on highway
= (78 ÷ 6.9) × 100 km
= 1130.434 78… km

Distance travelled in the city
= (78 ÷ 8.2) × 100 km
= 951.219 51… km

Difference = 1130.434 78… − 951.219 51…
= 179.215 27… km

On a full tank the car should travel about an extra 180 km on highways than it would in the city.

6 Heart rates ➲ p. 58

1 Beats per week = 81 × 60 × 24 × 7
= 816 480

Ruby's heart would beat 816 480 times in a week.

2 Maximum heart rate = 220 − 30
= 190

Now 50% of 190 = 0.5 × 190 = 95

and 70% of 190 = 0.7 × 190 = 133

So Shaun's target heart rate range is between 95 bpm and 133 bpm.

3 Maximum heart rate = 220 − 34
= 186 bpm

Heart rate reserve = 186 − 55
= 131 bpm

Minimum targeted heart rate = 131 × 0.35 + 55
= 100.85

Maximum targeted heart rate = 131 × 0.45 + 55
= 113.95

So Barney's targeted heart rate range is between about 101 bpm and 114 bpm.

4 Maximum heart rate = 220 − 50
= 170

Now $\frac{102}{170} \times 100\% = 60\%$

and $\frac{136}{170} \times 100\% = 80\%$

So Sam is aiming for between 60% and 80% of his maximum heart rate.

7 Blood pressure ➲ pp. 59–60

1 a **150 over 105**

b **The biggest difference was on day 5 and day 6.**

c The highest blood pressure was around day 6. **After that day his blood pressure on the whole steadily decreased.**

2 a **Make lifestyle changes to develop heart-healthy habits.**

b No, Jesse is not quite correct. The systolic reading would fall into the optimal category but the diastolic reading just falls into the category where he should make lifestyle changes. **Jesse should consider developing heart-healthy habits.**

c **Leanne should see a doctor as her diastolic reading is too high.**

3

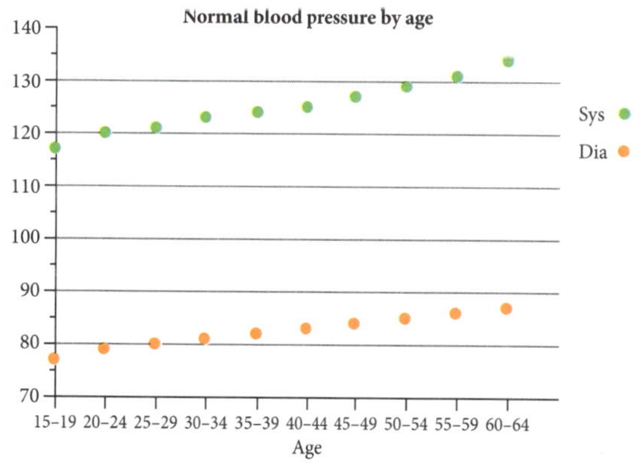

a **The normal blood pressure of a 23 year old is 120 over 79.**

b **The normal blood pressure is 129 over 85 for people in the 50–54 years age range.**

c Both the systolic and diastolic averages steadily increase as age increases. **The systolic readings increase slightly more quickly, meaning that the difference between the two readings also increases as age increases.**

8 Further applications of rates ➲ p. 61

1 Amount of the mix required
$= 25 \times 1.2$ kg
$= 30$ kg
Cost $= 30 \times \$2.40$
$= \mathbf{\$72.00}$

2 a Amount of alcohol = 13% of 750 mL
= **97.5 mL**

b Standard drinks per bottle $= 97.5 \div 12$
= **8.125**

3 a Total drops $= 30 \times 600$
$= 18\,000$

b Volume of fluid $= (18\,000 \div 20)$ mL
= **900 mL**

c 100 mL has 125 mg of medication.
So amount of medication in 900 mL
$= 9 \times 125$ mg
$= 1125$ mg
= **1.125 g**

4 a Total litres per day $= 20 + 125 + 40 + 60 + 30$
$= 275$
Number of days usage $= 33\,000 \div 275$
= **120**

b Saving per minute $= (8 - 5)$ L
$= 3$ L
Saving per day $= 3 \times 3$ L
$= 9$ L
Yearly saving $= 365 \times 9$ L
$= 3285$ L
$= 3.285$ kL
Jono would save 3.285 kilolitres in a year.

5 a 1 litre = 1000 mL
$= 200 \times 5$ mL
Amount of active ingredient $= 200 \times 100$ mg
$= 20\,000$ mg
= **20 g**

b 7.5×100 mg $= 750$ mg
So amount of solution $= 7.5 \times 5$ mL
= **37.5 mL**

HSC EXAM-TYPE QUESTIONS

Part A

1 A. Fuel consumption $= (42.5 \div 6.25)$ L/100 km
$= 6.8$ L/100 km

2 B. Number of days $= 25\,000 \div 320$
$= 78.125$
Now $78.125 \div 7 = 11.1607\ldots$
So the water will last 11 weeks.

3 C. Time taken $= (234 \div 72)$ h
$= 3.25$ h
$= 3$ h 15 min
Now 3 h and 15 min after 7:05 am is 10:20 am.
So James arrived at 10:20 am.

4 B. Mass required $= 46 \times 1.5$ mg
$= 69$ mg
Volume required $= \frac{69}{30} \times 5$ mL
$= 11.5$ mL

5 D. Maximum heart rate $= 220 - 56$
$= 164$ bpm
Low target heart rate $= 70 + 20$
$= 90$ bpm
High target heart rate $= 70 + 45$
$= 115$ bpm
Now $\frac{90}{164} \times 100\% = 54.878\ldots\%$
and $\frac{115}{164} \times 100\% = 70.121\ldots\%$
So Mick's target heart rate range is between 55% and 70%.

6 D. 4 L for \$14.80 = \$3.70/L
600 mL for \$2.25 = \$3.75/L
1.5 L for \$5.70 = \$3.80/L
2.5 L for \$9.00 = \$3.60/L
The cheapest price is \$3.60/L.
So 2.5 L for \$9.00 is the best buy.

7 A. Let x be the number of hours when the costs are the same.
So $90 + 80x = 40 + 90x$
$50 = 10x$
$5 = x$
The costs will be the same for 5 hours.
[Or try each option.]

8 C. Blood pressure of 156 over 99 is an example of mild hypertension.
[147 over 92 is mild hypertension; 121 over 88 is high normal blood pressure and 175 over 102 is moderate hypertension.]

Part B

9 a Cost per ha $= 0.8 \times \$82.50$
$= \mathbf{\$66.00}$ ✓

b Litres required $= 0.8 \times 15$
= **12 L** ✓

c Spray time $= 15 \times 12$ min
$= 180$ min
$= 3$ h
Cost to apply spray $= 3 \times \$93.00$
$= \$279.00$ ✓
Cost of spray $= 12 \times \$82.50$
$= \$990.00$
Total cost $= \$279 + \$990$
$= \$1269$
The total cost to spray the paddock is \$1269. ✓

10 a 6 h 40 min $= 6\frac{2}{3}$ hours
Average speed $= 465 \div 6\frac{2}{3}$
= **69.75 km/h** ✓

b Fuel used $= 4.65 \times 7$ L
$= 32.55$ L ✓
Cost $= 32.55 \times \$1.40$
$= \mathbf{\$45.57}$ ✓

11 Total hours charged = 7

Time cost = 7 × \$70
= \$490 ✓

Travel cost = 48 × \$5.00
= \$240

Total cost = \$30 + \$490 + \$240
= **\$760** ✓

12 a Number of hours = 400 ÷ 25
= 16

The infusion will take 16 hours. ✓

b Flow rate = 240 mL in 16 hours
= **15 mL/h** ✓

c 15 mL = (15 × 20) drops
= 300 drops ✓

Drip rate = 300 drops/h
= (300 ÷ 60) drops/min
= **5 drops/min** ✓

13 a 99 km in 1 h = 99 000 m in 60 min
= 1650 m in 60 s
= **27.5 m/s** ✓

b Time = 66 ÷ 27.5
= **2.4 s** ✓

c In 3 s, distance = 3 × 27.5 m
= 82.5 m

Total distance = 66 m + 82.5 m
= **148.5 m** ✓

14 a **121 over 79 is normal blood pressure.** ✓

b **Andrea's heart rate is 68 beats per minute.** ✓

c **Without knowing Andrea's age it isn't possible to determine her maximum heart rate, but a rate of 136 bpm would be within the target heart rate of most people.** ✓

15 a Run rate = 175 ÷ 20
= **8.75 runs/over** ✓

b Runs scored = 8 × 6.625 runs
= **53 runs** ✓

c To win the second team needs to score (at least) one more run than the first team.

Remaining overs = 12

Remaining runs required = 176 − 53
= 123

Required run rate = 123 ÷ 12
= 10.25

The second term needed to score at a rate of 10.25 runs/over for the remaining overs. ✓

CHAPTER 4 SCALE DRAWINGS

TOPIC: MEASUREMENT

1 Ratios

- **A ratio compares quantities of the same kind.**
- **Order is important in a ratio.** A ratio of 2 : 3 is different to a ratio of 3 : 2.
- **A ratio is expressed in simplest form** if expressed using whole numbers that have no common factors.

EXAMPLE 1

A fertiliser mix has 5.6 kg of phosphate and 16.8 kg of nitrogen. What is the ratio of phosphate to nitrogen in simplest form?

Phosphate : nitrogen = 5.6 : 16.8
= 1 : 3

The fraction button of a calculator can be used to simplify a ratio of two parts.

EXAMPLE 2

In a washing cycle, a machine spends 3 minutes washing and 20 seconds rinsing. What is the ratio of washing to rinsing in simplest form?

Ratio = 3 min : 20 s
= 180 s : 20 s
= 180 : 20
= 9 : 1

Quantities must be expressed in the same units.

EXAMPLE 3

A design uses tiles in three different colours. For every eight blue tiles, there are 12 yellow and 20 white tiles. What is the ratio of:

a **blue : white : yellow?**

b **yellow : blue and white?**

a blue : white : yellow = 8 : 20 : 12
= 2 : 5 : 3

4 is the highest common factor of 8, 20 and 12 so divide each by 4 to express the ratio in simplest form.

b yellow : blue and white = 12 : (8 + 20)
= 12 : 28
= 3 : 7

EXAMPLE 4

On a map a distance of 700 km is represented by a distance of 7 cm. What is the scale of the map?

Scale = 7 cm : 700 km
= 1 cm : 100 km
= 1 cm : 100 000 m
= 1 cm : 10 000 000 cm

So the scale is 1 : 10 000 000

A scale always puts the representative distance first and the real distance second.

EXAMPLE 5

In a recipe, $2\frac{1}{4}$ cups of flour and $\frac{1}{2}$ cup of coconut are used. What is the ratio of flour to coconut?

Ratio of flour to coconut $= 2\frac{1}{4} : \frac{1}{2}$

$= \frac{9}{4} : \frac{2}{4}$

$= 9 : 2$

Always give the answer to a ratio in simplest form. Simplest form means each part should be the simplest possible whole number.

➔ PRACTICE

1 **A fertiliser mix has 1.35 kg of phosphate and 3.15 kg of potassium. What is the ratio of phosphate to potassium in simplest form?**

2 **A worker spends 5 hours in his office and 25 minutes travelling. What is the ratio of office time to travel time in simplest form?**

3 **A bag holds marbles in three different colours. There are 18 blue, 12 green and 42 red marbles. What is the ratio of:**

a **blue : red : green?**

b **red : blue and green?**

4 On a map a distance of 90 km is represented by a distance of 9 cm. What is the scale of the map?

5 A recipe for lemon meringue pie uses $\frac{3}{4}$ cup sugar and $\frac{2}{3}$ cup of lemon juice. What is the ratio of sugar to lemon juice in simplest form?

6* A cleaning solution uses 120 mL of vinegar in 3.6 L of water. What is the ratio of water to vinegar in simplest form?

Answers ➲ p. 84

2 Dividing in a given ratio

- **To divide in a given ratio** means to separate an amount into different parts so that the parts form the required ratio.
- **When dividing in a given ratio**, the amount is first divided by the total number of parts in the ratio and then the required share of the amount is found for each component of the ratio.

EXAMPLE 1

Divide $50 in the ratio 3 : 2.

Total parts = 3 + 2
= 5
Each part = $50 ÷ 5
= $10
The first share is 3 × $10 = $30.
The second share is 2 × $10 = $20.

Check that the total shares add to the initial amount.

EXAMPLE 2

Three friends each contributed different amounts of money to begin a business. They agreed to share the profits in the ratio of their investments. Bill invested $20 000, Heath $15 000 and Amira $25 000. In the first year a profit of $28 800 was made. How much did Amira receive?

Ratio of contributions = 20 000 : 15 000 : 25 000
= 4 : 3 : 5
Amira has 5 of 12 shares.
Amount Amira receives = $\frac{5}{12}$ of $28 800
= $12 000

Bill would receive $9600 and Heath $7200.

→ PRACTICE

1 Divide $40 in the ratio 3 : 5.

2 Three friends each contributed different amounts of money to begin a business. They agreed to share the profits in the ratio of their investment. Jill invested $75 000, Heather $45 000 and Amir $90 000. In the first year a profit of $33 600 was made. How much did Jill receive?

3* Ted has 186 saplings that he is going to divide between three properties in the ratio 9 : 10 : 12. How many saplings will each property receive?

Answers ➲ p. 84

3 Solving problems involving ratios

- **The unitary method can be used with ratios.** First find the value for one item and then the value for the required number of items.
- **Ratios can be expressed as fractions and problems solved** using properties of fractions.
- **Given the ratio of all parts that make up a whole**, the percentage composition can be found. For example, if a mixture has red, blue and green in the ratio 5 : 3 : 2 then its percentage composition is 50% red, 30% blue and 20% green.

EXAMPLE 1

The ratio of water to cordial in a mixture is 9 : 2. If there is 405 mL of water, how much cordial should be used?

Nine parts is 405 mL.
So one part is (405 ÷ 9) mL or 45 mL.
Two parts is 2 × 45 mL or 90 mL.
So 90 mL of cordial should be used.

Water is mentioned first in the ratio so water is nine parts.

EXAMPLE 2

The ratio of blue to white tiles in a tiling pattern is 5 : 3.

a If there are 75 white tiles, how many blue tiles will there be?

b If there are 240 tiles altogether, how many will there be of each colour?

a Three parts is 75.
So one part is 75 ÷ 3 or 25.
5 parts = 5 × 25
= 125
So 125 blue tiles will be needed.

White tiles are mentioned second in the ratio so white tiles are three parts.

b Total parts = 5 + 3 = 8
Each part is 240 ÷ 8 or 30.
Now 5 × 30 = 150
and 3 × 30 = 90
So there will be 150 blue tiles and 90 white tiles.

EXAMPLE 3

Weeds in a paddock are to be sprayed at the rate of 0.8 L per hectare. A wetting agent is added to the spray to improve its effectiveness. The ratio of wetting agent to spray is 1 : 25. How many millilitres of wetting agent will be used on a 5 ha paddock?

Amount of spray required = 5 × 0.8 L
= 4 L
= 4000 mL
Now 25 parts is 4000 mL.
1 part is 4000 ÷ 25 or 160 mL.
So 160 mL of wetting agent will be needed.

Take care with the units.

EXAMPLE 4

A nut mixture has almonds, macadamias, walnuts and hazelnuts in the ratio 5 : 6 : 3 : 2. What is the percentage composition of the nut mixture?

Number of parts = 5 + 6 + 3 + 2
= 16

% of almonds = $\frac{5}{16} \times 100\%$
= 31.25%

% of macadamias = $\frac{6}{16} \times 100\%$
= 37.5%

% of walnuts = $\frac{3}{16} \times 100\%$
= 18.75%

% of hazelnuts = $\frac{2}{16} \times 100\%$
= 12.5%

So the percentage composition of the mixture is 31.25% almonds, 37.5% macadamias, 18.75% walnuts and 12.5% hazelnuts.

It is always a good idea to check that the percentage composition does add to 100%.

EXAMPLE 5

There were 120 people entered in a fun run. The ratio of children to adults was 3 : 5. After another 19 children and five adults enter the fun run what is the new ratio of children to adults?

Originally $\frac{3}{8}$ of the entrants were children.

Number of children entered = $\frac{3}{8}$ of 120
= 45
Number of adults entered = 120 − 45
= 75
New number of children = 45 + 19
= 64
New number of adults = 75 + 5 = 80
New ratio of children to adults = 64 : 80
= 4 : 5

Adding a number to each part of a ratio will not produce an equivalent ratio.

→ PRACTICE

1 **The ratio of petrol to oil in a fuel mixture is 25 : 1. If there is 1 L of petrol, how much oil should be added?**

2 **The ratio of red to yellow ribbons awarded at a sports event is 7 : 2.**
   a **If there are 112 yellow ribbons, how many red ribbons would be awarded?**
   b **If 396 red and yellow ribbons were awarded altogether, how many of each colour would there be?**

3 **Fertiliser is to be spread on a paddock at the rate of 125 kg per hectare. Nitrogen is one of the nutrients in the fertiliser. The ratio of nitrogen to fertiliser is 7 : 50. How many kilograms of nitrogen will be added to a 4 ha paddock?**

4 **An orchard has apricot, plum, peach and apple trees in the ratio 4 : 2 : 3 : 7. What is the percentage composition of the orchard?**

5 **There were 154 calves born on a farm in the first 11 months of a year. The ratio of male to female calves was 6 : 5. The next month another six male and 10 female calves were born. What is the ratio of male to female calves for the whole year?**

6* **The ratio of bricks to tiles is 18 : 7. If there are 1925 more bricks than tiles, how many tiles are there?**

Answers ➲ pp. 84–85

4 Similar figures

- **Similar figures are the same shape** but not necessarily the same size.
- **The corresponding angles of similar figures** are equal and the corresponding sides are in the same ratio.
- **Two similar figures will differ by a scale factor** (enlargement factor or reduction factor).

EXAMPLE 1

Jamil has this photo that he wants to enlarge.

The photo is currently 12 cm long and 8 cm wide. Jamil wants to make it 66 cm long. How wide will the enlarged photo be?

Enlargement factor = 66 ÷ 12
= 5.5

New width = 5.5 × 8 cm
= 44 cm

The enlarged photo will be 44 cm wide.

> The original (object) and the enlargement (image) are similar figures.

EXAMPLE 2

Sandi drew this star design but wanted to make it half as big again, so enlarged it.

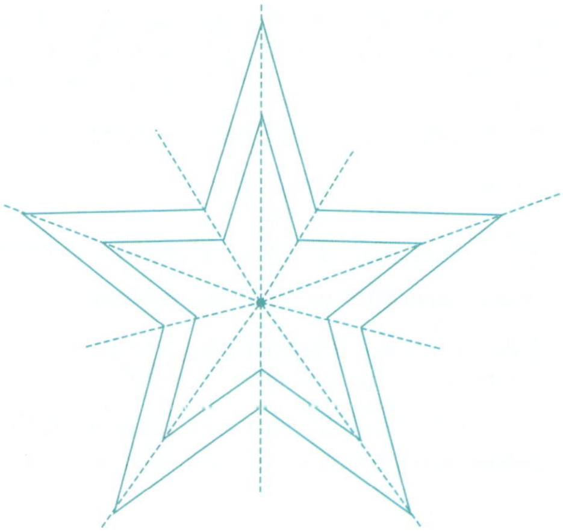

a **What is the enlargement factor?**

b **If each side of the original star was 13 cm long, what would be the perimeter of the enlarged star in metres?**

c **What reduction factor could be applied to the enlarged star to return it to the original size?**

a Enlargement factor = 1.5

> The image is half as big again so it is $1\frac{1}{2}$ times the original object.

b Each side of enlarged star = 1.5 × 13 cm
= 19.5 cm

Perimeter = 10 × 19.5 cm
= 195 cm
= 1.95 m

c Ratio of object to image = 1 : 1.5
= 2 : 3

So reduction factor = $\frac{2}{3}$.

EXAMPLE 3

A wallpaper pattern consists of different spots (circles). The smallest circle has diameter 4 cm.

a **Briefly explain why all circles are similar.**

b **Suppose the spots are given a letter from the smallest *A* to the largest *H*. What is the enlargement factor of spot *D* from spot *A* if its diameter is 10 cm?**

c **The scale factor of spot *F* (from *A*) is 3.5. What is the circumference of *F* to the nearest centimetre?**

d **Circle *G* is 1.6 times the size of spot *D*. What is its enlargement factor from *A*?**

e **Find the area of the largest circle, *H*, if the scale factor from *A* is 4.5.**

a The diameter and circumference of a circle depend on the radius. So any circle is similar to another according to the ratio of their radii.

> So there is a relationship between any two of the circles shown here.

b Enlargement factor = $\frac{10}{4}$
= 2.5

c Diameter of *F* = 3.5 × 4 cm
= 14 cm

Circumference = $\pi \times 14$ cm
= 43.982 2971… cm
= 44 cm (nearest cm)

d Diameter of $G = 1.6 \times 10$ cm
= 16 cm
= 4×4 cm

So spot G is 4 times spot A.

e H has diameter 4.5×4 cm or 18 cm.

So its radius is 9 cm.

Area = $\pi \times 9^2$
= 254.469 004…
= 254.5 (1 d.p.)

The area of spot H is 254.5 cm^2 to one decimal place.

EXAMPLE 4

These two quadrilaterals are similar.

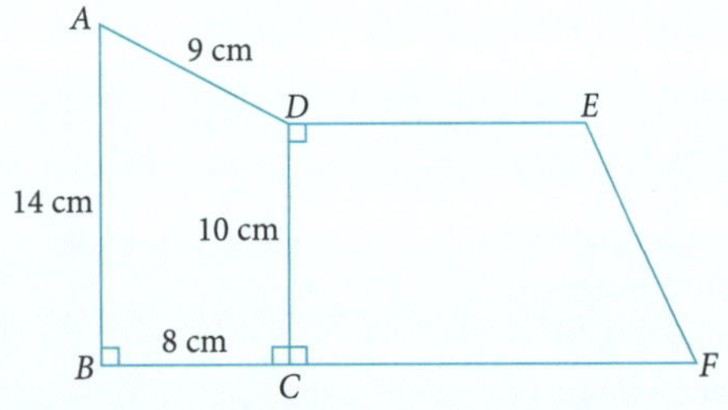

a Which side of $CDEF$ corresponds to side AB of $ABCD$?

b What is the scale factor?

c Find the length of BF.

a FC corresponds to AB.

b CD of $CDEF$ corresponds to BC of $ABCD$.

So enlargement factor = $\frac{10}{8}$
= 1.25

Quadrilateral *ABCD* has been enlarged, reflected in a vertical line and rotated 90° clockwise.

c $FC = 1.25 \times 14$ cm
= 17.5 cm

So $BF = (8 + 17.5)$ cm = 25.5 cm.

→ PRACTICE

1 Jamila has this photo that she wants to enlarge.

The photo is currently 10 cm long and 8 cm wide. Jamila wants to make it 75 cm long. How wide will the enlarged photo be?

2 Sandip drew this design but wanted to make it one-third as big again, so he enlarged it.

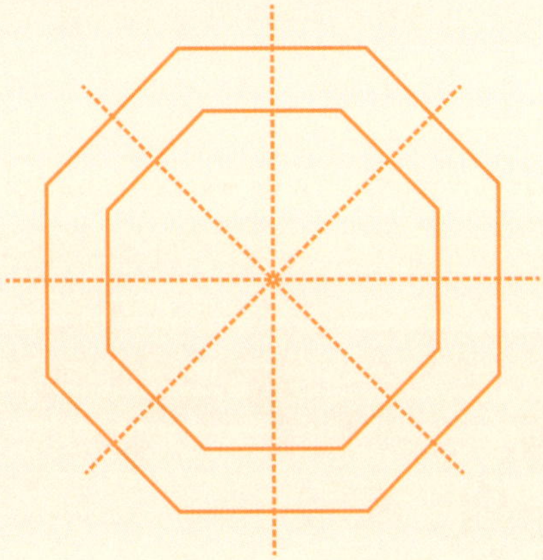

a What is the enlargement factor?

b If each side of the original octagon was 27 cm long, what would be the perimeter of the enlarged octagon in metres?

c What reduction factor could be applied to the enlarged octagon to return it to the original size?

3 A wallpaper pattern consists of different squares. The smallest square has side length 15 mm.

a Briefly explain why all squares are similar.

b Suppose the squares are given a letter from the smallest A to the largest G. What is the enlargement factor of square C from square A if its side length is 25 mm?

c The scale factor of square E (from A) is 3. What is the perimeter of E in centimetres?

d Square F is 1.4 times the size of square C. What is its enlargement factor from A?

e Find the area of the largest square, G, if the scale factor from A is 4.

4 These two quadrilaterals are similar.

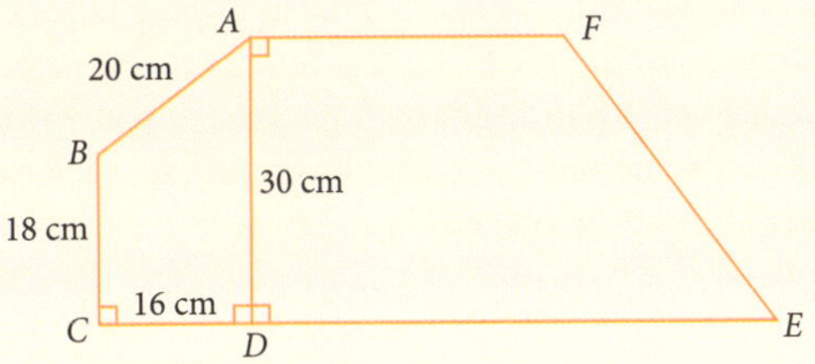

a Which side of *AFED* corresponds to side *AB* of *ABCD*?

b What is the scale factor?

c Find the length of *CE*.

5* Are these two quadrilaterals similar? Justify your answer.

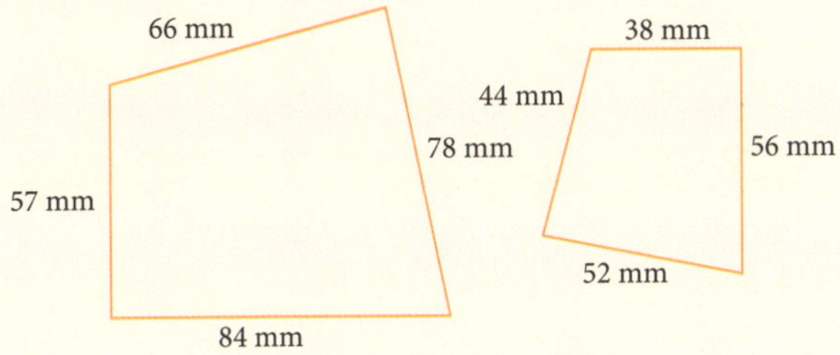

Answers ➲ p. 85

5 Similar triangles

- **Similar triangles are a special case of similar figures.**
- **If two triangles are similar,** then corresponding angles are equal and corresponding sides are in the same ratio.
- **Two triangles will be similar if any of the following occur:**
  - corresponding angles are equal
  - corresponding sides are in the same ratio
  - two pairs of corresponding sides are in the same ratio and the included angles are equal
  - the hypotenuse and another side of one right-angled triangle are in the same ratio as the hypotenuse and another side of another right-angled triangle.
- **There are many applications of similar triangles.**

EXAMPLE 1

A tree casts a shadow of length 38 m on level ground. At the same time a nearby post casts a shadow of 1.75 m. If the post is 1.2 m high, find the height of the tree.

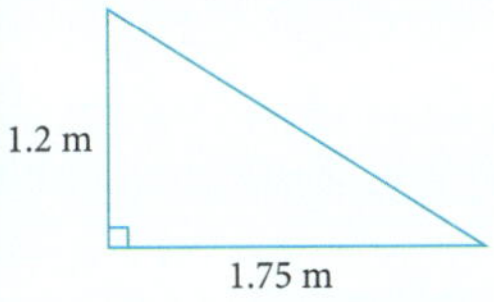

The triangles are similar (equiangular).
Let h m be the height of the tree.

We assume that the tree and post are both vertical and at right angles to the ground. The sun is in the same position so the angles of elevation are the same in both triangles.

So $\frac{h}{1.2} = \frac{38}{1.75}$

$h = \frac{38}{1.75} \times 1.2$

$= 26.057\,1428\ldots$

$= 26$ (nearest unit)

This shadow stick method can be used to find the height of objects.

The height of the tree is 26 m, to the nearest metre.

EXAMPLE 2

An underground pipeline has a steady fall as shown in the diagram. At the furthest point, 90 m from the start, the pipe is 16.5 m below ground.

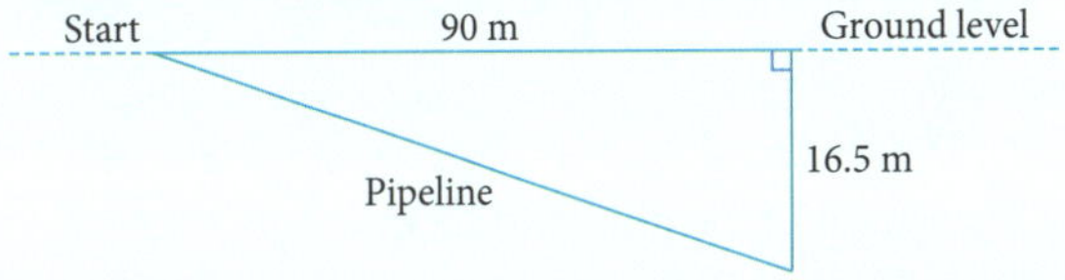

a **How far below ground is the pipe at 75 m from the start?**

b **How far from the start is the pipe 6.6 m below ground?**

c **How long is the pipe at 51 m from the start?**

a Let d m be the depth of the pipe at 75 m.

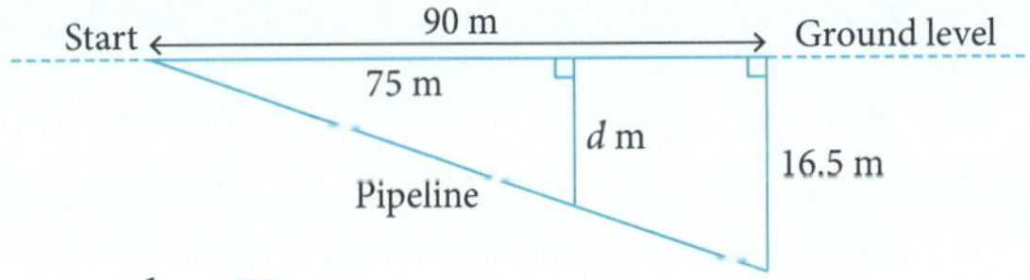

So $\frac{d}{16.5} = \frac{75}{90}$

$d = \frac{75}{90} \times 16.5$

$= 13.75$

The triangles are similar because both have a common angle and a right angle.

The depth of the pipe is 13.75 m

b Let x m be the distance from the start when the pipe is 6.6 m below the ground.

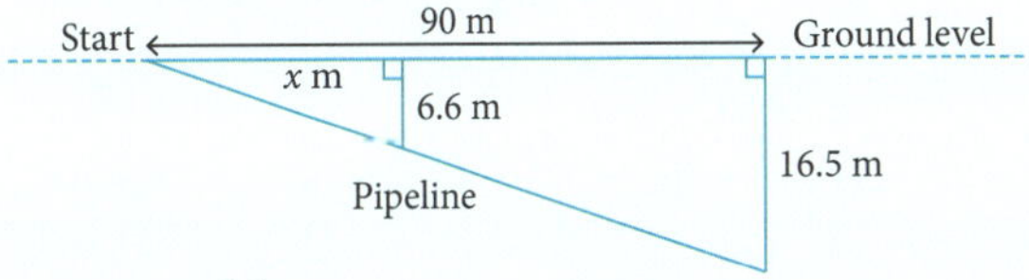

So $\frac{x}{90} = \frac{6.6}{16.5}$

It is not necessary to have the unknown at the top (in the numerator) of the first fraction, but it is easiest to do that.

$$x = \frac{6.6}{16.5} \times 90$$
$$= 36$$

The distance from the start is 36 m.

c Let h m be the depth of the pipe at 51 m.

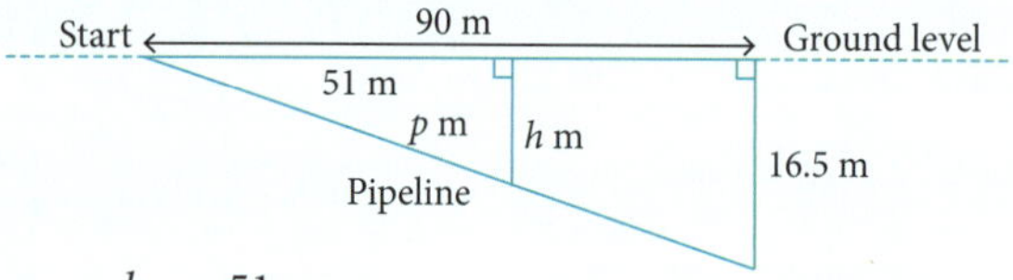

So $\frac{h}{16.5} = \frac{51}{90}$

$$h = \frac{51}{90} \times 16.5$$
$$= 9.35$$

The depth of the pipe is 9.35 m.

Let p m be the length of the pipe.

So $p^2 = 51^2 + 9.35^2$

$$= 2688.4225$$
$$p = \sqrt{2688.4225}$$
$$= 51.85$$

The length of the pipe is 51.85 m.

> We use Pythagoras' theorem. Alternatively, the length of the whole pipe could have been found and then similar triangles used.

EXAMPLE 3

A table has sets of legs in a triangular form as shown in the diagram.

a Briefly explain why the two triangles shown here are similar.

b Find the value of x.

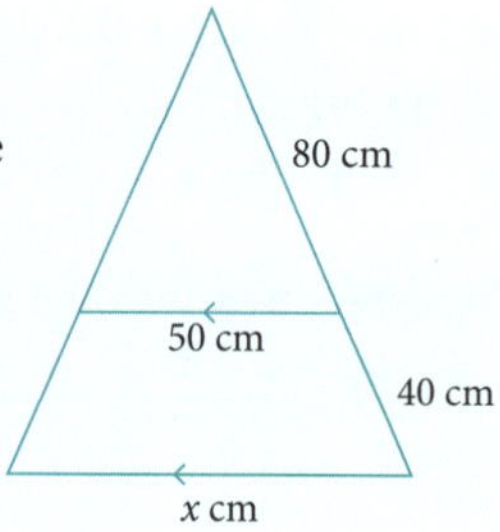

a The triangles have a common angle at the top.

The two base lines of the triangles are parallel, so the corresponding angles formed by the parallel lines are equal.

So the two triangles are equiangular and are similar.

> Be careful when talking about corresponding angles. We mean the matching angles in similar triangles, which is the same reason we have corresponding angles formed by parallel lines. However, in general when we talk about corresponding angles in similar figures we don't mean there are parallel lines.

b Now $\frac{x}{50} = \frac{120}{80}$

$$x = \frac{120}{80} \times 50$$
$$= 75$$

> x is the base of the large triangle. The side of the large triangle must be used as well, so its length is (80 + 40) cm or 120 cm.

EXAMPLE 4

A cable is attached to a point 5.6 m up a pole from a point 28 m from the base of the pole on level ground. A support post, 1.8 m high, is to be used at some point along the cable as shown in the diagram. How far from the pole should it be placed?

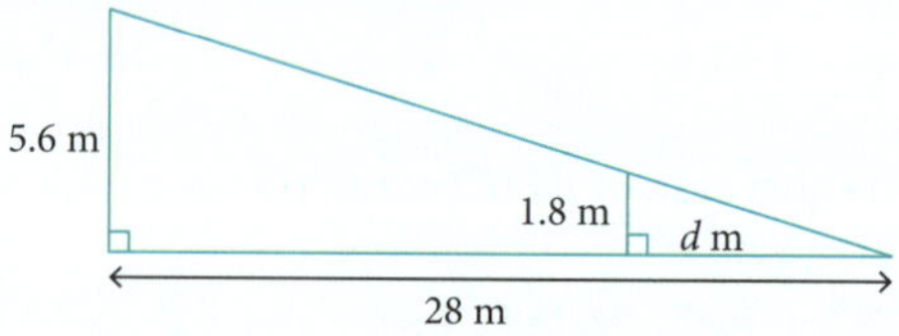

Let d m be the distance along the ground of the support post from the foot of the cable.

So $\frac{d}{28} = \frac{1.8}{5.6}$

$$d = \frac{1.8}{5.6} \times 28$$
$$= 9$$

> It is simpler to first find the distance from the foot. If x m was the distance to the pole then it would be necessary to solve an equation involving $(28 - x)$.

The distance from the foot of the cable is 9 m.

Distance from the pole = (28 − 9) m
= 19 m

The pole should be placed 19 metres from the foot of the pole.

EXAMPLE 5

There is to be an awning 1.3 m long at the top of a building and a second awning 0.8 m long 2 m from the top so that the ends of the awnings form a straight line with the base of the building as shown in the diagram. Find the height of the building.

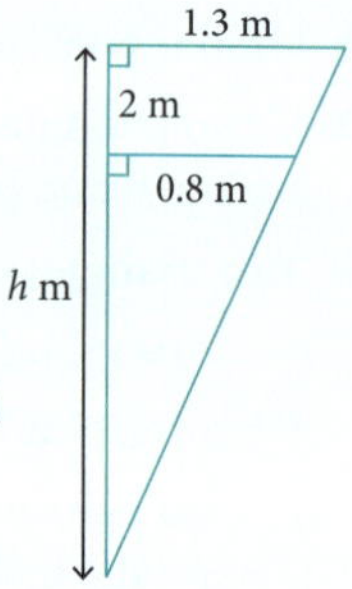

The triangles are similar (equiangular).

Let h m be the height of the building.

$\therefore (h - 2)$ m is the height of the lower awning.

So $\frac{h-2}{h} = \frac{0.8}{1.3}$

$$1.3h - 2.6 = 0.8h$$
$$0.5h = 2.6$$
$$h = 5.2$$

> We need to solve an equation because there are two heights we don't know, but they are related.

The height of the building is 5.2 m.

→ PRACTICE

1 A tree casts a shadow of length 47 m on level ground. At the same time a nearby post casts a shadow of 2 m. If the post is 1.4 m high, find the height of the tree.

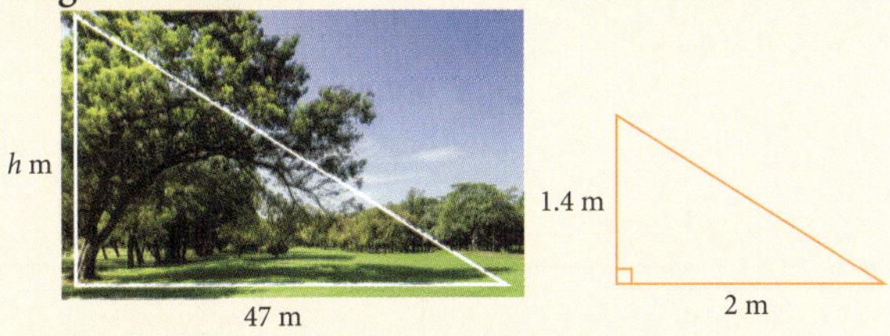

2 An underground cable has a steady fall as shown in the diagram. At the furthest point, 80 m from the start, the cable is 18 m below ground.

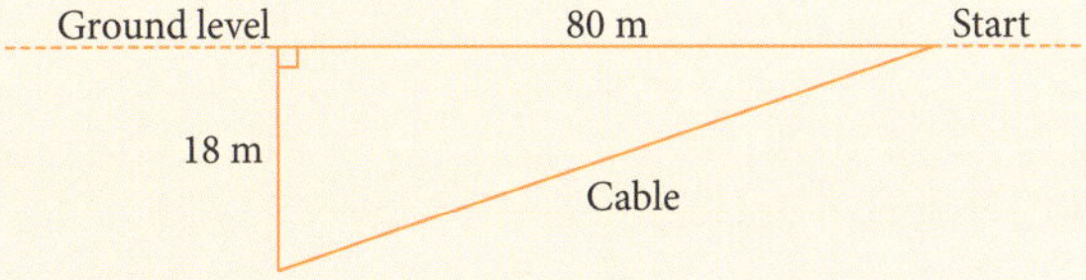

a How far below ground is the cable at 65 m from the start?

b How far from the start is the pipe 8.1 m below ground?

c How long is the pipe at 54 m from the start?

3 A hanging rack is in a triangular form as shown in the diagram.

a Briefly explain why the two triangles shown here are similar.

b Find the value of x.

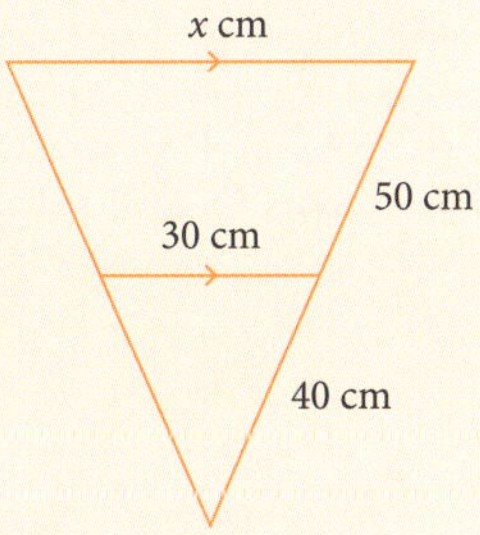

4 A rafter is attached to a point 2.6 m up a wall from a point 4 m from the base of the wall. A support post, 80 cm high, is to be used at some point along the rafter as shown in the diagram. How far from the wall should it be placed?

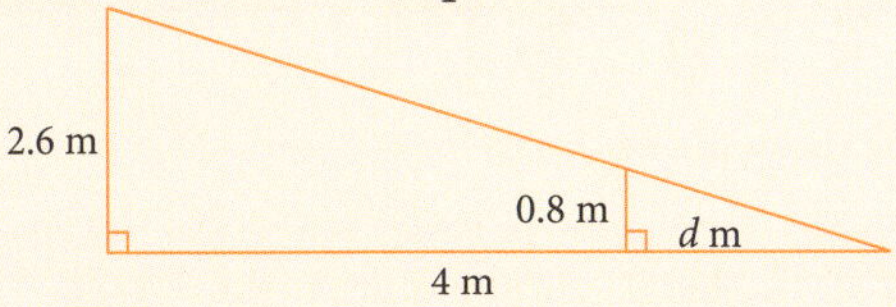

5 A ramp is to be constructed. It needs to be 1.7 m high at one end and 1.4 m at a point 1 m from that end as shown in the diagram. Find the length of the ramp along the ground.

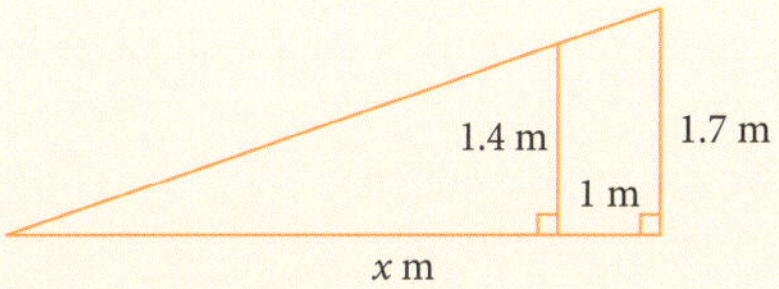

6* The shaded part of the diagram is a rectangle of width 5 m.

a Explain why the two smaller triangles are similar and why both are similar to the large triangle.

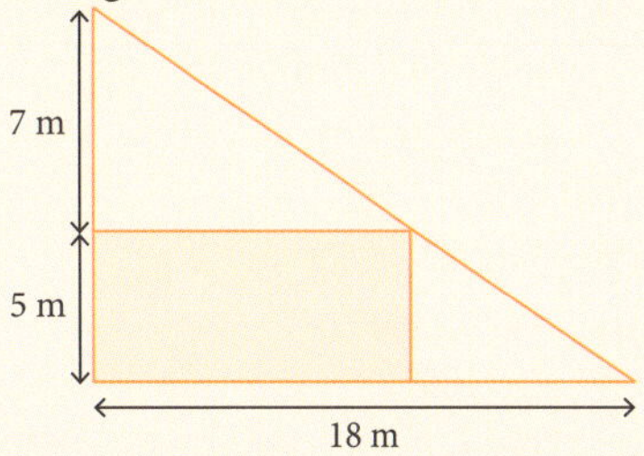

b Is the area of the rectangle more than, less than or the same as half the area of the large triangle?

Answers ⊃ pp. 85–86

6 Maps and plans

→ **A plan is said to be drawn to scale** if all measurements are in the same proportion as the original.

→ **If a diagram is drawn to scale**, and the scale is known, then by careful measurement and calculation all actual measurements can be found. Once dimensions are known, perimeter and area can be estimated or found by appropriate formulae.

→ **Aerial photographs are an example of scale diagrams.** It is possible to find distances by calculating the scale from features of a photograph.

→ **Tools in computer programs or online resources** can also provide information about distances and directions.

EXAMPLE 1

The scale on a map is given as 1 : 10 000.

a A distance of 3.5 cm on the map would be what actual distance?

b The distance between two towns is 6.3 km. How far apart would they be on the map?

c Arthur measured a rectangular piece of land shown on the map and found it to be 16 mm long and 9 mm wide. If fencing costs $15 per metre, how much will it cost to fence the entire boundary of the land?

a 3.5 cm on the map represents a real distance of $3.5 \times 10\,000$ cm.

Now $3.5 \times 10\,000$ cm $= 35\,000$ cm
$= 350$ m.

b 6.3 km = 6300 m
$= 630\,000$ cm

Using a map or photograph to find distances can be helpful in planning and decision-making.

Now 1 cm represents 10 000 cm.

So 63 cm would represent 630 000 cm.

On the map the two towns would be 63 cm apart.

c 9 mm represents 90 000 mm or 90 m.

16 mm represents 160 000 mm or 160 m.

Perimeter $= 2 \times (160 + 90)$ m
$= 500$ m

So the boundary would be 500 m long.

Cost of fencing $= \$15 \times 500$
$= \$7500$

It would cost \$7500 to fence the entire boundary.

EXAMPLE 2

This is a scale drawing of a hobby farm. It has been divided into three sections.

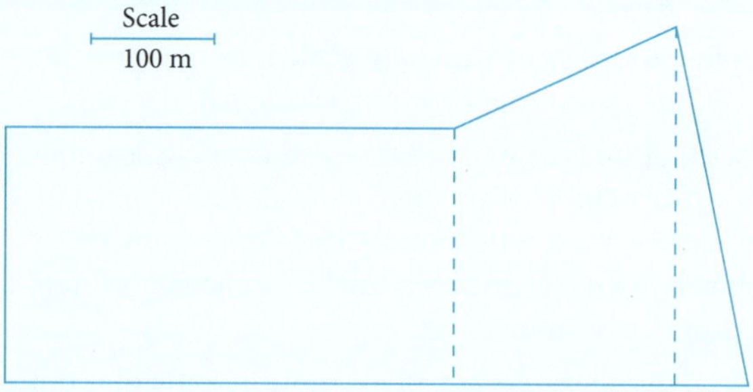

What is the area of the hobby farm in hectares to two decimal places?

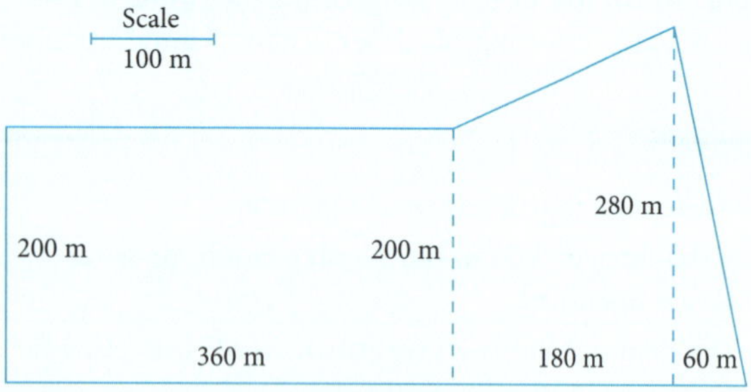

From the scale, 12.5 mm represents 100 m so the above measurements are found.

Area 1: rectangle

$A = lb$

$= 360 \times 200$

$= 72\,000 \text{ m}^2$

The required area has been divided into a rectangle, a trapezium and a triangle.

Area 2: trapezium

$A = \frac{1}{2}h(a + b)$

$= \frac{1}{2} \times 180 \times (200 + 280)$

$= 43\,200 \text{ m}^2$

Area 3: triangle

$A = \frac{1}{2}bh$

$= \frac{1}{2} \times 60 \times 280$

$= 8400 \text{ m}^2$

Total area $= (7200 + 43\,200 + 8400) \text{ m}^2$
$= 123\,600 \text{ m}^2$
$= 12.36$ ha

EXAMPLE 3

This diagram depicts an aerial view of a town. An Olympic-sized swimming pool is shown towards the bottom right of the diagram. Given that this pool is 50 m long, answer the following questions.

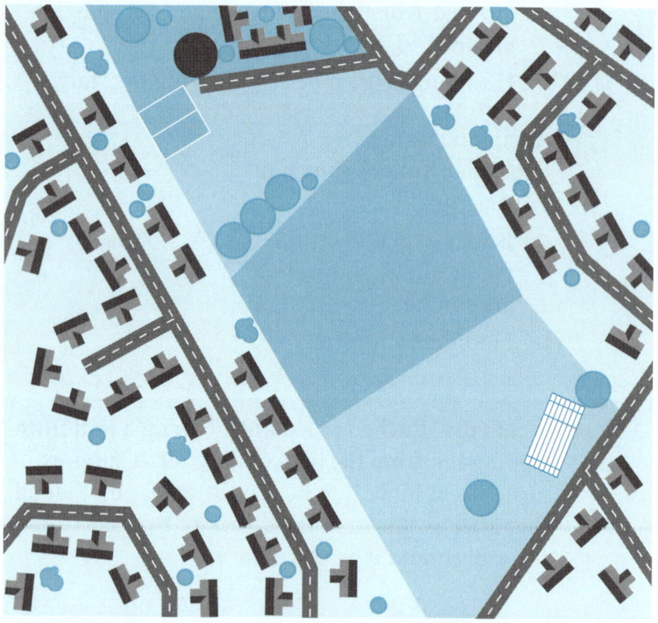

a **What is the scale of the map?**

b **The black circular object at the top left of the map is a hall at the local school. What is the actual diameter of the hall?**

c **Between the pool and the road at the top centre of the map is a trapezoidal cricket field. Estimate the area of the field, in square metres.**

a On the map the pool is 1 cm long. So the scale is 1 cm represents 50 m.

1 cm : 50 m = 1 cm : 5000 cm
= 1 : 5000

b The diameter on the photo is 6 mm.

Actual diameter = 5000 × 0.6 cm
= 3000 cm
= 30 m

The diameter of the hall is 30 m.

c The two parallel sides are 2.3 cm and 3.1 cm. So these represent distances of 115 m and 155 m respectively.

The distance between those parallel sides is 3.3 cm on the map which is a real distance of 165 m.

$$A = \frac{1}{2}h(a + b)$$
$$= \frac{1}{2} \times 165 \times (115 + 155)$$
$$= 22\,275 \text{ m}^2$$

The estimate should be reasonably accurate with careful measurement and calculation.

The area of the field is about 22 275 m².

EXAMPLE 4

From an image of her property on her computer a farmer decided to find the area of a triangular paddock on her farm. She used the ruler in a computer program to find that the distance from one corner of the paddock to a second corner was 280 m on a heading (bearing) of 137°. From that first corner to the third corner was 300 m on a heading of 226°. Find the approximate area of the paddock in hectares.

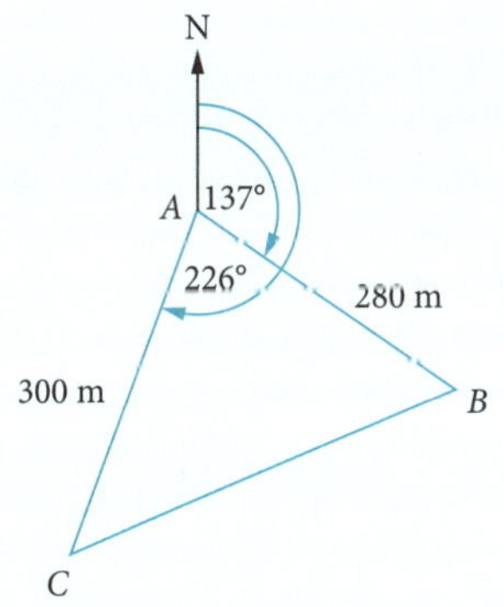

$\angle BAC = 226° - 137°$
$= 89°$

So the area can be approximated by finding the area of a right-angled triangle.

$$\text{Area} = \frac{1}{2}bh$$
$$= \frac{1}{2} \times 300 \times 280$$
$$= 42\,000 \text{ m}^2$$
$$= 4.2 \text{ ha}$$

The area could have been calculated using Heron's rule if the length of boundary *BC* was also found from the computer program.

The area of the paddock is approximately 4.2 ha.

→ PRACTICE

1 **The scale on a map is given as 1:100 000.**

a **What actual distance would a distance of 4.5 cm on the map be?**

b **The distance between two towns is 7.2 km. How far apart would they be on the map?**

c **Arlene measured a rectangular piece of land shown on the map and found it to be 18 mm long and 8 mm wide. If fencing costs $23 per metre, how much will it cost to fence the entire boundary of the land?**

2 **This is a scale drawing of a park. It has been divided into three sections.**

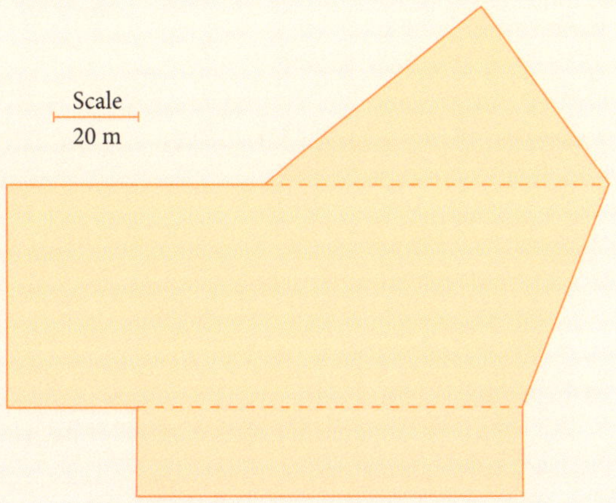

What is the area of the park in hectares to two decimal places?

3 **This is an aerial illustration of part of a town. The distance between the goalposts on the football field is 100 m.**

a What is the scale of the illustration?

b What is the distance between the centres of the two roads that lead off Thirroul Road?

c Between the football field and Thirroul Road is a trapezoidal field. Estimate the area of the field, in square metres.

4 From an image of his property on his computer a farmer decided to find the area of a triangular paddock on his farm. He used the ruler in a computer program to find that the distance from one corner of the paddock to a second corner was 372 m on a heading (bearing) of 148°. From that first corner to the third corner was 552 m on a heading of 239°. Find the approximate area of the paddock in hectares.

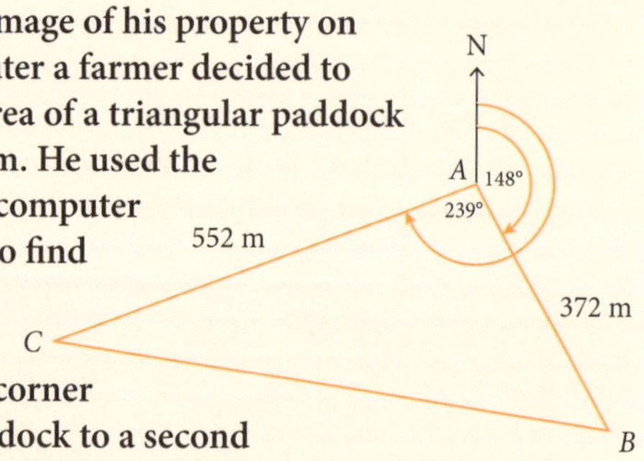

5* A paddock is bordered on one side by a river. A scale drawing was made of the paddock using the scale 1: 2000. The measurements on the scale drawing are shown on this sketch (which is not to scale). Find the approximate area of the paddock.

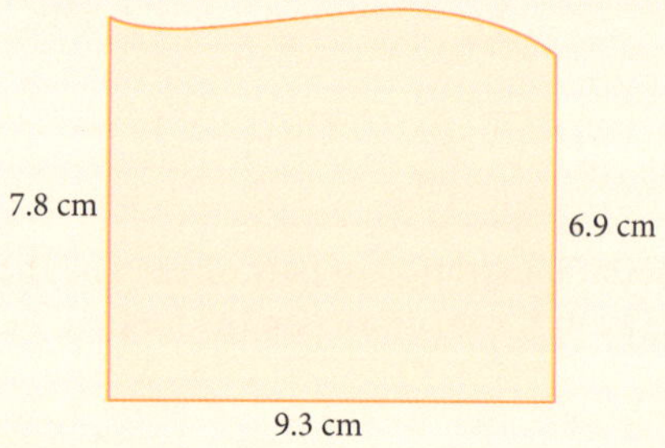

Answers ➲ pp. 86–87

7 Building plans

- **Floor plans for buildings show the layout of rooms and measurements.** All measurements are in millimetres.
- **An elevation view shows how the building will look.** Elevations might be numbered or named for the direction in which they face.
- **A site plan shows the position of any buildings and other features of the land.**

EXAMPLE 1

This is a plan of Summer's new house.

a What is the meaning of the abbreviations WIR, WC, Ens. and Ptry?

b What does the symbol  in the bathroom mean?

c The measurements of the study are given as 3300 by 2250. How many square metres is the room?

d Summer needs to replace the carpet in the study. The standard width of carpet is 3.66 m. Which way do you suggest Summer should run the carpet in the study? Justify your answer.

a WIR is walk-in robe. It is the wardrobe in the master suite.
WC stands for water closet.
It is the toilet.
Ens. is the ensuite or bathroom attached to the master bedroom.
Ptry is the pantry for storage in the kitchen.

You need to be familiar with common symbols and notations used in floor plans.

b The symbol  is the shower.

c Area of study = 3.3 m × 2.25 m
= 7.425 m^2

d Summer should use the width of the carpet for the length of the room. This would mean a leftover piece 0.36 m by 2.25 m. The alternative, which is trying to run it in the other direction, would mean either a much larger part left over, or lots of joins and the weave running in different directions.

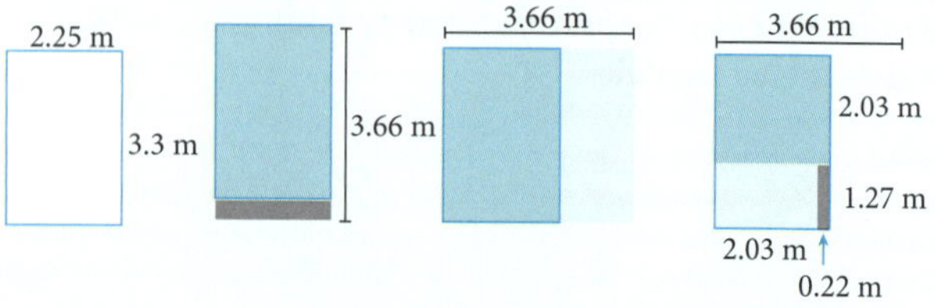

There is no specific answer to this type of question. As long as you justify your response with explanations or calculations, the answer should be correct.

EXAMPLE 2

Nell has drawn a floor plan for the house that she intends to build. She has also drawn to scale a plan of the block of land where she hopes to build the house and has drawn one elevation.

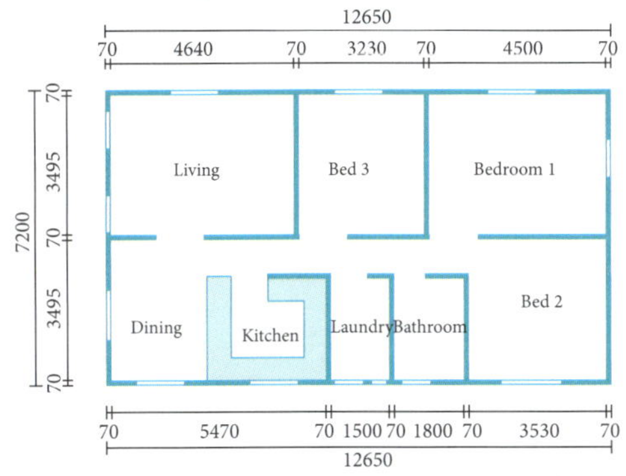

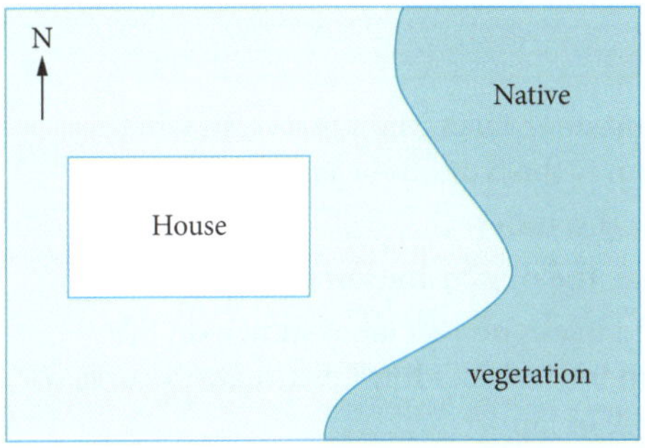

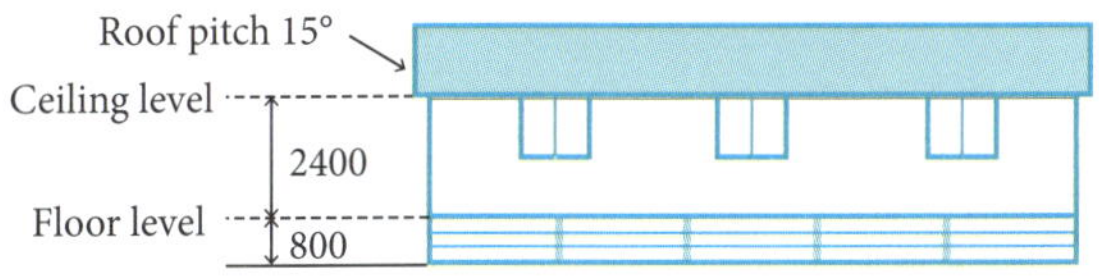

a **What are the dimensions of the living room in metres?**

b **How thick are the walls?**

c **The floor area is calculated as the area inside the exterior walls. What is the floor area of Nell's house in square metres?**

d **What is the scale that Nell has used on the site plan?**

e **What is the area of Nell's block of land?**

f **Bedroom 1 looks out onto the native vegetation at the back of the block. In what direction is the kitchen window facing?**

g **What direction name would be given to the elevation?**

h **What is the height of the ceilings above the floor in Nell's house?**

a The living room is 4640 mm or 4.64 m long and 3495 mm or 3.495 m wide.
So it is 4.64 m by 3.495 m.

b The walls are 70 mm thick.

c The house is 12 650 mm long.

Inner length = (12 650 – 2 × 70) mm
= 12 510 mm
= 12.51 m

The house is 7200 mm wide.

Inner width = (7200 – 2 × 70) mm
= 7060 mm
= 7.06 m

Floor area = (12.51 × 7.06) m^2
= 88.3206 m^2

d The house is 12 650 mm long.
By measurement, the length of the house on the scale drawing is 23 mm.
So the scale is 23 : 12 650 or 1 : 550.

e The measured length is 60 mm.

Actual length = 550 × 60 mm
= 33 000 mm
= 33 m

Using the scale found in part **d**

The measured width is 40 mm.

Actual width = 550 × 40 mm
= 22 000 mm
= 22 m

Area = 33 m × 22 m
= 726 m^2

f When looking out of the bedroom window towards the native vegetation Nell will be facing east. This means that the kitchen window will be facing south.

g The elevation shows three equally spaced windows, so it is the side of the house with the living room and bedroom 3 and bedroom 1 windows. They are facing north so it is the northern elevation.

h The ceiling is 2400 mm or 2.4 m above the floor.

→ PRACTICE

1 **This is a plan of Hugo's house.**

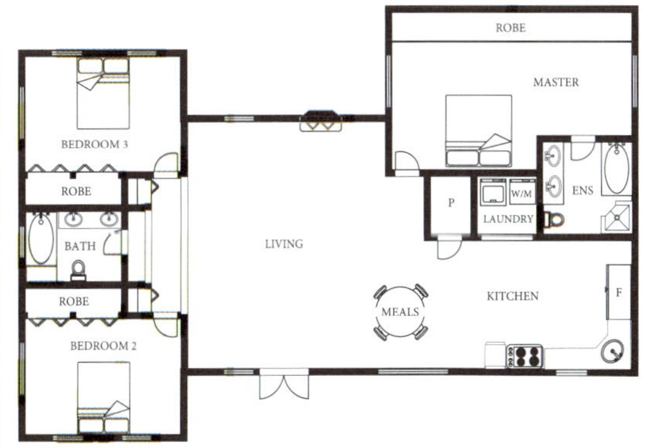

a What is the meaning of the abbreviations W/M, P and F?

b What does the symbol  found in the kitchen mean?

c There is one shelf in the robe in the master bedroom. Its measurements are given as 3850 by 550. How many square metres is the shelf?

d Hugo needs to line the shelf in the robe. He uses lining paper from a roll 2 m long and 50 cm wide. Which way do you suggest Hugo should lay the paper? Justify your answer.

2 Neil has drawn a floor plan for the house that he intends to build. He has also drawn to scale a plan of the block of land where he hopes to build the house and has drawn one elevation.

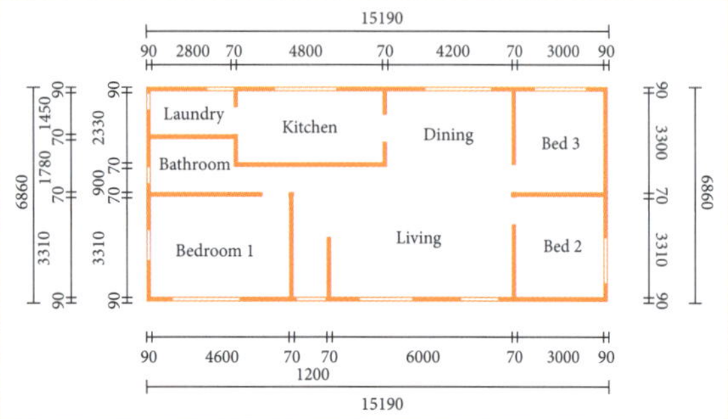

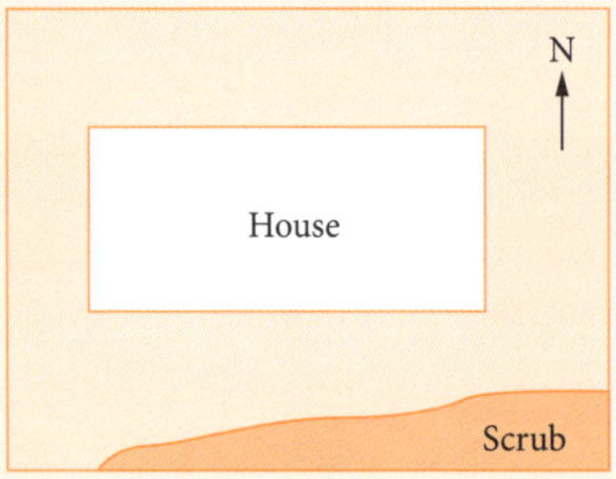

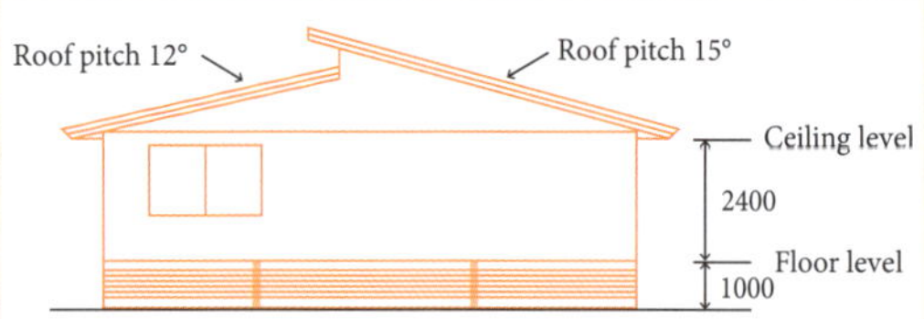

a What are the dimensions of bedroom 1?

b How much thicker are the exterior walls than the interior walls?

c The floor area is calculated as the area inside the exterior walls. What is the floor area of Neil's house in square metres?

d What is the scale that Neil has used on the site plan?

e What is the area of Neil's block of land? (Give the answer to the nearest square metre.)

f Bedroom 3 looks out onto the scrub at the back of the block. In what direction is the bathroom window facing?

g What direction name would be given to the elevation?

h What is the height of the ceiling above ground level in Neil's house?

3* This plan, which is drawn to scale, shows the position of a house on a block of land. The block of land has a 21.75 m frontage to Main Street and is 42 m deep on the northern side.

a The front door faces Main Street. In what direction is the front door facing?

b What is the scale?

c What are the dimensions of the house?

d The land backs on to a lake (shown on the diagram by dark shading). Estimate the area of the block of land.

Answers ➲ pp. 87–88

8 Further applications

- **Measurements from scale drawings can be used** in many applications, such as in painting, furnishing and landscaping.
- **Finding amounts of materials, likely costs or sustainability issues** can often be expedited by referring to maps and plans.

EXAMPLE 1

This is part of a floor plan of a house. The walls are 2.4 m high and the windows 1.25 m high. Find the cost of buying paint for the three walls of each bedroom if each receives two coats and the paint costs $62.50 for 4 L and covers 16 m²/L.

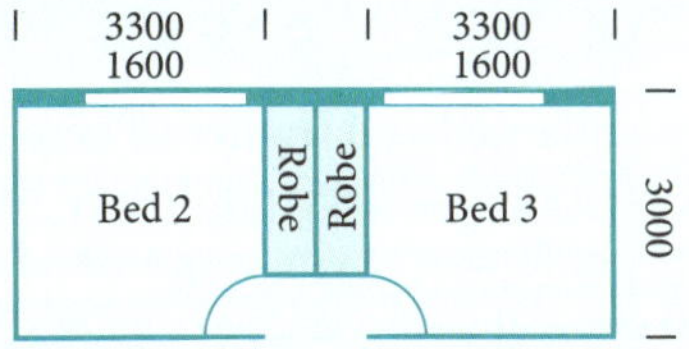

Both rooms are the same size: 3.3 m by 3 m.

Each has two walls 3.3 m by 2.4 m and one wall 3 m by 2.4 m to be painted, less the windows.

Now $2 \times (2 \times 3.3 \times 2.4 + 3 \times 2.4) = 46.08$

Area of two windows $= 2 \times 1.6 \times 1.25$

$= 4$

Area to be painted $= 42.08$ m²

Two coats of paint area $= 84.16$ m²

Number of litres required $= 84.16 \div 16$

$= 5.26$

So 5.26 L will be required.

This means two cans of the paint will be required.

Cost $= 2 \times \$62.50$

$= \$125$

The cost of the paint actually used is around $82.

EXAMPLE 2

This is an aerial photograph of one of Frank's paddocks. Frank knows that the shorter fence line at the top of the photo is 50 m long. Estimate the amount of superphosphate to apply to the paddock at the rate of 125 kg per hectare.

The paddock is roughly trapezoidal in shape.

The two parallel sides are about 110 m and 150 m in length.

$$A \approx \frac{h}{2}(d_f + d_l)$$

$$= \frac{50}{2}(110 + 150)$$

$= 6500$ m²

$= 0.65$ ha

Amount of superphosphate $= 0.65 \times 125$ kg

$= 81.25$ kg

So around 81 kg of superphosphate would be required.

Because the question asked for an estimate, different approaches could have been taken.

→ PRACTICE

1 **This is part of a floor plan of a house. The walls are 2.4 m high and the windows 1.2 m high. Find the cost of painting the three walls of each bedroom with two coats of paint if the paint costs $94.40 for 4 L and covers 15 m²/L.**

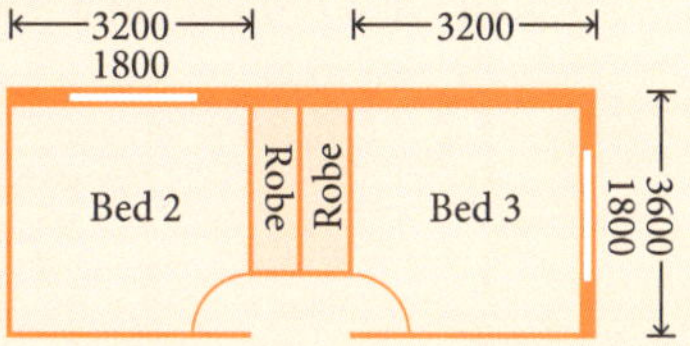

2 **This is an aerial photograph of Fran's paddock, which is bordered by a creek shown at the bottom of the photo. Fran knows that the shorter fence line at the top of the photo is 70 m long. Estimate the amount of fertiliser to apply to the paddock at the rate of 110 kg per hectare.**

3* **This is a scale drawing of an area that needs to be concreted.**

Scale 1 : 100

The concrete will be to a depth of 75 mm.

A 20 kg bag of cement will cover 1 m³.

Find the amount of sand and gravel that are required if the ratio of cement to sand to gravel is 1 : 2 : 3.

Answers ➲ p. 88

HSC EXAM-TYPE QUESTIONS

Total marks: 30

PART A

(Suggested time: 12 minutes)
(1 mark each)

1 Sue, Leon and Mo share $18 000 in the ratio 3:5:4. What is Leon's share?

A $4500 **B** $6000 **C** $7500 **D** $8000

2 The scale on a building plan is 1:500. How long will a wall that is 56 mm long on the plan actually be?

A 1.12 m **B** 2.8 m **C** 11.2 m **D** 28 m

3 A tree casts a shadow 29 m long. At the same time a 1 m stick casts a shadow 80 cm long. About how tall is the tree?

A 36 m **B** 33 m **C** 31 m **D** 23 m

4 This is a floor plan and elevation of a house.

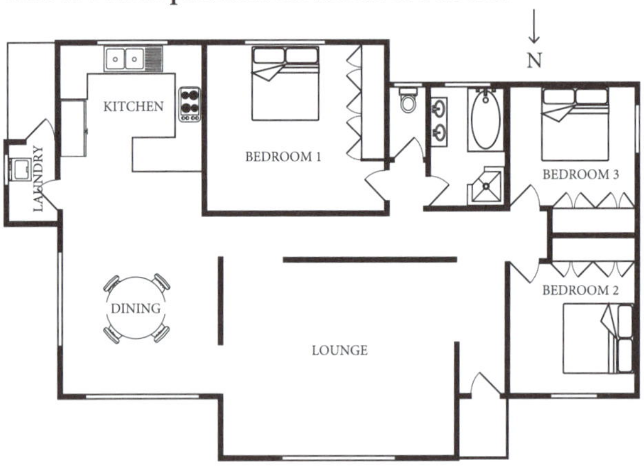

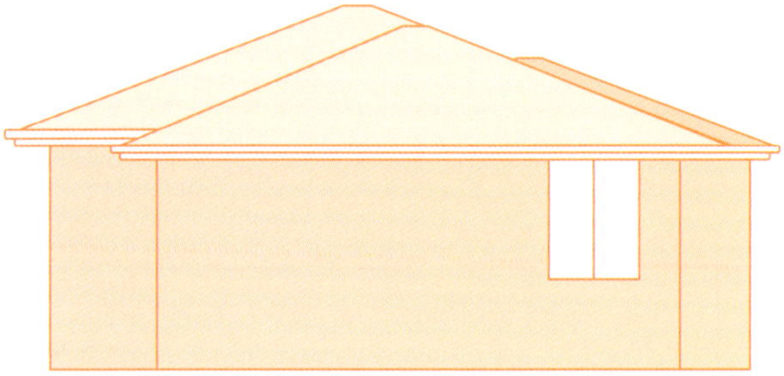

Which elevation is shown?

A North **B** South **C** East **D** West

5 The ratio of red to blue marbles in a bag is 4:3. If there are 17 fewer blue marbles than red, how many red marbles are there?

A 51 **B** 68 **C** 34 **D** 85

6 A recipe uses $2\frac{1}{2}$ cups of flour and $\frac{3}{4}$ cup of sugar. What is the ratio of flour to sugar in simplest form?

A 3:1 **B** 5:2 **C** 10:3 **D** 13:4

7 The ratio of phosphorus to sulphur in superphosphate is 4:5. How many kg per ha of sulphur are applied if, when superphosphate is applied at the rate of 125 kg/ha, 11 kg/ha of phosphorus is applied?

A 13.75 **B** 25 **C** 31.25 **D** 69.45

8 A photo was enlarged and framed. There is a 3 cm wide border between the enlarged photo and the frame. If the original photo was 12 cm long and 7.5 cm wide and it is now 90 cm wide, find the perimeter of the frame.

A 4.5 m **B** 4.68 m **C** 4.8 m **D** 4.92 m

PART B

(Suggested time: 33 minutes)

Show all working.

9 On a map a distance of 17.6 km is represented by a distance of 88 mm.

a What scale is used on the map? (1 mark)

b What actual distance would be represented by a distance on the map of 3.5 cm? (1 mark)

10 A safety expert recommends that the ratio of the distance from the wall of a foot of a ladder to the height the ladder reaches up the wall should be 4:15.

a How far from a wall should the foot of a ladder be if it reaches 12 m up the wall? (1 mark)

b How far up the wall should a ladder reach if its foot is 50 cm from the wall? (1 mark)

c Would it be safe to place a 4 m long ladder so that it reaches 3.2 m up a wall? Justify your answer with appropriate calculations. (2 marks)

11 Fertiliser was added to a rose garden at the rate of 100 g/m^2.

a How many square metres does a 5 kg bag of fertiliser cover? (1 mark)

b The fertiliser consists of nitrogen, phosphorus and potassium in the ratio 9:4:12. How many grams of each element is that per square metre? (3 marks)

12 This is a scale drawing of a paddock. The scale used is 1 : 20 000.

What is the area of the paddock in hectares? (3 marks)

13 This is a partly completed plan of a house.

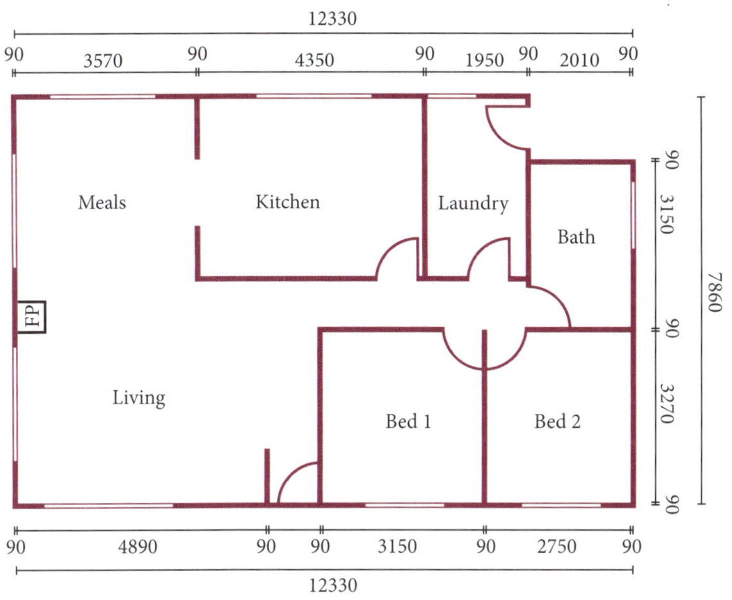

a What is the feature marked FP on the plan? (1 mark)

b How many windows does the house have? (1 mark)

c What would it cost to carpet the two bedrooms at $\$115/m^2$? (Give the answer to the nearest \$10.) (3 marks)

d What is the total area of the house? (4 marks)

ANSWERS

PRACTICE

1 Ratios ➲ pp. 68–69

1 Phosphate : potassium = 1.35 : 3.15
= **3 : 7**

2 Ratio = 5 h : 25 min
= 300 min : 25 min
= 300 : 25
= **12 : 1**

3 a Blue : red : green = 18 : 42 : 12
= **3 : 7 : 2**

b Red : blue and green = 42 : (18 + 12)
= 42 : 30
= **7 : 5**

4 Scale = 9 cm : 90 km
= 1 cm : 10 km
= 1 cm : 10 000 m
= 1 cm : 1 000 000 cm

So the scale is 1 : 1 000 000.

5 Ratio of sugar to lemon juice

$= \frac{3}{4} : \frac{2}{3}$

$= \frac{9}{12} : \frac{8}{12}$

= **9 : 8**

6 Ratio of water to vinegar
= 3.6 L : 120 mL
= 3600 mL : 120 mL
= 3600 : 120
= **30 : 1**

2 Dividing in a given ratio p. 69

1 Total parts = 3 + 5
= 8

Each part = \$40 ÷ 8
= \$5

The first share is 3 × \$5 = \$15.

The second share is 5 × \$5 = \$25.

2 Ratio of contributions = 75 000 : 45 000 : 90 000
= 5 : 3 : 6

Jill has 5 of 14 shares.

Amount Jill receives = $\frac{5}{14}$ of \$33 600
= **\$12 000**

3 Total parts = 9 + 10 + 12
= 31

Each part = 186 ÷ 31
= 6

The first property will receive 9 × 6 or 54 saplings.

The second property will receive 10 × 6 or 60 saplings.

The third property will receive 12 × 6 or 72 saplings.

3 Solving problems involving ratios p. 70

1 25 parts is 1 L or 1000 mL.

So 1 part is (1000 ÷ 25) mL or 40 mL.

So 40 mL of oil should be added.

2 a 2 parts is 112

So 1 part is 112 ÷ 2 or 56

7 parts = 7 × 56
= 392

So 392 red ribbons would be awarded.

b Total parts = 7 + 2 = 9

Each part is 396 ÷ 9 or 44

Now 7 × 44 = 308

and 2 × 44 = 88

So there will be 308 red ribbons and 88 yellow ribbons.

3 Amount of fertiliser required = 4 × 125 kg
= 500 kg

Now 50 parts is 500 kg.

So 1 part is 500 ÷ 50 or 10 kg

7 parts is 7 × 10 kg = 70 kg

So 70 kg of nitrogen will be added.

4 Number of parts = 4 + 2 + 3 + 7
= 16

% of apricots $= \frac{4}{16} \times 100\%$
= 25%

% of plums $= \frac{2}{16} \times 100\%$
= 12.5%

% of peaches $= \frac{3}{16} \times 100\%$
= 18.75%

% of apples $= \frac{7}{16} \times 100\%$
= 43.75%

So the percentage composition of the orchard is 25% apricots, 12.5% plums, 18.75% peaches and 43.75% apples.

5 Originally $\frac{6}{11}$ of the calves were male.

Number of male calves = $\frac{6}{11}$ of 154
= 84

Number of female calves = 154 − 84
= 70

New number of males = 84 + 6
= 90

New number of females = 70 + 10 = 80

New ratio of male to female = 90 : 80
= **9 : 8**

6 Difference in parts = 18 – 7
= 11

So 11 parts is 1925.

$\therefore$ 1 part is 175.

7 parts is 175 × 7 = 1225.

So there are 1225 tiles.

4 Similar figures pp. 72–73

1 Enlargement factor = 75 ÷ 10
= 7.5

New width = 7.5 × 8 cm
= 60 cm

The enlarged photo will be 60 cm wide.

2 a Enlargement factor = $\mathbf{1\frac{1}{3}}$ **or** $\mathbf{\frac{4}{3}}$ **or** $\mathbf{1.\dot{3}}$.

b Each side of enlarged octagon = $\frac{4}{3} \times 27$ cm
= 36 cm

Perimeter = 8 × 36 cm
= 288 cm
= **2.88 m**

c Ratio of object to image = $1:\frac{4}{3}$
= 3:4

So reduction factor = $\mathbf{\frac{3}{4}}$.

3 a **The length and width of a square are always equal. So any square is similar to another because there is only one variable and so it can always be in proportion to another.**

b Enlargement factor = $\frac{25}{15}$
= $\mathbf{\frac{5}{3}}$

c Side length of E = 3 × 15 mm
= 45 mm

Perimeter = 4 × 45 mm
= 180 mm
= **18 cm**

d Side length of F = 1.4 × 25 mm
= 35 mm

Enlargement factor from A = $\frac{35}{15}$
= $\mathbf{\frac{7}{3}}$

e G has side length 4 × 15 mm or 60 mm or 6 cm.

Area = 6^2
= 36

The area of square G is 36 cm^2.

4 a ***EF* corresponds to *AB*.**

b *AD* of *AFED* corresponds to *CD* of *ABCD*.

So enlargement factor = $\frac{30}{16}$
= **1.875**

c *ED* of *AFED* corresponds to *AD* of *ABCD*.

ED = 1.875 × 30 cm
= 56.25 cm

So *CE* = (16 + 56.25) cm
= 72.25 cm.

5 Ratio of the shortest sides = 38 : 57
= 2 : 3

Ratio of the next shortest sides = 44 : 66
= 2 : 3

Ratio of second longest sides = 52 : 78
= 2 : 3

Ratio of longest sides = 56 : 84
= 2 : 3

So the quadrilaterals are similar as the corresponding sides are in the same ratio.

5 Similar triangles p. 75

1 The triangles are similar.

Let h m be the height of the tree.

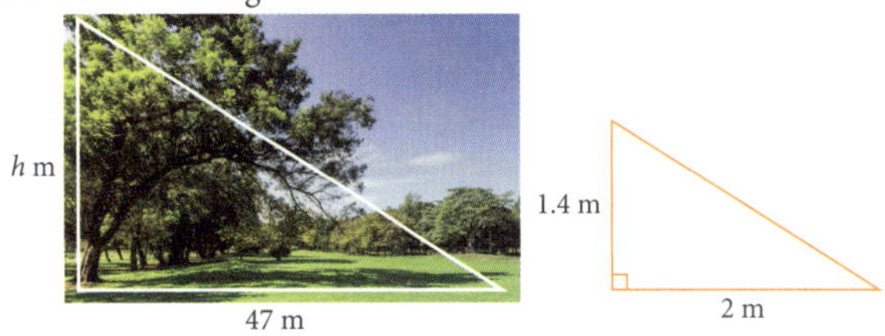

So $\frac{h}{1.4} = \frac{47}{2}$

$h = \frac{47}{2} \times 1.4$
= 32.9
= 33 (nearest unit)

The height of the tree is 33 m, to the nearest metre.

2 a Let d m be the depth of the cable at 65 m.

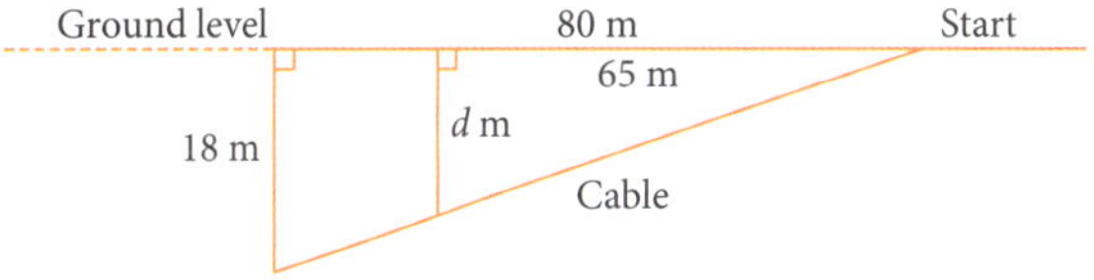

So $\frac{d}{18} = \frac{65}{80}$

$d = \frac{65}{80} \times 18$
= 14.625

The depth of the cable is 14.625 m.

b Let x m be the distance from the start when the cable is 8.1 m below the ground.

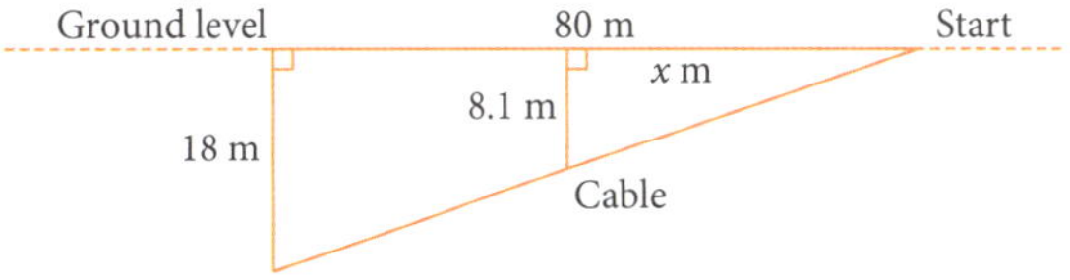

So $\frac{x}{80} = \frac{8.1}{18}$

$x = \frac{8.1}{18} \times 80$
= 36

The distance from the start is 36 m.

c Let h m be the depth of the cable at 54 m.

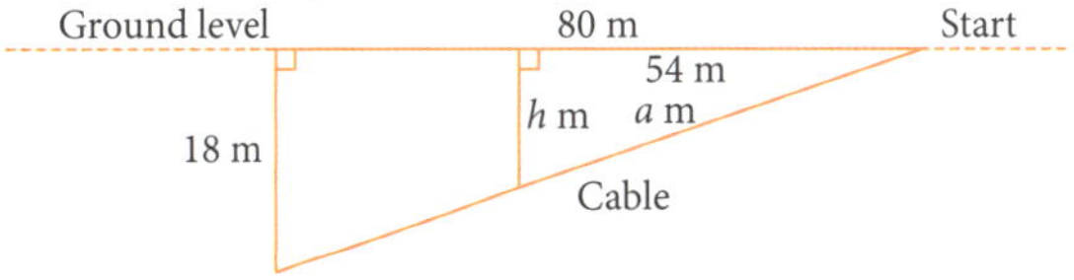

So $\frac{h}{18} = \frac{54}{80}$

$h = \frac{54}{80} \times 18$

$= 12.15$

The depth of the cable is 12.15 m.

Let a m be the length of the cable.

So $a^2 = 54^2 + 12.15^2$

$= 3063.6225$

$a = \sqrt{3063.6225}$

$= 55.35$

The length of the cable is 55.35 m.

3 a The triangles have a common angle at the bottom.

The two top lines of the triangles are parallel, so the corresponding angles formed by the parallel lines are equal.

So the two triangles are equiangular and so are similar.

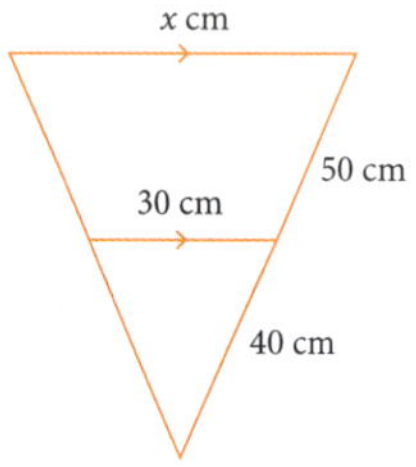

b Now $\frac{x}{30} = \frac{90}{40}$

$x = \frac{90}{40} \times 30$

$= \mathbf{67.5}$

4 Let d m be the distance of the support post from the foot of the rafter.

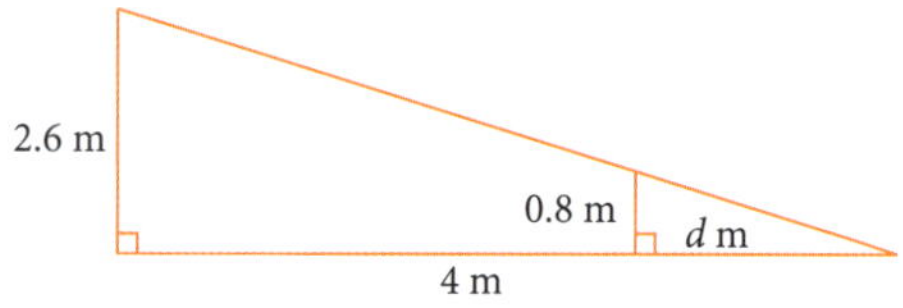

So $\frac{d}{4} = \frac{0.8}{2.6}$

$d = \frac{0.8}{2.6} \times 4$

$= 1.23076923\ldots$

The distance from the foot of the rafter is 1.23 m to two decimal places.

Distance from the wall = (4 − 1.23) m

= 2.77 m

The post is 2.77 m from the wall, to the nearest centimetre.

5 Let x m be the length of the ramp along the ground.

$\therefore (x - 1)$ m is the length to known point.

So $\frac{x-1}{x} = \frac{1.4}{1.7}$

$1.7x - 1.7 = 1.4x$

$0.3x = 1.7$

$x = 5.666666\ldots$

The length of the ramp is 5.67 m, to the nearest centimetre.

6 a The triangles are all similar because they are equiangular.

The shaded part of the diagram is a rectangle.

So each of the triangles has a right angle.

The large triangle has a common angle with each of the smaller triangles, so is similar to both.

Therefore the two smaller triangles must also be similar.

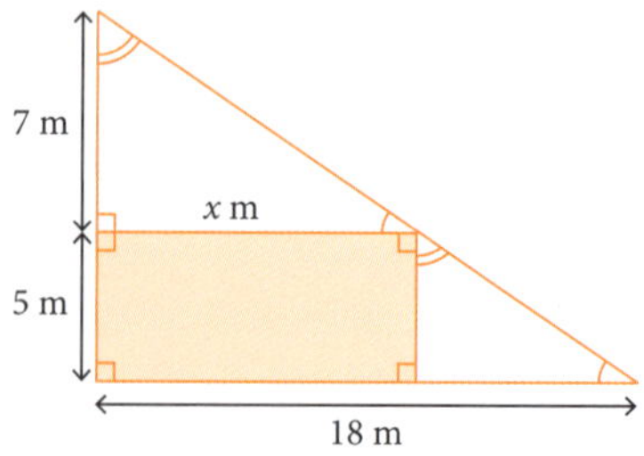

b Let x m be the length of the rectangle.

So $\frac{x}{18} = \frac{7}{12}$

$x = \frac{7}{12} \times 18$

$= 10.5$

Area of rectangle = 10.5×5

$= 52.5 \text{ m}^2$

Area of large triangle = $\frac{1}{2} \times 18 \times 12$

$= 108 \text{ m}^2$

Now $\frac{1}{2}$ of 108 is 54, so the area of the rectangle is slightly less than half that of the big triangle.

6 Maps and plans pp. 77–78

1 a 4.5 cm on the map represents a real distance of $4.5 \times 100\,000$ cm.

Now $4.5 \times 100\,000$ cm = 450 000 cm

= 4500 m

= **4.5 km.**

b 1 cm represents 100 000 cm.

100 000 cm = 1000 m = 1 km

So 1 cm represents 1 km.

So 7.2 km would be represented by 7.2 cm.

c 18 mm is 1.8 cm so represents 1.8 km or 1800 m.

8 mm is 0.8 cm so represents 0.8 km or 800 m.

Perimeter = $2 \times (1800 + 800)$ m

= 5200 m

So the boundary would be 5200 m long.

Cost of = $\$23 \times 5200$

= \$119 600

It would cost \$119 600 to fence the entire boundary.

2 Using the scale to determine the distances:

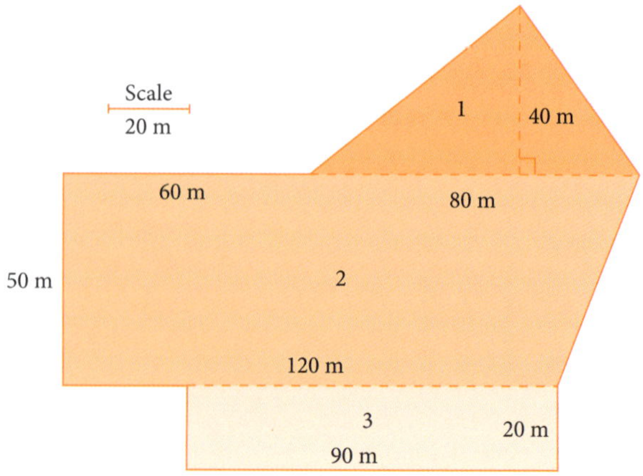

Area 1: triangle

$A = \frac{1}{2}bh$

$= \frac{1}{2} \times 80 \times 40$

$= 1600\ \text{m}^2$

Area 2: trapezium

$A = \frac{1}{2}h(a + b)$

$= \frac{1}{2} \times 50 \times (140 + 120)$

$= 6500\ \text{m}^2$

Area 3: rectangle

$A = lb$

$= 90 \times 20$

$= 1800\ \text{m}^2$

Total area $= (1600 + 6500 + 1800)\ \text{m}^2$

$= 9900\ \text{m}^2$

$=$ **0.99 ha**

3 a On the map the distance between the goalposts is 50 mm.

The real distance is 100 m.

So 50 mm represents 100 m.

1 mm represents 2 m.

But 2 m = 2000 mm.

So the scale is 1 : 2000.

b On the map the distance between the two roads is 39 mm.

Real distance $= 39 \times 2000$ mm

$= 78\,000$ mm

$= 78$ m

The real distance between the two roads is **78 m**.

c The distance between the parallel sides is 16 mm on the map. The parallel sides are 15 mm and 17 mm.

Each mm represents 2 m so the real measurements are:

$A = \frac{1}{2}h(a + b)$

$= \frac{1}{2} \times 32 \times (30 + 34)$

$=$ **1024 m²**

So the area of the field is about 1024 m².

4 $\angle BAC = 239° - 148°$

$= 91°$

So the area can be approximated by finding the area of a right-angled triangle.

Area $= \frac{1}{2}bh$

$= \frac{1}{2} \times 552 \times 372$

$= 102\,672\ \text{m}^2$

$= 10.2672$ ha

The area of the paddock is approximately 10.3 ha.

5 7.8 cm represents 2000×7.8 cm or 15 600 cm or 156 m.

6.9 cm represents 2000×6.9 cm or 13 800 cm or 138 m.

9.3 cm represents 2000×9.3 cm or 18 600 cm or 186 m.

$A \approx \frac{h}{2}(d_f + d_l)$

$= \frac{186}{2}(156 + 138)$

$= 27\,342$

The area of the paddock is about 27 000 m² or 2.7 ha.

7 Building plans pp. 79–80

1 a W/M stands for **washing machine**. P is the **pantry** and F the **refrigerator**.

b The symbol  is the **kitchen sink**.

c Area of shelf $= 3.85\ \text{m} \times 0.55\ \text{m}$

$=$ **2.1175 m²**

d Hugo should run the paper **along the length of the shelf**. Two rolls of paper would cover most of the shelf. He could then use offcuts to complete some of the strip at the back of the shelf, if he so desired. (Although he would need three rolls to do the shelf properly as the area of each roll is 1 m².)

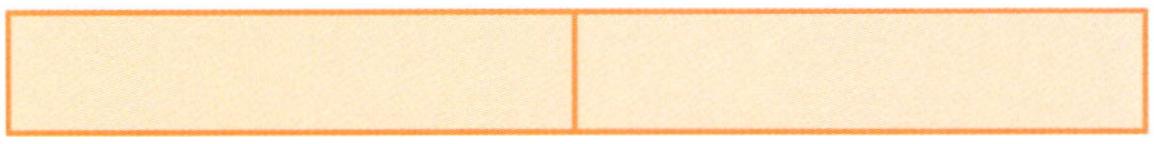

He could run the paper the other way but it would mean lots of joins.

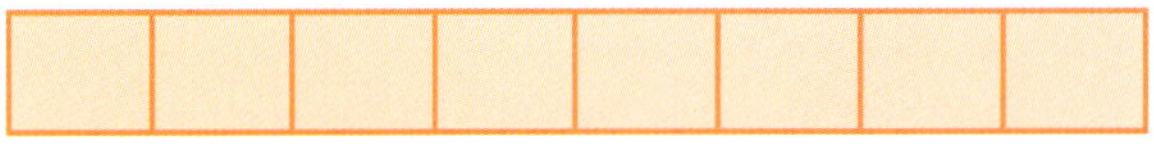

2 a Bedroom 1 is 4600 mm long and 3310 mm wide.

So it is **4.6 m by 3.31 m**.

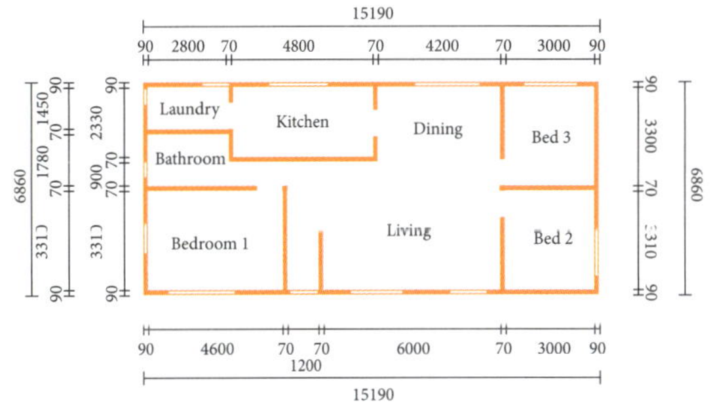

b The exterior walls are 90 mm thick and the interior walls are 70 mm thick. **So the exterior walls are 20 mm thicker than the interior walls.**

c Total length = 15 190 mm

Inside length $= (15\,190 - 2 \times 90)$ mm

$= 15\,010$ mm

$= 15.01$ m

Total width = 6860 mm

Inside width $= (6860 - 2 \times 90)$ mm

$= 6680$ mm

$= 6.68$ m

Floor area $= (15.01 \times 6.68)\ \text{m}^2$

$= 100.2668\ \text{m}^2$

The floor area is 100 m² to the nearest square metre.

d The house is 15 190 mm long.

By measurement, the length of the house on the scale drawing is 31 mm.

So the scale is 31 : 15 190 or **1 : 490**.

e The measured length is 47 mm.

Actual length = 490 × 47 mm
= 23 030 mm
= 23.03 m

The measured width is 35 mm.

Actual width = 490 × 35 mm
= 17 150 mm
= 17.15 m

Area = 23.03 m × 17.15 m
= 394.9645 m^2
= **395 m^2** (nearest square metre)

f When looking out of the bedroom window towards the scrub, Neil will be facing south. This means the bathroom window will be facing **east**.

g The elevation shows just one window, so it is the side of the house with bedroom 2 and bedroom 3. That will be the western side so it is the **western elevation**.

h The ceiling is 2400 mm or 2.4 m above the floor, which is 1000 mm or 1 m above the ground. So the ceiling will be 3400 or **3.4 m** above ground level.

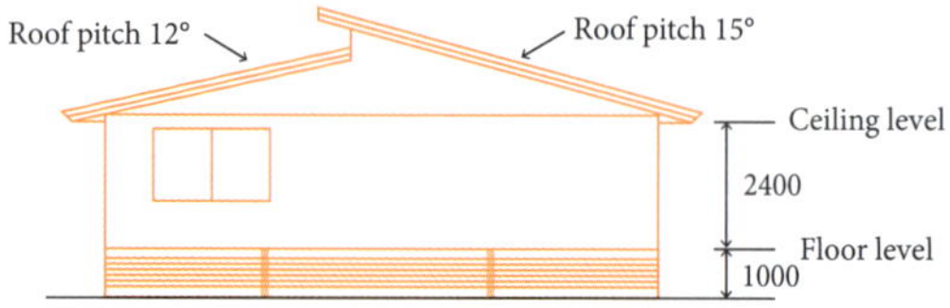

3 a The front door will be facing **east**.

b The block is 42 m deep.

On the map this measurement is 56 mm.

So 56 mm represents 42 m.

56 mm represents 42 000 mm.

Scale = 56 : 42 000
= **1 : 750**

c On the map the house is 26 mm long and 12 mm wide.

750 × 26 mm = 19 500 mm
= 19.5 m

750 × 12 mm = 9000 mm
= 9 m

So the house is 19.5 m long and 9 m wide.

d Each millimetre on the map represents 750 mm or 0.75 m.

By measurement and calculation these measurements, in metres, are found:

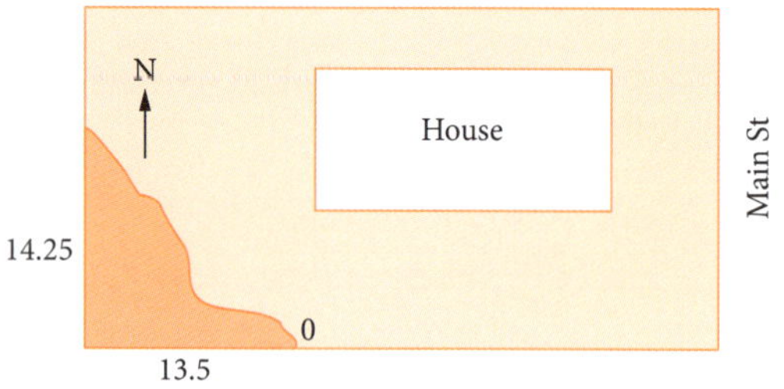

Area of lake $\approx \frac{1}{2} \times 13.5 \times 14.25$
= 96.1875

So the area taken up by the lake is approximately 96 m^2.

Area of land ≈ 42 × 21.75 − 96
= 817.5

So the area of the block of land is approximately 818 m^2.

8 Further applications p. 81

1 Both rooms are the same size: 3.6 m by 3.2 m.

Each has 2 walls 3.2 m by 2.4 m and 1 wall 3.6 m by 2.4 m to be painted, less the windows.

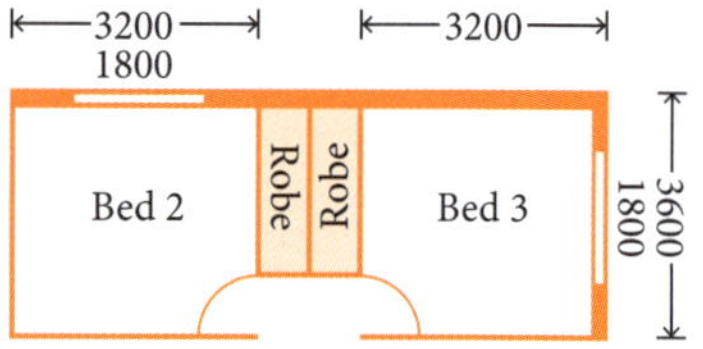

Now 2 × (2 × 3.2 × 2.4 + 3.6 × 2.4) = 48.

Area of 2 windows = 2 × 1.8 × 1.2
= 4.32

Area to be painted = 48 − 4.32
= 43.68 m^2

Two coats of paint area = 87.36 m^2

Number of litres required = 87.36 ÷ 15
= 5.824

So 5.824 L will be required.

This means 2 cans of the paint will be required.

Cost = 2 × \$94.40
= **\$188.80**

2 The paddock is roughly trapezoidal in shape.

The two parallel sides are about 170 m and 200 m in length.

$A \approx \frac{h}{2}(d_f + d_l)$

$= \frac{70}{2}(170 + 200)$

= 12 950 m^2
= 1.295 ha

Amount of fertiliser = 1.295 × 110 kg
= 142.45 kg

So around 142 kg of fertiliser would be required.

3 Using the scale 1 cm represents 100 cm so 1 cm represents 1 m.

Now the scale drawing measures 6 cm by 2.5 cm.

So the actual area would be 6 m by 2.5 m.

The depth of concrete is 75 mm or 0.075 m.

Volume of concrete = 6 × 2.5 × 0.075
= 1.125 m^3

Amount of cement required = 1.125 × 20 kg
= 22.5 kg

Amount of sand = 2 × 22.5 kg
= 45 kg

Amount of gravel = 3 × 22.5
= 67.5 kg

So 22.5 kg of cement, 45 kg of sand and 67.5 kg of gravel would be needed.

HSC EXAM-TYPE QUESTIONS

Part A

1 C. Total shares = 3 + 5 + 4
= 12

Leon's share = $\frac{5}{12} \times \$18\,000$
= \$7500

2 D. 56 mm represents 500 × 56 mm
= 28 000 mm
= 28 m

3 **A.** Let h m be the height of the tree.

$$\text{So } \frac{h}{1} = \frac{29}{0.8}$$

$$h = 36.25$$

So the tree is about 36 m high.

4 **D.** The elevation shown is the side showing the walls of bedrooms 2 and 3.

The northern elevation includes the front door.

So the elevation shown is west.

5 **B.** The marbles in the bag are 4 parts red and 3 blue.

The difference is 17 so 1 part is 17 marbles.

Number of red marbles = 4 × 17
= 68

6 **C.** Flour to sugar $= 2\frac{1}{2} : \frac{3}{4}$
$= 10:3$

7 **A.** 4 units = 11 kg

1 unit = 11 ÷ 4 = 2.75 kg

5 units = 2.75 × 5 = 13.75 kg

8 **D.** Enlargement factor = 90 ÷ 7.5
= 12

New length = 12 × 12 cm
= 144 cm

Length of frame = (144 + 6) cm
= 150 cm
= 1.5 m

Width of frame = (90 + 6) cm
= 96 cm
= 0.96 m

Total = 2 × (1.5 + 0.96) m
= 4.92 m

Part B

9 **a** 88 mm represents 17.6 km.

Now 17.6 km = 17 600 m
= 17 600 000 mm

So the scale is 88 : 17 600 000 = **1 : 200 000** ✓

b 3.5 cm represents 3.5 × 200 000 cm or 700 000 cm.

Now 700 000 cm = 7000 m
= 7 km

So the actual distance is 7 km. ✓

10 **a** Let x m be the distance of the foot from the wall.

$$\text{So } \frac{x}{4} = \frac{12}{15}$$

$$x = \frac{12}{15} \times 4$$

$$= 3.2$$

The foot should be 3.2 m from the wall. ✓

b Let h m be the distance up the wall.

$$\text{So } \frac{h}{15} = \frac{0.5}{4}$$

$$h = \frac{0.5}{4} \times 15$$

$$= 1.875$$

The ladder should reach 1.875 m up the wall. ✓

c Let d m be the distance from the wall when the ladder reaches 3.2 m up the wall.

$$\text{So } \frac{d}{4} = \frac{3.2}{15}$$

$$d = \frac{3.2}{15} \times 4$$

$$= 0.853\,333\ldots$$

So the foot should be around 0.85 m from the wall. ✓

If the ladder is 4 m long, then the distance from the wall will be:

$$d^2 = 4^2 - 3.2^2$$

$$= 5.76$$

$$d = \sqrt{5.76}$$

$$= 2.4$$

The foot would actually be 2.4 m from the wall. ✓

So it would not be safe.

11 **a** 5 kg = 5000 g

Now 100 g covers 1 m^2.

So 5000 g covers 50 m^2.

A 5 kg bag covers **50 square metres.** ✓

b Number of parts = 9 + 4 + 12
= 25

Amount of nitrogen $= \frac{9}{25} \times 100$ g
= 36 g ✓

Amount of phosphorus $= \frac{4}{25} \times 100$ g
= 16 g ✓

Amount of potassium $= \frac{12}{25} \times 100$ g
= 48 g ✓

12 Scale is 1 : 20 000.

So 1 mm represents 20 000 mm.

But 20 000 mm = 20 m.

So each millimetre on the diagram is a real distance of 20 m. ✓

The top boundary is 55 mm long.

55 × 20 m = 1100 m

The bottom boundary is 33 mm.

33 × 20 m = 660 m

The distance between the top and bottom boundaries is 19 mm.

19 × 20 m = 380 m ✓

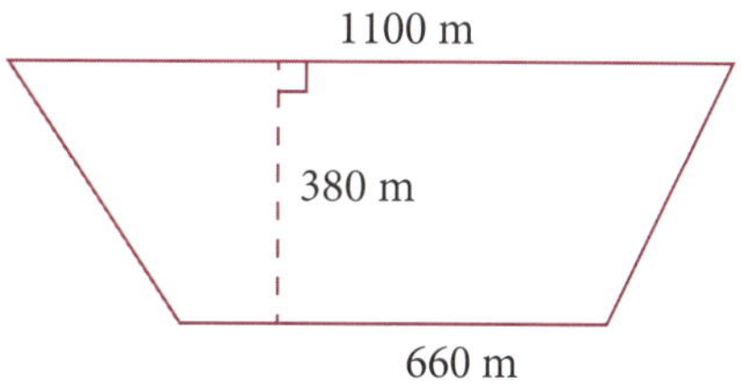

$$A = \frac{h}{2}(a + b)$$

$$= \frac{380}{2} \times (1100 + 660)$$

$$= 334\,400 \text{ m}^2$$

$$= 33.44 \text{ ha}$$

The area of the paddock is 33.44 hectares. ✓

13 a FP is a **fireplace**. ✓

b **The house has 9 windows.** ✓

c Both bedrooms are 3.27 m long.

Together the 2 bedrooms are (3.15 + 2.75) m wide.

Total width = 5.9 m ✓

Total area = 3.27×5.9

$= 19.293\text{ m}^2$ ✓

Cost of carpet = $19.293 \times \$115$

= \$2218.695

= **\$2220** (nearest \$10) ✓

d The house is 12 330 mm or 12.33 m long and 7860 mm or 7.86 m wide.

Now 90 + 3270 + 90 + 3150 + 90 = 6690.

So one side is 6690 mm or 6.69 m long. ✓

Width of exterior wall with laundry door

= (7.86 − 6.69) m

= 1.17 m ✓

2010 + 90 = 2100

So one wall of the bathroom is 2100 mm or 2.1 m long. ✓

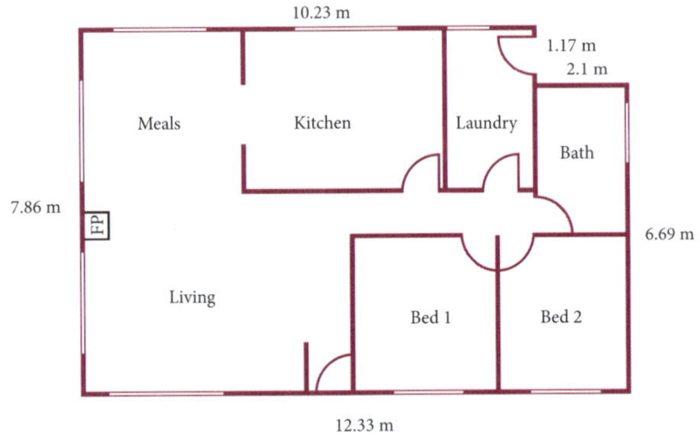

$A = 12.33 \times 7.86 - 2.1 \times 1.17$

$= 94.4568$

$= 94.5$ (1 d.p.)

The area of the house is approximately 94.5 square metres. ✓

CHAPTER 5 INVESTMENT

TOPIC: FINANCIAL MATHEMATICS

1 Simple interest

- **Interest is the amount paid for the use of money**; for example, the amount someone pays to borrow money.
- **Interest is also the amount earned on an investment**.
- **Simple interest is where the amount paid at regular intervals** is the same over the whole period of the investment or loan.
- **Simple interest is calculated using the formula $I = Prn$** where P is the principal or initial amount of money, r is the interest rate per period of time expressed as a decimal and n is the number of time periods.

EXAMPLE 1

If \$8000 is invested at 6% p.a. simple interest over 3 years, find the amount of interest earned.

$P = \$8000$, $r = 0.06$, $n = 3$

$I = Prn$

$= \$8000 \times 0.06 \times 3$

$= \$1440$

Remember that p.a. means 'per annum' or per year.

EXAMPLE 2

Find the simple interest paid in 18 months on an amount of \$1725 invested at 4% p.a.

$P = \$1725, \quad r = 0.04$

Now 18 months is $1\frac{1}{2}$ years so $n = 1.5$.

$I = Prn$

$= \$1725 \times 0.04 \times 1.5$

$= \$103.50$

The amount of interest paid is \$103.50.

The period of time must always match the interest rate.

EXAMPLE 3

Sara places \$9000 in an account earning 1.2% per quarter simple interest. How much interest does Sara earn in 3 years?

$P = \$9000$, $r = 0.012$, $n = 12$

$I = Prn$

$= \$9000 \times 0.012 \times 12$

$= \$1296$

Sara earns \$1296 interest.

A quarter is a fourth of a year. 3 years = 3 × 4 quarters

EXAMPLE 4

Find the amount invested if \$2100 interest was earned in 2 years at 8% p.a. simple interest.

$I = \$2100$, $r = 0.08$, $n = 2$

$I = Prn$

$\$2100 = P \times 0.08 \times 2$

$\$2100 = 0.16P$

$P = \$2100 \div 0.16$

$= \$13\,125$

The amount invested was \$13 125.

If the amount of interest is known, solving an equation after substituting into the simple interest formula allows us to find either the principal, interest rate or period of an investment.

EXAMPLE 5

Find the rate of simple interest paid per year if an amount of \$13 500 earned \$3240 in interest over 6 years.

Interest earned per year $= \$3240 \div 6$

$= \$540$

$\text{Rate} = \frac{\$540}{\$13\,500} \times 100\%$

$= 4\%$

The rate of interest is 4% p.a.

Alternatively substitute into the formula and solve the resulting equation for r. r will be a decimal and will need to be converted to a percentage.

→ PRACTICE

1 **Find the amount of simple interest earned if \$7250 is invested for 5 years at $6\frac{1}{2}$% p.a.**

2 **Find the simple interest paid in 6 months on an amount of \$2350 invested at 3% p.a.**

3 Courtney places \$6200 in an account earning 0.9% per quarter simple interest. How much interest does Courtney earn in 4 years?

4 Find the amount invested if \$4334.40 interest was earned in 3 years at 7% p.a. simple interest.

5 Find the rate of simple interest paid per year if an amount of \$16 000 earned \$11 520 in interest over 9 years.

6* Find the annual rate of simple interest if \$24.75 interest was earned in 6 months on an investment of \$900.

Answers ➲ p. 101

2 Compound interest

- **In an investment earning compound interest**, each amount of interest is added to the investment so that the amount upon which interest is earned increases over time. As a result, the amount of compound interest earned will also increase with each period of time.
- **The final value of an investment is called the future value (FV).** The principal or initial amount is called the present value (PV).
- **The formula $FV = PV(1 + r)^n$ gives the value of an amount earning compound interest** for n periods where r is the interest rate expressed as a decimal.
- **The amount of compound interest earned** is the difference between the future value and the present value: $I = FV - PV$.

EXAMPLE 1

Find the future value, to the nearest dollar, if an amount of \$17 000 is invested for 3 years at 4.5% p.a.

$PV = \$17\,000,\ r = 0.045,\ n = 3$

$FV = PV(1 + r)^n$

$= \$17\,000(1 + 0.045)^3$

$= \$19\,399.8241\ldots$

$= \$19\,400$ (nearest dollar)

The future value is \$19 400 to the nearest dollar.

The formula as it is used here is called the future value formula. It is commonly used in the financial world.

EXAMPLE 2

Find the amount of interest earned if \$5500 is invested at 5% p.a. compound interest for 2 years.

$PV = \$5500,\quad r = 0.05,\quad n = 2$

$FV = PV(1 + r)^n$

$= \$5500(1 + 0.05)^2$

$= \$6063.75$

$I = FV - PV$

$= \$6063.75 - \$5500$

$= \$563.75$

An amount of \$563.75 is earned in interest.

The formula $A = P(1 + r)^n$ is an equivalent formula, often called the compound interest formula.

EXAMPLE 3

An amount of \$4500 is invested at 6% p.a. interest compounded monthly for 4 years. Find the future value.

$PV = \$4500,\ r = \dfrac{0.06}{12},\ n = 48$

$FV = PV(1 + r)^n$

$= \$4500\left(1+\dfrac{0.06}{12}\right)^{48}$

$= \$5717.201\,22\ldots$

$= \$5717$ (nearest dollar)

The interest rate must be expressed per the compounding period and the period must match the interest rate. So here the rate must be the monthly rate and the period must be the number of months.

EXAMPLE 4

Find the amount of interest earned if \$16 250 is invested for 3 years at 9% p.a. interest, compounded quarterly.

$PV = \$16\,250,\ r = \dfrac{0.09}{4},\ n = 3 \times 4$

$FV = PV(1 + r)^n$

$= \$16\,250\left(1+\dfrac{0.09}{4}\right)^{3\times 4}$

$= \$21\,223.3123\ldots$

$= \$21\,223$ (nearest dollar)

$I = \$21\,223 - \$16\,250$

$= \$4973$

The compound interest earned is \$4973, to the nearest dollar.

As $\dfrac{0.09}{4} = 0.0225$ the interest rate could have been expressed as 0.0225 in the formula.

EXAMPLE 5

David is to invest \$23 000 at 2.75% per half year compound interest. He wants to have \$25 000 in 18 months time. Will he achieve that goal?

$PV = \$23\,000,\quad r = 0.0275,\quad n = 3$

$FV = PV(1 + r)^n$

$= \$23\,000(1 + 0.0275)^3$

$= \$24\,950.1595\ldots$

$= \$24\,950$ (nearest dollar)

So David will be about $50 short of his goal.

18 months is a year and a half so 3 half-years.

→ PRACTICE

1 Find the future value, to the nearest dollar, if an amount of $15 000 is invested for 5 years at 6.25% p.a.

2 Find the amount of interest earned if $9400 is invested at 6% p.a. compound interest for 2 years.

3 An amount of $24 000 is invested at 7.5% p.a. interest, compounded monthly for 3 years. Find the future value.

4 Find the amount of interest earned if $23 400 is invested for 4 years at 8% p.a. interest, compounded quarterly.

5 Davina is to invest $27 500 at 1.95% per half year compound interest. She wants to have $30 000 in 30 months time. Will she achieve that goal?

6* $60 000 is invested for 5 years at 8% p.a. interest. How much greater is the future value if the interest is compounded monthly rather than half-yearly?

Answers ➲ p. 101

3 Finding the present value

→ **The present value of an investment is the single amount of money** that could be invested now to give the known future value.

→ **The formula** $PV = \dfrac{FV}{(1+r)^n}$ **gives the present value** when the future value, interest rate and period are known. This formula is just a rearrangement of the future value formula.

→ **The present value, or the interest rate, could also be found** by substituting into the future value formula and solving the resulting equation.

EXAMPLE 1

Find the amount of money that needs to be invested now to give a future value of $15 150 in 4 years time if invested at 6% p.a.

$FV = \$15\,150,\ r = 0.06,\ n = 4$

$FV = PV(1 + r)^n$

$\$15\,150 = PV(1 + 0.06)^4$

$PV = \dfrac{\$15\,150}{(1+0.06)^4}$

$= \$12\,000.218\ldots$

$= \$12\,000$ (nearest dollar)

Substituting into the future value formula and rearranging gives the alternative formula for finding the present value.

EXAMPLE 2

An amount of money was invested at 8% p.a. interest compounded quarterly for 6 years. If the future value is $11 259, find the present value.

$FV = \$11\,259,\ r = \dfrac{0.08}{4},\ n = 6 \times 4$

$PV = \dfrac{FV}{(1+r)^n}$

$= \dfrac{\$11\,259}{\left(1+\dfrac{0.08}{4}\right)^{6\times 4}}$

$= \$6999.962\,23\ldots$

$= \$7000$ (nearest dollar)

So $7000 invested now would grow to $11 259 in 6 years.

EXAMPLE 3

Rose wants to have $20 000 for an overseas trip that she intends to take in 4 years time. What single sum of money does she need to invest now at 4.5% p.a. compounded monthly?

$FV = \$20\,000,\ r = 0.045 \div 12 = 0.003\,75,\ n = 4 \times 12 = 48$

$PV = \dfrac{FV}{(1+r)^n}$

$= \dfrac{\$20\,000}{(1+0.003\,75)^{48}}$

$= \$16\,711.0291\ldots$

$= \$16\,711$ (nearest dollar)

Rose needs to invest $16 711 now.

The interest is compounded monthly so the rate must be per month and the period must be the number of months.

EXAMPLE 4

What annual rate of compound interest is paid if $24 000 accrues to $27 822.58 in 5 years?

$PV = \$24\,000,\ FV = \$27\,822.58,\ n = 5$

$FV = PV(1 + r)^n$

$\$27\,822.58 = \$24\,000(1 + r)^5$

$(1 + r)^5 = 1.159\,274\,16\ldots$

$1 + r = 1.030\,000\,01\ldots$

Divide both sides by $24 000.

Take the fifth root of both sides.

$= 1.03$ (2 d.p.)

$r = 0.03$

So the annual rate of interest is 3%.

Subtract 1 from both sides.

EXAMPLE 5

Find the annual rate of compound interest paid if \$60 000 increases to \$76 256.67 in 10 years, interest compounded monthly.

$PV = \$60\,000$, $FV = \$76\,256.67$, $n = 10 \times 12 = 120$

$FV = PV(1 + r)^n$

$\$76\,256.67 = \$60\,000(1 + r)^{120}$

$(1 + r)^{120} = 1.270\,944\ldots$

$1 + r = 1.002$

$r = 0.002$

So the monthly rate of interest is 0.2%.

The annual rate of interest is $12 \times 0.2\% = 2.4\%$.

Take care to answer the question. Although the interest is compounded monthly, the required interest rate is the annual rate.

→ PRACTICE

1 **Find the amount of money that needs to be invested now to give a future value of \$47 000 in 12 years time if invested at 4.5% p.a. interest.**

2 **An amount of money was invested at 9% p.a. interest, compounded monthly, for 5 years. If the future value is \$20 000, find the present value.**

3 **Ross wants to have \$28 000 for a new vehicle that he intends to buy in 5 years time. What single sum of money does he need to invest now at 4.2% p.a. compounded quarterly?**

4 **What annual rate of compound interest is paid if \$30 000 accrues to \$40 202.87 in 6 years?**

5 **Find the annual rate of compound interest paid if \$55 000 increases to \$63 889 in 5 years, interest compounded monthly.**

6* **Find the annual interest rate if \$13 000 increases to \$15 543 in 3 years, interest compounded quarterly.**

Answers ➲ p. 102

4 Tables of compounded values

- **Often tables are available** that can be used to find the amounts to which an investment grows.
- **If a table gives compounded values of \$1**, multiplying the given factor for the appropriate interest rate and period by the amount involved gives the future value.

Future value of \$1

Period	1%	1.5%	2%	2.5%	3%	4%	5%	6%
1	1.0100	1.0150	1.0200	1.0250	1.0300	1.0400	1.0500	1.0600
2	1.0201	1.0302	1.0404	1.0506	1.0609	1.0816	1.1025	1.1236
3	1.0303	1.0457	1.0612	1.0769	1.0927	1.1249	1.1576	1.1910
4	1.0406	1.0614	1.0824	1.1038	1.1255	1.1699	1.2155	1.2625
5	1.0510	1.0773	1.1041	1.1314	1.1593	1.2167	1.2763	1.3382
6	1.0615	1.0934	1.1262	1.1597	1.1941	1.2653	1.3401	1.4185
12	1.1268	1.1956	1.2682	1.3449	1.4258	1.6010	1.7959	2.0122
18	1.1961	1.3073	1.4282	1.5597	1.7024	2.0258	2.4066	2.8543
24	1.2697	1.4295	1.6084	1.8087	2.0328	2.5633	3.2251	4.0489
30	1.3478	1.5631	1.8114	2.0976	2.4273	3.2434	4.3219	5.7435
36	1.4308	1.7091	2.0399	2.4325	2.8983	4.1039	5.7918	8.1473

EXAMPLE 1

Find the value of an investment of $6500 at 6% p.a., compounded annually for 5 years.

Future value = $6500 × 1.3382
= $8698.30

The interest rate is 6% and the period is 5 years so the factor from the table is 1.3382.

EXAMPLE 2

$8000 is invested for 3 years at 10% p.a. interest, compounded quarterly. What is the future value?

Interest rate = 10% ÷ 4 = 2.5%
Period = 3 × 4 quarters = 12 quarters
Future value = $8000 × 1.3449
= $10 759.20

The interest is compounded quarterly so the interest rate must be the rate per quarter and the period must be the number of quarters.

EXAMPLE 3

An amount of $15 000 invested at 12% p.a. interest, compounded monthly, grows to $19 045.50. For how many years was the money invested?

Interest rate = 12% p.a.
= 1% per month
Now $15 000 × factor = $19 045.50
factor = $19 045.50 ÷ $15 000
= 1.2697
From the table this is the factor for 24 periods.
Now 24 months = 2 years.
So the money was invested for 2 years.

Look in the table for the required factor using the appropriate interest rate.

➔ PRACTICE

1 **Find the value of an investment of $7500 at 5% p.a., compounded annually for 6 years.**

2 **$11 000 is invested for 6 years at 8% p.a., interest compounded quarterly. What is the future value?**

3 **An amount of $9000 invested at 18% p.a., interest compounded monthly, grows to $15 381.90. For how many years was the money invested?**

4* a **Find the annual interest rate if $16 400 amounts to $19 953.88 in $2\frac{1}{2}$ years, interest compounded half-yearly.**

b **How much interest would be earned if the original amount was invested for 15 years at the same interest rate (and same compounding period)?**

Answers ➲ p. 102

5 Comparing investments

➔ **When considering investment opportunities** things to take into account include the type of interest paid, the interest rate, and the period and compounding period.

➔ **Graphs of future values can be used in comparisons.** Such graphs are often examples of exponential growth.

EXAMPLE 1

a **Hannah is considering investing $10 000 for 4 years in an account earning 6.5% simple interest. What will be the future value of Hannah's investment?**

b **An alternative for Hannah is to invest, for the same amount of time, in an account earning 0.5% per month interest, compounded monthly. In which account would you recommend Hannah should invest? Justify your answer with appropriate calculations.**

a $P = \$10\,000$, $r = 0.065$, $n = 4$
$I = Prn$
$= \$10\,000 \times 0.065 \times 4$
$= \$2600$
Future value = $10 000 + $2600
= $12 600

The term 'future value' is not limited to compound interest investments.

b $PV = \$10\,000$, $r = 0.005$, $n = 4 \times 12 = 48$
$FV = PV(1 + r)^n$
$= \$10\,000(1 + 0.005)^{48}$
$= \$12\,704.8916\ldots$
$= \$12\,704.89$ (nearest cent)

Hannah should take the second option because it gives a greater future value.

EXAMPLE 2

$15 000 is invested for 5 years at 4.8% p.a. interest, compounded monthly. About how much extra would need to be invested at 4.8% p.a. compounded quarterly to produce the same result in 4 years?

$PV = \$15\,000$, $r = 0.048 \div 12 = 0.004$, $n = 12 \times 5 = 60$
$FV = PV(1 + r)^n$
$= \$15\,000(1 + 0.004)^{60}$
$= \$19\,059.6107\ldots$
$= \$19\,060$ (nearest dollar)

$FV = \$19060$, $r = 0.048 \div 4 = 0.012$, $n = 4 \times 4 = 16$

$$PV = \frac{FV}{(1+r)^n}$$

$$= \frac{\$19\,060}{(1+0.012)^{16}}$$

$$= \$15\,748.3368\ldots$$

$$\approx \$15\,750$$

About an extra $750 would need to be invested.

Use common sense when rounding off, particularly when the question says 'about'.

EXAMPLE 3

Ben decides to invest $2500 and considers some different options.

a **One option is to invest the money in an account earning $225 simple interest every year. Write a formula for the amount of money ($*A*) Ben will have in the account after *n* years if he chooses this option.**

b **A second option is to invest the money to earn compound interest. Ben's bank supplies him with a graph to show the amount he will have in the account over several years.**

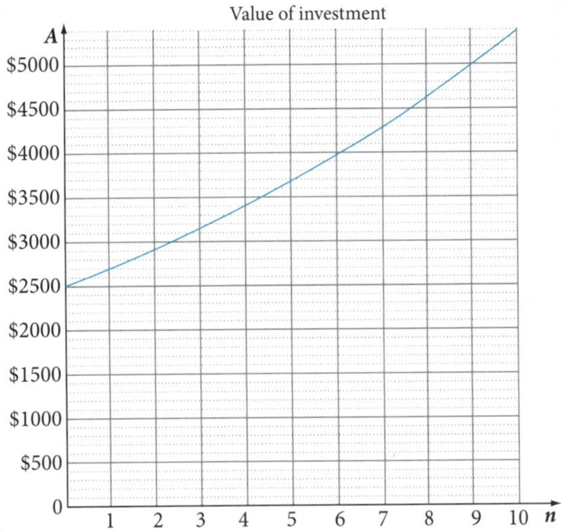

Copy the graph and on the same diagram draw the graph of the amount Ben will have if he chooses the first option.

c **After how many years will the amounts be the same?**

d **After how many years will the amount double if Ben chooses the compound interest option?**

e **Use the formula from part a to find the number of years it will take the simple interest option to double.**

f **Which option do you think Ben should choose? Justify your answer.**

a $A = 2500 + 225n$

He will have the original amount ($2500) plus $225 for every year.

b

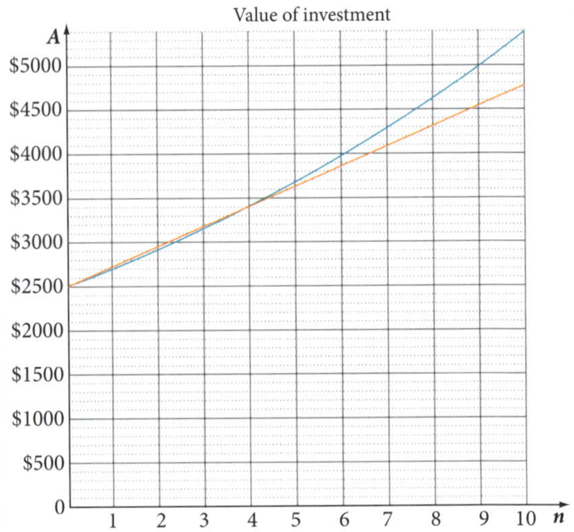

c 4 years

d 9 years

When $A = 5000$

e $A = 2500 + 225n$

When the amount has doubled, $A = 5000$.

$5000 = 2500 + 225n$

$2500 = 225n$

$n = 2500 \div 225$

$= 11.111\,11\ldots.$

The simple interest option doubles in just over 11 years.

f If Ben intends to invest for less than 4 years he should choose the simple interest option but if he intends to invest the money for more than 4 years he should choose the compound interest option. If he doesn't know how long the investment might be, the compound interest option is better. This is because, although the simple interest produces a slightly better result for a small number of years, the compound interest produces a significantly better result after a few years.

→ PRACTICE

1 a Henry is considering investing \$20 000 for 3 years in an account earning 5.9% simple interest. What will be the future value of Henry's investment?

b An alternative for Henry is to invest for 3 years in an account earning 0.45% per month interest, compounded monthly. In which account would you recommend Henry should invest? Justify your answer with appropriate calculations.

2 \$25 000 is invested for 6 years at 5.4% p.a. interest, compounded quarterly. About how much extra would need to be invested at 5.4% p.a. compounded monthly to produce the same result in 5 years?

3 Shannon decides to invest \$1000 and considers some different options.

a One option is to invest the money in an account earning \$80 simple interest every year. Write a formula for the amount of money (\$$A$) Shannon will have in the account after n years if he chooses this option.

b A second option is to invest the money to earn compound interest. Shannon's financial institution supplies him with a graph to show the amount he will have in the account over several years.

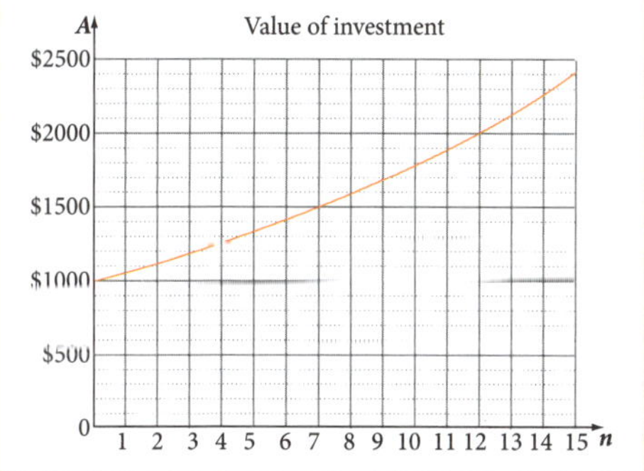

Copy the graph and on the same diagram draw the graph of the amount Shannon will have if he chooses the first option.

c After how many years will the amounts be the same?

d After how many years will the amount double if Shannon chooses the compound interest option?

e Use the formula from part a to find the number of years it will take that option to double.

f Which option do you think Shannon should choose? Justify your answer.

4* \$125 000 is to be invested for 2 years. Which would give the best return: 6% p.a. compounded annually, 5.95% compounded half-yearly, 5.9% compounded quarterly or 5.85% compounded monthly?

Answers ➲ pp. 102–103

6 Practical applications

- ➔ **If an asset increases in value over time** it is said to appreciate in value.
- ➔ **The future value formula can be used** to find the appreciated value.
- ➔ **The rate of inflation can also be determined** by considering the present and future values.

EXAMPLE 1

A ring currently valued at \$7200 is expected to appreciate at the rate of 4% p.a. What is its expected value in 10 years time? Give the answer to the nearest 100 dollars.

$PV = \$7200,\ r = 0.04,\ n = 10$

$$\begin{aligned} FV &= PV(1+r)^n \\ &= \$7200(1+0.04)^{10} \\ &= \$10\,657.7588\ldots \\ &= \$10\,700 \quad \text{(nearest \$100)} \end{aligned}$$

We just need to find the future value using the value of the ring as the present value.

The expected value of the ring in 10 years time is \$10 700 to the nearest 100 dollars.

EXAMPLE 2

A painting was bought for \$23 000 in December 2008 and sold for \$57 000 in December 2019.

a What was the annual rate of appreciation? (Give the answer as a percentage correct to one decimal place.)

b If the painting continues to appreciate at that rate, find its approximate worth at the end of 2025.

a $FV = \$57\,000,\ PV = \$23\,000,\ n = 11$

$$\begin{aligned} FV &= PV(1+r)^n \\ \$57\,000 &= \$23\,000(1+r)^{11} \\ (1+r)^{11} &= \frac{57\,000}{23\,000} \\ 1+r &= \sqrt[11]{\frac{57\,000}{23\,000}} \\ &= 1.086\,0043\ldots \\ r &= 0.086 \quad \text{(3 d.p.)} \end{aligned}$$

We still call the first amount the present value and the second the future value even though both might be in the past!

So the annual rate of appreciation is 8.6%.

b In 2025, $n = 6$

$FV = \$57\,000(1 + 0.086)^6$
$= \$93\,509.0849\ldots$

The painting will be worth approximately $93 500 at the end of 2025.

EXAMPLE 3

A gemstone purchased for $2780 is predicted to appreciate at a rate of 5% p.a. Use a trial and error method to determine how many years it will take to be worth $5000.

$FV = \$5000, PV = \$2780, r = 0.05$

$FV = PV(1 + r)^n$

$\$5000 = \$2780(1 + 0.05)^n$

So $1.05^n = \dfrac{5000}{2780}$

$= 1.798\,561\,15\ldots$

$n \approx 12$

It will take around 12 years for the gemstone to be worth $5000.

Begin with any value for n. For example, try 10. $1.05^{10} = 1.628\ldots$ which is too low so the value must be higher. Try $n = 15$. $1.05^{15} = 2.078\ldots$ which is too high so the value of n is between 10 and 15.

EXAMPLE 4

Lara currently pays $420 per week rent. Assuming a constant inflation rate of 2.3% p.a., and assuming Lara's rent is adjusted each year with the inflation rate, how much will she need to pay in three years time?

$PV = \$420, r = 0.023, n = 3$

$FV = PV(1 + r)^n$

$= \$420(1 + 0.023)^3$

$= \$449.651\,650\ldots$

$= \$450$ (nearest dollar)

Lara will need to pay $450 per week in 3 years time.

Many wages and expenses are adjusted periodically with the inflation rate.

EXAMPLE 5

If the quarterly rate of inflation is 0.75%, find the percentage by which prices will increase in 3 years.

$PV = \$100, r = 0.0075, n = 3 \times 4 = 12$

$FV = PV(1 + r)^n$

$= \$100(1 + 0.0075)^{12}$

$= \$109.380\,689\ldots$

So the percentage increase in prices is approximately 9.4%.

We are not given a present value in the question so could use any amount and determine the percentage increase. The simplest option is to use $100 as the present value because there won't be any further calculations needed.

→ PRACTICE

1 A necklace currently valued at $3500 is expected to appreciate at the rate of 6% p.a. What is its expected value in 5 years time? Give the answer to the nearest 100 dollars.

2 A vase was bought for $6900 in July 2006 and sold for $16 000 in July 2020.

a What was the annual rate of appreciation? (Give the answer as a percentage correct to one decimal place.)

b If the vase continues to appreciate at that rate, find its approximate worth at the end of 2027.

3 A painting purchased for $7910 is predicted to appreciate at a rate of 4% p.a. Use a trial and error method to determine how many years it will take to be worth $10 000.

4 Larry currently earns $1700 per week. Assuming a constant inflation rate of 2.4% p.a. and assuming Larry's wage is adjusted each year with the inflation rate, how much will he earn per week in 4 years time?

5 If the quarterly rate of inflation is 0.85%, find the percentage by which prices will increase in 7 years.

6* A house increased in value from $650 000 in January 2012 to $1.1 million in January 2018. What was the annual rate of appreciation?

Answers ➲ p. 103

HSC EXAM-TYPE QUESTIONS

Total marks: 30

PART A

(Suggested time: 12 minutes)
(1 mark each)

1 Find the amount of simple interest earned if $8000 is invested at 9% per year for 18 months.

A $1080 B $1104 C $1296 D $4960

2 $2000 is placed in an account earning 8.4% p.a. interest, compounded monthly. Find the value (to the nearest dollar) of the investment after 8 years.

A $3813 B $3907 C $3344 D $4465

3 An artwork valued at $125 000 is predicted to appreciate at the rate of 6% p.a. over 5 years. Which is closest to its predicted value then?

A $164 000 B $162 500 C $167 500 D $166 000

4 How much interest (to the nearest dollar) is earned if $9000 is placed in an account earning 15% p.a. interest, compounded quarterly for 7 years?

A $9450 B $11 646 C $14 940 D $16 230

5 Which will be worth the most at the end of 3 years?

A $8000 cash kept under the mattress
B $6750 earning 7% p.a. simple interest
C $6500 earning 7% interest compounded annually
D $6250 earning 0.7% per month compound interest

6 A certain amount is to be invested into an account earning 5.5% p.a. interest, compounded half-yearly. At the end of 3 years the future value will be $4295. What is the present value?

A $3650 B $3750 C $3850 D $3950

7 What rate of compound interest per quarter will mean that an amount of $6000 will accumulate to around $6500 in 2 years?

A 4.1% B 2.5% C 1% D 0.25%

8 $10 000 is invested at 0.6% per month compound interest for 5 years. What annual rate of simple interest will produce the same result?

A 2.7% B 6.8% C 7.2% D 8.6%

PART B

(Suggested time: 33 minutes)

Show all working.

9 Find the future value if an amount of $7600 is invested at 7.5% p.a. interest, compounded monthly for 3 years. (2 marks)

10 Find the amount of interest earned if $24 350 is invested for 5 years at 8% p.a. interest, compounded quarterly. (2 marks)

11 An amount of money was invested at 6% p.a. interest, compounded monthly, for 4 years. If the future value is $32 700, find the present value. (2 marks)

12 What annual rate of compound interest is paid if $65 000 accrues to $87 260.60 in 8 years? (2 marks)

13 A diamond, bought in July 2004 for $6200, appreciated at the rate of 12% p.a. What was its value when it was sold in July 2018? (2 marks)

14 The table shows the future value of $1 for certain interest rates and periods.

Future value of $1				
Period	1.5%	2%	3%	4%
2	1.0302	1.0404	1.0609	1.0816
3	1.0457	1.0612	1.0927	1.1249
4	1.0614	1.0824	1.1255	1.1699
6	1.0934	1.1262	1.1941	1.2653
12	1.1956	1.2682	1.4258	1.6010

An amount of $6500 is invested at 4% p.a. interest, compounded half-yearly. Use the table to find the value of the investment after 6 years. (2 marks)

15 James wants $15 000 in 3 years time to replace a piece of machinery.

a If he invests $12 000 at 4% p.a. interest, compounded monthly now, by how much, to the nearest dollar, will he fall short of his goal? (2 marks)

b How much more would James need to invest now at that interest rate to reach his goal? (2 marks)

c At what monthly interest rate would James need to invest the $12 000 to have $15 000 after 3 years? (2 marks)

16 If the quarterly rate of inflation is 0.7%, find the price in 4 years time of a trolley of groceries that currently costs $300. (2 marks)

17 The graph shows the value of an amount of $2000 invested at 2.5% p.a. or at 5% p.a. for different numbers of years.

a After 25 years, what is the approximate difference in value? (1 mark)

b About how many years longer does the investment take to double at 2.5% p.a. than at 5% p.a.? (1 mark)

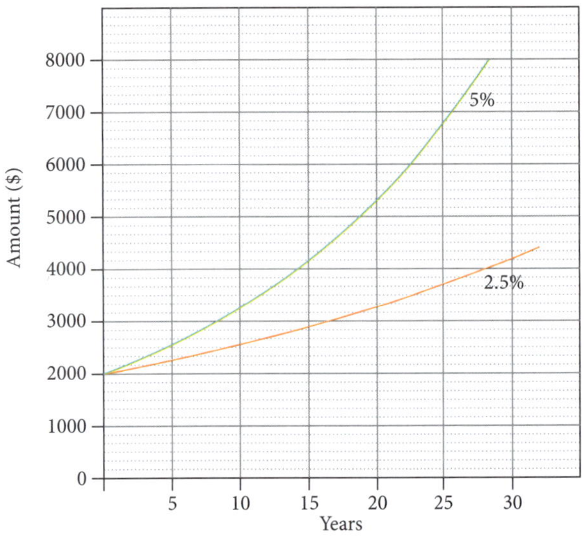

ANSWERS

PRACTICE

1 Simple interest ➲ pp. 91–92

1 $P = \$7250$, $r = 0.065$, $n = 5$

$I = Prn$
$= \$7250 \times 0.065 \times 5$
$= \mathbf{\$2356.25}$

2 $P = \$2350$, $r = 0.03$

Now 6 months is $\frac{1}{2}$ of a year so $n = 0.5$.

$I = Prn$
$= \$2350 \times 0.03 \times 0.5$
$= \$35.25$

The amount of interest paid is $35.25.

3 $P = \$6200$, $r = 0.009$, $n = 16$

$I = Prn$
$= \$6200 \times 0.009 \times 16$
$= \$892.80$

Courtney earns $892.80 interest.

4 $I = \$4334.40$, $r = 0.07$, $n = 3$

$I = Prn$
$\$4334.40 = P \times 0.07 \times 3$
$\$4334.40 = 0.21P$
$P = \$4334.40 \div 0.21$
$= \$20\,640$

The amount invested was $20 640.

5 $P = \$16\,000$, $I = \$11\,520$, $n = 9$

$I = Prn$
$\$11\,520 = \$16\,000 \times r \times 9$
$\$11\,520 = \$144\,000r$
$r = \$11\,520 \div \$144\,000$
$= 0.08$

So the interest rate is 8% p.a.

6 $I = \$24.75$, $P = \$900$, $n = 0.5$

$I = Prn$
$\$24.75 = \$900 \times r \times 0.5$
$24.75 = 450r$
$r = \frac{24.75}{450}$
$= 0.055$

So the rate of interest was 5.5% p.a.

2 Compound interest ➲ p. 93

1 $PV = \$15\,000$, $r = 0.0625$, $n = 5$

$FV = PV(1 + r)^n$
$= \$15\,000(1 + 0.0625)^5$
$= \$20\,311.2173\ldots$
$= \$20\,311$ (nearest dollar)

The future value is $20 311, to the nearest dollar.

2 $PV = \$9400$, $r = 0.06$, $n = 2$

$FV = PV(1 + r)^n$
$= \$9400(1 + 0.06)^2$
$= \$10\,561.84$
$I = FV - PV$
$= \$10\,561.84 - \$9400$
$= \$1161.84$

An amount of $1161.84 is earned in interest.

3 $PV = \$24\,000$, $r = \frac{0.075}{12}$, $n = 3 \times 12$

$FV = PV(1 + r)^n$
$= \$24\,000\left(1 + \frac{0.075}{12}\right)^{36}$
$= \$30\,034.7072\ldots$
$= \mathbf{\$30\,034.71}$ (nearest cent)

4 $PV = \$23\,400$, $r = 0.02$, $n = 16$

$FV = PV(1 + r)^n$
$= \$23\,400(1.02)^{16}$
$= \$32\,123.185\ldots$
$= \$32\,123$ (nearest dollar)
$I = \$32\,123 - \$23\,400$
$= \$8723$

The compound interest earned is $8723, to the nearest dollar.

5 $PV = \$27\,500$, $r = 0.0195$

30 months = 5 × 6 months
So $n = 5$

$FV = PV(1 + r)^n$
$= \$27\,500(1 + 0.0195)^5$
$= \$30\,287.877\ldots$
$= \$30\,288$ (nearest dollar)

So Davina will exceed her goal by about $288.

6 Monthly: $PV = \$60\,000$, $r = \frac{0.08}{12}$, $n = 5 \times 12$

$FV = PV(1 + r)^n$
$= \$60\,000\left(1 + \frac{0.08}{12}\right)^{5\times12}$
$= \$89\,390.742\ldots$
$= \$89\,390.74$ (nearest cent)

Half-yearly: $PV = \$60\,000$, $r = \frac{0.08}{2}$, $n = 5 \times 2$

$FV = PV(1 + r)^n$
$= \$60\,000\left(1 + \frac{0.08}{2}\right)^{5\times2}$
$= \$88\,814.657\ldots$
$= \$88\,814.66$ (nearest cent)

Difference $= \$89\,390.74 - \$88\,814.66$
$= \$576.08$

The future value is about $576 greater when the interest is compounded monthly rather than half-yearly.

3 Finding the present value ➲ p. 94

1 $FV = \$47\,000$, $r = 0.045$, $n = 12$

$$FV = PV(1+r)^n$$
$$\$47\,000 = PV(1+0.045)^{12}$$
$$PV = \frac{\$47\,000}{(1.045)^{12}}$$
$$= \$27\,714.2016\ldots$$
$$= \$27\,714 \text{ (nearest dollar)}$$

An amount of $27 714, to the nearest dollar, needs to be invested now.

2 $FV = \$20\,000$, $r = \frac{0.09}{12}$, $n = 5 \times 12$

$$PV = \frac{FV}{(1+r)^n}$$
$$= \frac{\$20\,000}{\left(1+\frac{0.09}{12}\right)^{5\times 12}}$$
$$= \$12\,773.9939\ldots$$
$$= \mathbf{\$12\,774} \text{ (nearest dollar)}$$

3 $FV = \$28\,000$, $r = 0.042 \div 4 = 0.0105$, $n = 5 \times 4 = 20$

$$PV = \frac{FV}{(1+r)^n}$$
$$= \frac{\$28\,000}{(1+0.0105)^{20}}$$
$$= \$22\,721.2214\ldots$$
$$= \$22\,721 \text{ (nearest dollar)}$$

Ross needs to invest $22 721 now.

4 $PV = \$30\,000$, $FV = \$40\,202.87$, $n = 6$

$$FV = PV(1+r)^n$$
$$\$40\,202.87 = \$30\,000(1+r)^6$$
$$(1+r)^6 = 1.340\,095\,66\ldots$$
$$1 + r = 1.050\,0000\ldots$$
$$r = 0.05 \text{ (2 d.p.)}$$

So the annual rate of interest is 5%.

5 $PV = \$55\,000$, $FV = \$63\,889$, $n = 5 \times 12 = 60$

$$FV = PV(1+r)^n$$
$$\$63\,889 = \$55\,000(1+r)^{60}$$
$$(1+r)^{60} = 1.161\,618\,18\ldots$$
$$1 + r = 1.002\,5000\ldots$$
$$- 1.0025 \text{ (4 d.p.)}$$
$$r = 0.0025$$

So the monthly rate of interest is 0.25%.

The annual rate of interest is 12 × 0.25% = 3%.

6 $PV = \$13\,000$, $FV = \$15\,543$, $n = 3 \times 4 = 12$

$$FV = PV(1+r)^n$$
$$\$15\,543 = \$13\,000(1+r)^{12}$$
$$(1+r)^{12} = 1.195\,615\,38\ldots$$
$$1 + r = 1.014\,9998\ldots$$
$$= 1.015 \text{ (3 d.p.)}$$
$$r = 0.015$$

So the quarterly rate of interest is 1.5%.

The annual rate of interest is 4 × 1.5% = 6%.

4 Tables of compounded values ➲ p. 95

1 Future value = $7500 × 1.3401
= **$10 050.75**

2 Interest rate = 8% ÷ 4
= 2%

Period = 6 × 4 quarters
= 24 quarters

Future value = $11 000 × 1.6084
= **$17 692.40**

3 Interest rate = 18% p.a.
= 1.5% per month

Now $9000 × factor = $15 381.90
factor = $15 381.90 ÷ $9000
= 1.7091

From the table, at 1.5% per period, this is the factor for 36 periods.

Now 36 months = 3 years.

So the money was invested for 3 years.

4 a 2.5 years = 5 half-years

Now $16 400 × factor = $19 953.88
factor = $19 953.88 ÷ $16 400
= 1.2167

From the table, for 5 periods, this is the factor for 4% per period.

So the rate is 4% per half year or 8% p.a.

b 15 years = 30 half-years

$FV = \$16\,400 \times 3.2434$
$= \$53\,191.76$

Interest = $53 191.76 – $16 400
= **$36 791.76**

5 Comparing investments ➲ p. 97

1 a $P = \$20\,000$, $r = 0.059$, $n = 3$

$$I = Prn$$
$$= \$20\,000 \times 0.059 \times 3$$
$$= \$3540$$

Future value = $20 000 + $3540
= **$23 540**

b $PV = \$20\,000$, $r = 0.0045$, $n = 3 \times 12 = 36$

$$FV = PV(1+r)^n$$
$$= \$20\,000(1+0.0045)^{36}$$
$$= \$23\,508.6599\ldots.$$
$$= \$23\,508.66 \text{ (nearest cent)}$$

Henry should take the first option because it gives a greater future value.

2 $PV = \$25\,000$, $r = 0.054 \div 4 = 0.0135$, $n = 6 \times 4 = 24$

$$FV = PV(1+r)^n$$
$$= \$25\,000(1+0.0135)^{24}$$
$$= \$34\,491.3411\ldots$$
$$= \$34\,491 \text{ (nearest dollar)}$$

$FV = \$34\,491$, $r = 0.054 \div 12 = 0.0045$, $n = 5 \times 12 = 60$

$$PV = \frac{FV}{(1+r)^n}$$
$$= \frac{\$34\,491}{(1+0.0045)^{60}}$$
$$= \$26\,345.6744\ldots$$
$$= \$26\,350 \text{ (nearest \$10)}$$

About an extra $1350 would need to be invested.

3 a **$A = 1000 + 80n$**

[The original amount plus \$80 per year.]

b

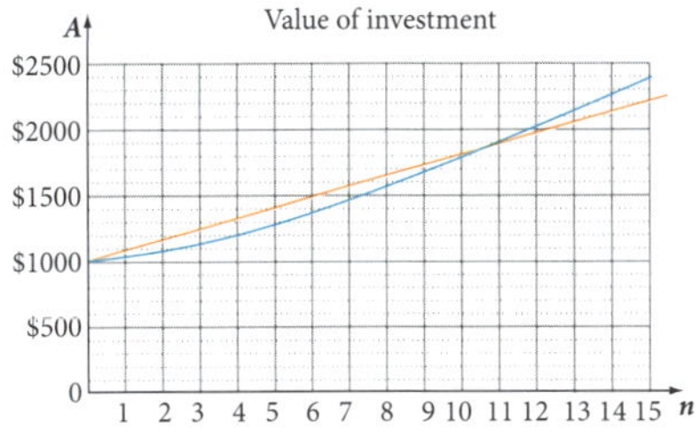

c **$10\frac{1}{2}$ years**

d **12 years** [When $A = 2000$]

e $A = 1000 + 80n$

When amount has doubled

$A = 2000$

$2000 = 1000 + 80n$

$1000 = 80n$

$n = 1000 \div 80$

$= 12.5$

The simple interest option doubles in $12\frac{1}{2}$ years.

f **If Shannon intends to invest for less than 10 years he should choose the simple interest option. Otherwise he should use the compound interest option. Although the simple interest option produces a slightly better result for the first 10 years, the compound interest produces a significantly better result after 10 years.**

4 $PV = \$125\,000$, $r = 0.06$, $n = 2$

$FV = PV(1 + r)^n$

$= \$125\,000 \times (1 + 0.06)^2$

$= \$140\,450$

$PV = \$125\,000$, $r = 0.0595 \div 2 = 0.029\,75$, $n = 2 \times 2 = 4$

$FV = PV(1 + r)^n$

$= \$125\,000 \times (1 + 0.029\,75)^4$

$= \$140\,552.06$ (nearest cent)

$PV = \$125\,000$, $r = 0.059 \div 4 = 0.014\,75$, $n = 2 \times 4 = 8$

$FV = PV(1 + r)^n$

$= \$125\,000 \times (1 + 0.014\,75)^8$

$= \$140\,534.35$ (nearest cent)

$PV = \$125\,000$, $r = 0.0585 \div 12 = 0.004\,875$, $n = 2 \times 12 = 24$

$FV = PV(1 + r)^n$

$= \$125\,000 \times (1 + 0.004\,875)^{24}$

$= \$140\,474.99$ (nearest cent)

5.95% compounded half-yearly will give the best return.

6 Practical applications ➲ p. 98

1 $PV = \$3500$, $r = 0.06$, $n = 5$

$FV = PV(1 + r)^n$

$= \$3500(1 + 0.06)^5$

$= \$4683.789\,52\ldots$

$= \$4700$ (nearest \$100)

The expected value of the necklace in 5 years time is \$4700 to the nearest 100 dollars.

2 a $FV = \$16\,000$, $PV = \$6900$, $n = 14$

$FV = PV(1 + r)^n$

$\$16\,000 = \$6900(1 + r)^{14}$

$(1 + r)^{14} = \dfrac{16\,000}{6900}$

$1 + r = \sqrt[14]{\dfrac{16\,000}{6900}}$

$= 1.061\,917\ldots$

$r = 0.062$ (3 d.p.)

So the annual rate of appreciation is 6.2%.

b In 2027, $n = 7$

$FV = \$16\,000(1 + 0.062)^7$

$= \$24\,377.6366\ldots$

The vase will be worth approximately \$24 400 at the end of 2027.

3 $FV = \$10\,000$, $PV = \$7910$, $r = 0.04$

$FV = PV(1 + r)^n$

$\$10\,000 = \$7910(1 + 0.04)^n$

So $1.04^n = \dfrac{10\,000}{7910}$

$= 1.264\,222\,50\ldots$

$n \approx 6$

It will take around 6 years for the painting to be worth \$10 000.

4 $PV = \$1700$, $r = 0.024$, $n = 4$

$FV = PV(1 + r)^n$

$= \$1700(1 + 0.024)^4$

$= \$1869.169\,76\ldots$

$= \$1869$ (nearest dollar)

Larry will earn about \$1869 per week in 4 years time.

5 $PV = \$100$, $r = 0.0085$, $n = 7 \times 4 = 28$

$FV = PV(1 + r)^n$

$= \$100(1 + 0.0085)^{28}$

$= \$126.743\,376\ldots$

So the percentage increase in prices is approximately 26.7%.

6 $PV = \$650\,000$, $FV = \$1\,100\,000$, $n = 6$

$FV = PV(1 + r)^n$

$\$1\,100\,000 = \$650\,000(1 + r)^6$

$(1 + r)^6 = \$1\,100\,000 \div \$650\,000$

$= 1.692\,307\,69\ldots$

$1 + r = 1.091\,641\,12\ldots$

$r = 0.091\,641\,12\ldots$

$= 0.09$ (2 d.p.)

The annual rate of appreciation is about 9%.

HSC EXAM-TYPE QUESTIONS

Part A

1 **A.** $P = \$8000$, $r = 0.09$, $n = 1.5$

$I = Prn$

$= \$8000 \times 0.09 \times 1.5$

$= \$1080$

2 **B.** $PV = \$2000$, $r = 0.007$, $n = 96$

$FV = PV(1 + r)^n$

$= \$2000(1.007)^{96}$

$= \$3907$ (nearest dollar)

3 **C.** $PV = \$125\,000$, $r = 0.06$, $n = 5$

$FV = PV(1 + r)^n$

$= \$125\,000(1 + 0.06)^5$

$= \$167\,278.197\ldots$

The value is closest to \$167 500.

4 **D.** $PV = \$9000$, $r = 0.0375$, $n = 28$

$FV = PV(1 + r)^n$

$= \$9000(1.0375)^{28}$

$= \$25\,230$ (nearest dollar)

Interest = \$25 230 − \$9000

= \$16 230

5 **B.** Try each option:

A: \$8000

B: $I = \$6750 \times 0.07 \times 3 = \$1417.50$

$FV = \$6750 + \$1417.50 = \$8167.50$

C: $FV = \$6500(1 + 0.07)^3 = \$7962.78$

D: $FV = \$6250(1 + 0.007)^{36} = \$8034.17$

\$6750 earning 7% p.a. simple interest is worth the most.

6 **A.** $FV = \$4295$, $r = 0.055 \div 2 = 0.0275$, $n = 3 \times 2 = 6$

$$PV = \frac{FV}{(1+r)^n}$$

$$= \frac{\$4295}{(1+0.0275)^6}$$

$= \$3649.826\,20\ldots$

$= \$3650$ (nearest dollar)

7 **C.** $FV = \$6500$, $PV = \$6000$, $n = 2 \times 4 = 8$

$FV = PV(1 + r)^n$

$\$6500 = \$6000(1 + r)^8$

$$\text{So } (1 + r)^8 = \frac{6500}{6000}$$

$$1 + r = \sqrt[8]{\frac{6500}{6000}}$$

$= 1.010\,055\,55\ldots$

$r = 0.01$ (2 d.p.)

The quarterly rate is 1%.

8 **D.** $PV = \$10\,000$, $r = 0.006$, $n = 5 \times 12 = 60$

$FV = PV(1 + r)^n$

$= \$10\,000(1 + 0.006)^{60}$

$= \$14\,317.8841\ldots$

$= \$14\,318$ (nearest dollar)

$I = \$14\,318 - \$10\,000$

$= \$4318$

Yearly interest = \$4318 ÷ 5

= \$863.60

$$\text{Rate} = \frac{\$863.60}{\$10\,000} \times 100\%$$

$= 8.636\%$

Part B

9 $PV = \$7600$ $r = 0.075 \div 12 = 0.006\,25$ $n = 3 \times 12 = 36$

$FV = PV(1 + r)^n$

$= \$7600(1 + 0.006\,25)^{36}$ ✓

$= \$9510.9906\ldots$

= **\$9511** (nearest dollar) ✓

10 $PV = \$24\,350$, $r = 0.08 \div 4 = 0.02$ $n = 5 \times 4 = 20$

$FV = PV(1 + r)^n$

$= \$24\,350(1 + 0.02)^{20}$

$= \$36\,182.8190\ldots$

$= \$36\,182.82$ (nearest cent) ✓

$I = \$36\,182.82 - \$24\,350$

$= \$11\,832.82$

The amount of interest earned is \$11 832.82, to the nearest cent. ✓

11 $FV = \$32\,700$, $r = 0.06 \div 12 = 0.005$, $n = 4 \times 12 = 48$

$$PV = \frac{FV}{(1+r)^n}$$

$$= \frac{\$32\,700}{(1+0.005)^{48}}$$ ✓

$= \$25\,738.1180\ldots$

= **\$25 738** (nearest dollar) ✓

12 $PV = \$65\,000$, $FV = \$87\,260.60$, $n = 8$

$FV = PV(1 + r)^n$

$\$87\,260.60 = \$65\,000(1 + r)^8$

$(1 + r)^8 = 1.342\,470\,76\ldots$ ✓

$1 + r = 1.037\,499\,99\ldots$

$= 1.0375$ (4 d.p.)

$r = 0.0375$

So the annual rate of interest is $3\frac{3}{4}$%. ✓

13 $PV = \$6200$, $r = 0.12$, $n = 14$

$FV = PV(1 + r)^n$

$= \$6200(1 + 0.12)^{14}$ ✓

$= \$30\,300.0961\ldots$

$= \$30\,300$ (nearest dollar)

The value of the diamond was \$30 300. ✓

14 4% p.a. = 2% per half-year

6 years = 12 half-years

So the required factor is 1.2682 ✓

$FV = 1.2682 \times \$6500$

$= \$8243.30$

So the value of the investment will be \$8243.30. ✓

15 a $PV = \$12\,000$, $r = \dfrac{0.04}{12}$, $n = 3 \times 12 = 36$

$$FV = PV(1 + r)^n$$
$$= \$12\,000\left(1+\frac{0.04}{12}\right)^{36} \quad ✓$$
$$= \$13\,527.2624\ldots$$
$$= \$13\,527 \quad \text{(nearest dollar)}$$

Shortfall = $15 000 − $13 527
= $1473

James will fall short of his goal by $1473. ✓

b $FV = \$15\,000$, $r = \dfrac{0.04}{12}$, $n = 3 \times 12 = 36$

$$PV = \frac{FV}{(1+r)^n}$$
$$= \frac{\$15000}{\left(1+\frac{0.04}{12}\right)^{36}} \quad ✓$$
$$= \$13\,306.4616\ldots$$
$$= \$13\,306 \quad \text{(nearest dollar)}$$

So James would need to invest an extra $1306. ✓

c $FV = \$15\,000$, $PV = \$12\,000$, $n = 36$

$$FV = PV(1 + r)^n$$
$$\$15\,000 = \$12\,000(1 + r)^{36}$$
$$\text{So } (1 + r)^{36} = \frac{15\,000}{12\,000}$$
$$1 + r = \sqrt[36]{\frac{15\,000}{12\,000}} \quad ✓$$
$$= 1.006\,217\,68\ldots$$
$$r = 0.006\,217\,68\ldots$$

So James would need to invest at 0.62% per month. ✓

16 $PV = \$300$, $r = 0.007$, $n = 4 \times 4 = 16$

$$FV = PV(1 + r)^n$$
$$= \$300(1 + 0.007)^{16} \quad ✓$$
$$= \$335.422\,957\ldots$$
$$= \$335 \quad \text{(nearest dollar)}$$

In 4 years time the trolley of groceries would cost $335. ✓

17 a After 25 years, at 5% the value is about $6800 while at 2.5% the value is about $3700. **So the approximate difference in value is about $3100.** ✓

b At 5% the amount doubles in about 14 years. At 2.5% it doubles in about 28 years. **So it takes about 14 years longer to double at 2.5% p.a. than at 5% p.a.** ✓

TOPIC: FINANCIAL MATHEMATICS

CHAPTER 6 DEPRECIATION AND LOANS

1 Depreciation

- **Many assets decrease in value over time.** This decrease in value is called depreciation and is a cost to businesses and individuals. Depreciation needs to be considered in many financial situations.
- **The straight-line method is one way of determining the depreciation** of an asset over time. Using this method we assume that the asset decreases in value by a fixed amount each period.
- **We can apply a formula to find the value of an asset** when it is being depreciated by a fixed amount every period. $S = V_0 - Dn$ where S is the salvage value of the asset after n periods, V_0 is the original value of the asset, D is the amount of depreciation allowed per period and n is the number of periods.

EXAMPLE 1

Some office furniture was purchased for $4800 and is being depreciated by $400 per year. What is the value of the furniture after 6 years?

$V_0 = \$4800, D = \$400, n = 6$
$S = V_0 - Dn$
$= \$4800 - \$400 \times 6$
$= \$2400$
After 6 years the value of the furniture will be $2400.

The subscript 0 in V_0 is a way of writing the value when time is zero. V_1 would be the value after 1 period, and so on.

EXAMPLE 2

A machine, originally purchased for $27 500, is being depreciated by the straight-line method. Find the amount of depreciation allowed per year if the salvage value of the machine is $16 300 after 7 years.

$S = \$16\,300, V_0 = \$27\,500, n = 7$
$S = V_0 - Dn$
$\$16\,300 = \$27\,500 - D \times 7$
$7D = \$27\,500 - \$16\,300$
$= \$11\,200$
$D = \$1600$
The amount of depreciation allowed per year is $1600.

Or, logically, just find the difference between the original value and the current value and divide that amount by 7 to find the depreciation each year.

PRACTICE

1 **A forklift was purchased for $26 400 and is being depreciated by $1850 per year. What is the value of the forklift after 8 years?**

2 **A machine, originally purchased for $24 500, is being depreciated by the straight-line method. Find the amount of depreciation allowed per year if the salvage value of the machine is $15 700 after 8 years.**

3* **A 6-year-old machine is currently valued at $4625. If it has been depreciated by $675 per year, find its purchase price.**

Answers ➲ p. 126

2 The declining-balance method of determining depreciation

- **In the declining-balance method of depreciation**, the depreciation per period is a fixed percentage of the value of the asset at the end of the previous period.
- **The future value formula, with a negative value for r,** allows us to find the current value of an asset that is being depreciated using the declining-balance method. $S = V_0(1 - r)^n$ where S is the salvage value of the asset at the end of n periods, V_0 is the original value of the asset, r is the interest rate per period expressed as a decimal and n is the number of periods.
- **The declining-balance formula gives the salvage value of the asset.** The depreciation after n years is the difference between the salvage value and the original value.

EXAMPLE 1

A tractor is being depreciated using the declining-balance method, at 8% p.a. If its purchase price was $65 000, find its value to the nearest 100 dollars, after 12 years.

$V_0 = \$65\,000$, $r = 0.08$, $n = 12$

$$
\begin{aligned}
S &= V_0(1 - r)^n \\
&= \$65\,000(1 - 0.08)^{12} \\
&= \$23\,898.315\ldots \\
&= \$23\,900 \text{ (nearest hundred dollars)}
\end{aligned}
$$

After 12 years the tractor is worth approximately \$23 900.

Under the declining-balance method, the depreciation is greatest when the asset is new.

EXAMPLE 2

An overlocker was bought for \$1100 five years ago. If it is being depreciated at 10% p.a., how much has it depreciated in value over that time?

$V_0 = \$1100$, $r = 0.1$, $n = 5$

$$
\begin{aligned}
S &= V_0(1 - r)^n \\
&= \$1100(1 - 0.1)^5 \\
&= \$649.539 \\
&= \$650 \text{ (nearest dollar)}
\end{aligned}
$$

$$
\begin{aligned}
\text{Depreciation} &= \$1100 - \$650 \\
&= \$450
\end{aligned}
$$

The overlocker has depreciated by \$450, to the nearest dollar, in that time.

The depreciation is 10% of the value each time, so in the first year it is \$110 but in the second year it is 10% of the decreased value (\$1100 – \$110 = \$990) or \$99, and so on.

EXAMPLE 3

A machine is bought for \$93 000. It is depreciated using the declining-balance method by 15% p.a.

a **Find the salvage value, to the nearest dollar, at the end of 4 years.**

b **Find the amount of depreciation allowed in the fifth year.**

a $V_0 = \$93\,000$, $r = 0.15$, $n = 4$

$$
\begin{aligned}
S &= V_0(1 - r)^n \\
&= \$93\,000(1 - 0.15)^4 \\
&= \$48\,546.581\,25\ldots \\
&= \$48\,547 \text{ (nearest dollar)}
\end{aligned}
$$

b Depreciation in fifth year

= 15% of \$48 547

= \$7282 (nearest dollar)

Each year the amount of depreciation allowed is 15% of the value at the end of the previous year.

EXAMPLE 4

The current value of a truck is \$19 000. Find the purchase price of the truck if it has been depreciated at 11% p.a. and it is 9 years old.

$S = \$19\,000$, $r = 0.11$, $n = 9$

$$
\begin{aligned}
S &= V_0(1 - r)^n \\
\$19\,000 &= V_0(1 - 0.11)^9 \\
\$19\,000 &= V_0 \times 0.350\,356\,403\ldots \\
V_0 &= \$19\,000 \div 0.350\,356\,403\ldots \\
&= \$54\,230.4915\ldots \\
&= \$54\,230 \text{ (nearest dollar)}
\end{aligned}
$$

The purchase price of the truck was \$54 230.

The current value is the salvage value at the current time.

EXAMPLE 5

A machine bought for \$37 000 three years ago is currently valued at \$28 810. Find the annual rate of depreciation if the machine has been depreciated using the declining-balance method.

$S = \$28\,810$, $V_0 = \$37\,000$, $n = 3$

$$
\begin{aligned}
S &= V_0(1 - r)^n \\
\$28\,810 &= \$37\,000(1 - r)^3 \\
(1 - r)^3 &= \frac{\$28\,810}{\$37\,000} \\
&= 0.778\,648\,648\ldots. \\
1 - r &= \sqrt[3]{0.778\,648\,648\ldots} \\
&= 0.919\,984\,502\ldots \\
&= 0.92 \text{ (2 d.p.)} \\
r &= 1 - 0.92 \\
&= 0.08
\end{aligned}
$$

The machine has been depreciated at 8% p.a.

It is not necessary to evaluate the expressions at each step.

$1 - r = \sqrt[3]{\frac{\$28\,810}{\$37\,000}}$

→ PRACTICE

1 A car is being depreciated using the declining-balance method, at 12% p.a. If its purchase price was \$42 000, find its value to the nearest 100 dollars, after 7 years.

2 A cement mixer was bought for \$1400 three years ago. If it is being depreciated at 8% p.a., how much has it depreciated in value over that time?

3 A machine was bought for \$76 000. It is being depreciated using the declining-balance method by 12.5% p.a.

a Find the salvage value, to the nearest dollar, at the end of 5 years.

b Find the amount of depreciation allowed in the sixth year.

4 The current value of a set of tools is \$3638. Find the purchase price of the tools if they have been depreciated at 8% p.a. and they are 6 years old.

5 A machine, bought for \$43 000 three years ago, is currently valued at \$26 400. Find the rate of depreciation allowed if the machine has been depreciated using the declining-balance method.

6* Find the amount of depreciation allowed for the fourth year if a bus is being depreciated, using the declining-balance method, at 9% per year and the purchase price was \$56 000.

Answers ➲ p. 126

3 Comparing methods of depreciation

- **The graph of the salvage value of an asset,** when the straight-line method of depreciation is used, is a straight line.
- **When using the declining-balance method**, the graph of the value of the asset over time is a curve. It is an example of exponential decay.
- **In the early stages an asset will be worth more if depreciated under the straight-line method** than under the declining-balance method. With the declining-balance method the amount of depreciation will decrease as the value of the asset decreases and so after a while the asset will be worth more under the declining-balance method.

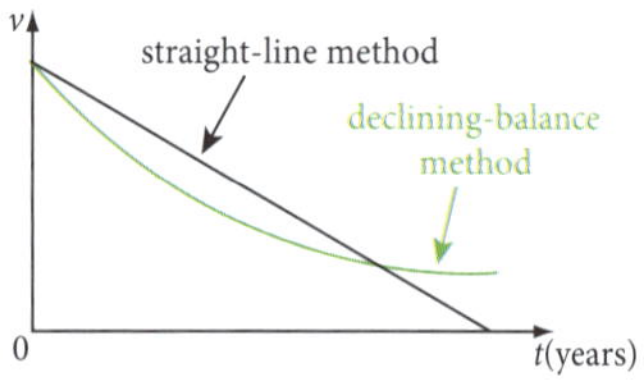

EXAMPLE 1

The graph shows the value of an asset, which is being depreciated using the declining-balance method, over the first 10 years.

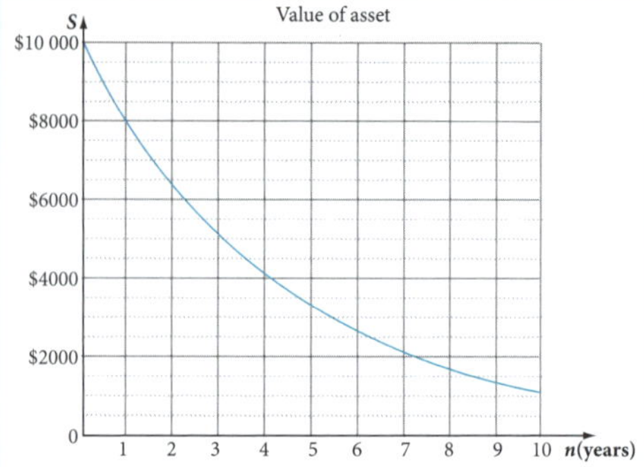

a What is the value after 4 years?

b When is the value of the asset \$2100?

c By what percentage is the value decreasing each year?

a After 4 years the value is about \$4100.

b The value of the asset is \$2100 after 7 years.

c When $t = 0$, $S = \$10\,000$.

When $t = 1$, $S = \$8000$.

$$\text{Depreciation} = \$10\,000 - \$8000 = \$2000$$

$$\%\text{ depreciation} = \frac{\$2000}{\$10\,000} \times 100\% = 20\%$$

The asset is decreasing in value by 20% per year.

So $S = 10\,000(0.8)^n$.

EXAMPLE 2

The graph shows the value of a machine over time, if it is depreciated at different rates.

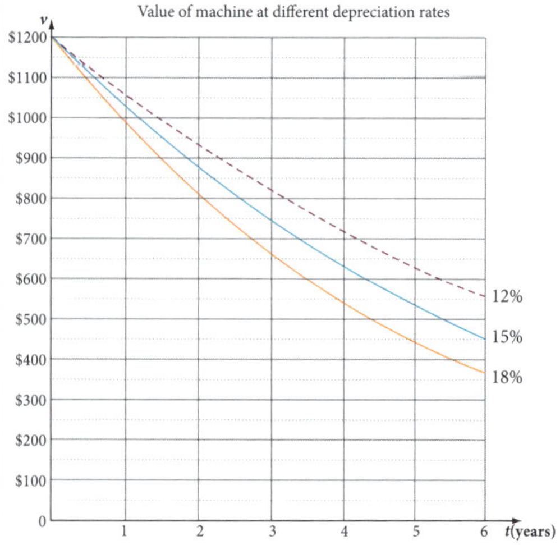

a What was the purchase price of the machine?

b What is the value of the machine after 3 years if it is depreciated at 18%?

c If the machine is depreciated at 15%, after how many years will it halve in value?

d If it is depreciated at 12%, how much did the machine lose in value in the fifth year?

e What is the difference in salvage value after 4 years if the machine is depreciated at 15% or 18%?

a The purchase price was $1200.

b After 3 years depreciation at 18% the machine is worth about $660.

> Values read from the graph can only be approximate.

c At 15% depreciation the machine halves in value after about 4 years and 4 months.

d At 12% depreciation the machine lost about $90 in the fifth year.

e After 4 years there is about $90 difference in salvage value if depreciated at 15% or at 18%.

EXAMPLE 3

A car is bought for $34 500.

a **If it is depreciated by the straight-line method with depreciation of $2800 allowed per year, what will the car be worth in 6 years?**

b **If it is depreciated using the declining-balance method at 11% p.a., what will it be worth in 6 years? (Give the answer to the nearest 100 dollars.)**

c **By which method of depreciation will the salvage value of the car be higher after 7 years? Justify your answer.**

a $V_0 = \$34\,500$, $D = \$2800$, $n = 6$

$S = V_0 - Dn$

$= \$34\,500 - \$2800 \times 6$

$= \$17\,700$

If depreciated by the straight-line method the car will be worth $17 700 after 6 years.

b $V_0 = \$34\,500$, $r = 0.11$, $n = 6$

$S = V_0(1 - r)^n$

$= \$34\,500(1 - 0.11)^6$

$= \$17\,145.8545$

$= \$17\,100$ (nearest $100)

c After a further year under the straight-line method of depreciation, the car will lose another $2800 in value.

Salvage value $= \$17\,700 - \$2800$

$= \$14\,900$

Under the declining-balance method, the car will lose a further 11% of its value.

Depreciation $= 0.11 \times \$17\,100$

$= \$1881$

Salvage value $= \$17\,100 - \$1881$

$= \$15\,219$

> Or simply use the formulae to find the values after 7 years.

The car will be worth more after 7 years using the declining-balance method.

→ PRACTICE

1 **The graph shows the value of an asset, which is being depreciated using the declining-balance method, over the first 10 years.**

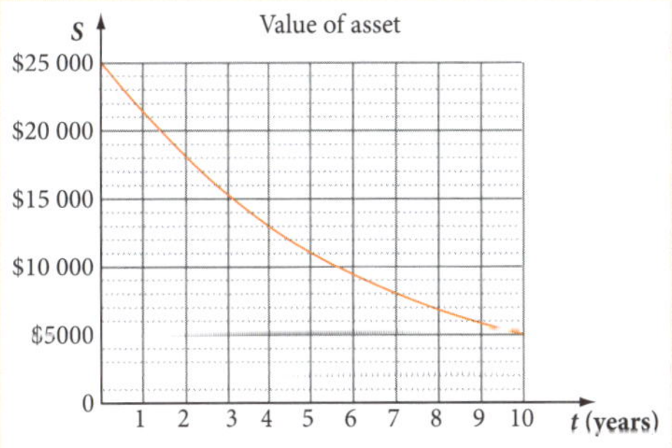

a **What is the value after 3 years?**

b **When is the value of the asset $8000?**

c **By what percentage is the value decreasing in each period?**

2 **The graph shows the value of a machine over time, if it is depreciated at different rates.**

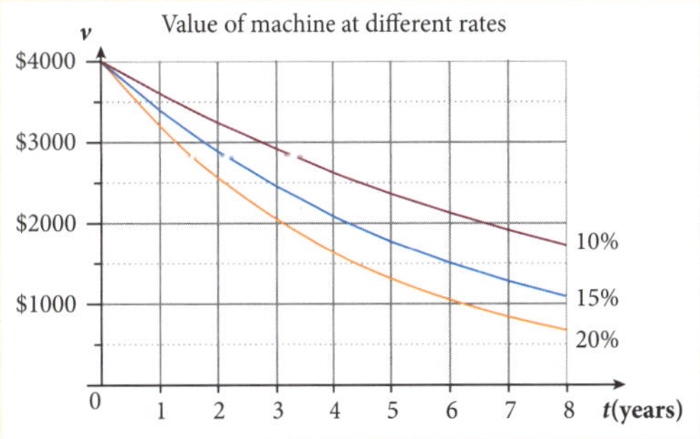

a What was the purchase price of the machine?
b What is the value of the machine after 4 years if it is depreciated at 20%?
c If the machine is depreciated at 10%, after how many years will it halve in value?
d If it is depreciated at 15%, how much did the machine lose in value in the second year?
e What is the difference in salvage value after 6 years if the machine is depreciated at 15% or 20%?

3 A van is bought for $32 500.
a If it is depreciated by the straight-line method with depreciation of $2700 allowed per year, what will the van be worth in 5 years?
b If it is depreciated using the declining-balance method at 9% p.a., what will it be worth in 5 years? (Give the answer to the nearest 100 dollars.)
c By which method of depreciation will the salvage value of the van be higher after 6 years? Justify your answer.

4* Some office furniture was bought for $7600.
a If it is depreciated using the declining-balance method at 10% p.a., find its value after 10 years to the nearest dollar.
b If the value of the furniture would be the same after 10 years if the straight-line method of depreciation was used, find the amount of depreciation allowed per year.
c Draw a rough sketch showing the value of the furniture over the first 12 years under both methods.

Answers ➲ pp. 126–127

4 Flat-rate loans

- A flat-rate loan is one where the interest charged is **simple interest**. The amount of interest charged is the same over the whole period of the loan.
- **The principal is the amount borrowed.**
- **The total to be repaid is the sum of the interest and principal.**
- **The amount of each regular repayment is the total to be repaid divided by the number of instalments.**

EXAMPLE 1

Aaron borrows $7200 at 0.9% per month simple interest. He repays the loan over 4 years.
a How much interest does Aaron pay?
b What is the total amount that Aaron must repay?

a $P = \$7200$, $r = 0.009$, $n = 48$
$I = Prn$
$= \$7200 \times 0.009 \times 48$
$= \$3110.40$
Aaron pays $3110.40 in interest.

Here the monthly interest rate and the number of months were used in the simple interest formula. The annual interest rate and number of years could have been used instead.

b Total to be repaid = $7200 + $3110.40
= $10 310.40
Aaron must repay a total of $10 310.40.

EXAMPLE 2

Zac borrows $9000 to buy a boat. He agrees to repay the loan over 4 years with equal monthly instalments of $232.50.
a How much does Zac repay?
b How much interest does he pay?
c What is the annual rate of simple interest charged?

a Total repaid = $232.50 × 48
= $11 160
b Interest = $11 160 – $9000
= $2160

The difference between the amount borrowed and the amount repaid is the interest.

c Total interest = $2160
Interest per year = $2160 ÷ 4
= $540
$\text{Interest rate} = \frac{\$540}{\$9000} \times 100\%$
= 6%
The rate of simple interest charged is 6% p.a.

Alternatively, substitute into the simple interest formula and solve the resulting equation.

EXAMPLE 3

A loan of $25 000 is taken out and is to be repaid over 7 years with equal monthly instalments. A flat rate of 5% p.a. interest is charged on the loan. Find, to the nearest cent, the amount of each instalment.

$P = \$25\,000$, $r = 0.05$, $n = 7$
$I = Prn$
$= \$25\,000 \times 0.05 \times 7$
$= \$8750$
Total to be repaid = $25 000 + $8750
= $33 750
Number of instalments = 7 × 12
= 84
Each instalment = $33 750 ÷ 84
= $401.785 714…
= $401.79 (nearest cent)

Usually loans are repaid with equal instalments. Instalments might be weekly, fortnightly, monthly or over some other time period.

EXAMPLE 4

Tabitha decides to buy a car priced at $18 000. She pays $2000 deposit and borrows the remainder at $7\frac{1}{2}$% p.a. simple interest over 5 years. Find:

- a **the amount of interest charged**
- b **the amount of each monthly repayment**
- c **the total amount Tabitha pays for the car.**

a Price = $18 000, deposit = $2000

Principal = $18 000 − $2000
= $16 000

$r = 0.075$, $n = 5$

$I = Prn$
= $16 000 × 0.075 × 5
= $6000

Interest is only paid on the amount that is borrowed. If a deposit is paid when an item is bought, for example, then that amount is subtracted from the purchase price to find the amount borrowed.

b Total to be repaid = $16 000 + $6000
= $22 000

Number of instalments = 5 × 12
= 60

Each instalment = $22 000 ÷ 60
= $366.666 66…
= $366.67 (nearest cent)

c Total paid = total to be repaid + deposit
= $22 000 + $2000
= $24 000

Alternatively, the total paid is the price plus the interest:
$18 000 + $6000 = $24 000.

→ PRACTICE

1 **Jordan borrows $6100 at 0.8% per month simple interest. He repays the loan over 5 years.**
- a **How much interest does Jordan pay?**
- b **How much must Jordan repay?**

2 **Carmen borrows $11 000 to buy a car. She agrees to repay the loan over 5 years with equal monthly instalments of $247.50.**
- a **How much does Carmen repay?**
- b **How much interest does she pay?**
- c **What rate of simple interest is charged?**

3 **A loan of $32 000 is to be repaid over 6 years. A flat rate of 5.5% p.a. interest is charged on the loan. Find to the nearest cent the amount of each monthly instalment.**

4 **Juliet decides to buy a car priced at $21 000. She pays $6000 deposit and borrows the remainder at 8% p.a. simple interest over 4 years. Find:**
- a **the amount of interest charged**
- b **the amount of each monthly repayment**
- c **the total amount Juliet pays for the car.**

5* **Zane paid $7000 deposit and 60 monthly instalments of $800.00 when he bought his ute.**
- a **What is the total amount Zane paid?**
- b **Find the interest rate if the cash price of the ute was $37 000.**

Answers ➲ p. 127

5 Reducing-balance loans

➔ **Most home loans are not flat-rate loans but reducing-balance loans**. In a reducing-balance loan (or reducible-interest loan) the rate of interest charged remains the same, but interest is only paid on the amount owing at a particular time. Because the amount owing decreases as amounts are paid off the loan, the amount of interest also decreases over time.

➔ **Tables of loan repayments can be drawn up to show the amount owing at particular time periods**. The principal is the amount owing at the beginning of the time period. The interest is calculated on that principal. The interest is added to the principal and then the repayment is subtracted to give the amount owing at the end of the period of time. The amount owing at the end of one period is the amount owing at the beginning of the following period.

EXAMPLE 1

A table has been drawn up to show the amount owing for the first 4 months of a loan.

Amount borrowed: $20 000
Reducible-interest rate: 0.9% per month
Monthly repayments: $378.64

Month	Principal	Interest	$P + I$	$P + I - R$
1	$20 000.00	$180.00	$20 180.00	$19 801.36
2	$19 801.36	$178.21	$19 979.57	$19 600.93
3	$19 600.93	$176.41	$19 777.34	$19 398.70
4	$19 398.70	$174.59	$19 573.29	$19 194.65

- a **How much was owing at the beginning of the third month?**
- b **How much interest was paid in the second month?**
- c **By how much had the loan reduced after 4 months?**
- d **What is the total of the repayments made after 4 months?**

e Explain the difference between the answers to parts c and d.

a \$19 600.93

reading straight from the table

b \$178.21

c After 4 months the amount owing is \$19 194.65.

You can see from the table that the amount of interest reduces each month—hence the term reducible-interest loan.

Reduction = \$20 000 – \$19 194.65
= \$805.35

The loan had reduced by \$805.35.

d Total repaid = 4 × \$378.64
= \$1514.56

e The difference is the amount of interest paid.

difference = \$1514.56 – \$805.35
= \$709.21
total interest paid
= \$180 + \$178.21 + \$176.41 + \$174.59
= \$709.21

EXAMPLE 2

Robyn borrowed \$18 000 at a reducible interest rate of 0.8% per month. The monthly repayments were \$295. Draw up a repayment table for the first 6 months.

Mth	Principal	Interest	$P + I$	$P + I - R$
1	\$18 000.00	\$144.00	\$18 144.00	\$17 849.00
2	\$17 849.00	\$142.79	\$17 991.79	\$17 696.79
3	\$17 696.79	\$141.57	\$17 838.36	\$17 543.36
4	\$17 543.36	\$140.35	\$17 683.71	\$17 388.71
5	\$17 388.71	\$139.11	\$17 527.82	\$17 232.82
6	\$17 232.82	\$137.86	\$17 370.68	\$17 075.68

The principal at the beginning of a new period is always the same as the last entry for the previous period. The interest each month is found by multiplying the appropriate principal by 0.008.

EXAMPLE 3

The following table shows the principal and interest over the first 4 months of a loan.

Mth	Principal	Interest	$P + I$	$P + I - R$
1	\$80 000.00	\$533.33	\$80 533.33	\$79 914.33
2	\$79 914.33	\$532.76	\$80 447.09	\$79 828.09
3	\$79 828.09	\$532.19	\$80 360.28	\$79 741.28
4	\$79 741.28	\$531.61	\$80 272.89	\$79 653.89

a **How much is each repayment?**

b **What interest rate per month is charged?**

c **What will be the principal in month 6?**

a $R = \$80\,533.33 - \$79\,914.33$
$= \$619$

or \$80 447.09 – \$79 828.09
or \$80 360.28 – \$79 741.28
or \$80 272.89 – \$79 653.89

b Interest rate $= \dfrac{\$533.33}{\$80\,000} \times 100\%$
$= 0.6666\ldots\ \%$

The interest rate per month is $0.\dot{6}\%$.

c Month 5: $P = \$79\,653.89$

Interest $= 0.006\,666\ldots \times \$79\,653.89$
$= \$531.03$ (nearest cent)

$P + I = \$79\,653.89 + \$531.03$
$= \$80\,184.92$

$P + I - R = \$80\,184.92 - \$619$
$= \$79\,565.92$

The principal in month 6 will be \$79 565.92.

If the interest rate and amount of each repayment are not given, they can easily be found from the entries in the table.

EXAMPLE 4

Jayden borrows \$4000 and agrees to repay it over 3 years. Interest of 9% p.a. is calculated annually and added to the loan. Jayden agrees to repay \$1500 per year at the end of the first 2 years and the balance owing at the end of the third year. Find:

a **the interest to be charged at the end of the first year**

b **the amount that will be owing at the beginning of the second year**

c **the balance that Jayden must pay at the end of the third year.**

a Interest = 0.09 × \$4000
= \$360

b $P + I - R = \$4000 + \$360 - \$1500$
$= \$2860$

∴ the amount owing at the beginning of the second year is \$2860.

c Interest 2nd year = 0.09 × \$2860
= \$257.40

Amount owing at beginning of 3rd year:
\$2860 + \$257.40 – \$1500 = \$1617.40

Interest 3rd year = 0.09 × \$1617.40
= \$145.57 (nearest cent)

$P + I = \$1617.40 + \$145.57$
$= \$1762.97$

Jayden must pay \$1762.97 to pay out the loan at the end of the third year.

The final interest must be calculated immediately before the loan is paid out.

EXAMPLE 5

Ben wants to borrow $14 000 and intends to repay the loan in annual instalments over 5 years. He has the option of choosing a flat-rate loan of 6% p.a. interest or a reducing-balance loan with an annual interest rate of 9%.

a How much are the equal annual instalments if Ben chooses the flat-rate loan?

b If Ben chooses the reducing-balance loan, the first four instalments will be $3600. By drawing up a table, find the amount of the final instalment.

c With which loan will Ben pay back the most and by how much?

a $P = \$14000$, $r = 0.06$, $n = 5$

$I = Prn$

$= \$14000 \times 0.06 \times 5$

$= \$4200$

Total to repay = $14 000 + $4200

= $18 200

Each instalment = $18 200 ÷ 5

= $3640

If Ben chooses the flat-rate loan, each instalment will be $3640.

b $P = \$14000$, $r = 9\%$, $R = \$3600$

	Principal	Interest	$P + I$	$P + I - R$
1	$14 000.00	$1260.00	$15 260.00	$11 660.00
2	$11 660.00	$1049.40	$12 709.40	$9109.40
3	$9109.40	$819.85	$9929.25	$6329.25
4	$6329.25	$569.63	$6898.88	$3298.88
5	$3298.88	$296.90	$3595.78	–

The final instalment will be $3595.78.

c With the flat-rate loan Ben will repay $18 200.

With the reducing-balance loan:

Total repaid = 4 × $3600 + $3595.78

= $17 995.78

Difference = $18 200 – $17 995.78

= $204.22

Ben will pay $204.22 more with the flat-rate loan.

The flexibility of a loan can be important. Some loans allow the borrower to make extra payments or repay more with each instalment. By choosing to repay a little more each time the borrower can save thousands of dollars over the term of the loan because the interest is reduced.

→ PRACTICE

1 A table has been drawn up for the first 3 months of a loan.

Amount borrowed:	$25 000
Reducible-interest rate:	0.8% per month
Monthly repayments:	$409.88

Mth	Principal	Interest	$P + I$	$P + I - R$
1	$25 000.00	$200.00	$25 200.00	$24 790.12
2	$24 790.12	$198.32	$24 988.44	$24 578.56
3	$24 578.56	$196.63	$24 775.19	$24 365.31

a How much was owing at the beginning of the second month?

b How much interest was paid in the third month?

c By how much had the loan reduced after 3 months?

d What is the total of the repayments made after 3 months?

e How much is the difference between the answers to parts c and d? What does this represent?

2 Roslyn borrowed $22 000 at a reducible-interest rate of 1% per month. The monthly repayments were $489.38. Draw up a repayment table for the first 4 months.

3 The table below covers the first 5 months of a loan.

a How much is each repayment?

b What monthly interest rate is charged?

c What will be the principal in month 7?

	Principal	Interest	$P + I$	$P + I - R$
1	$60 000.00	$420.00	$60 420.00	$59 472.83
2	$59 472.83	$416.31	$59 889.14	$58 941.97
3	$58 941.97	$412.59	$59 354.56	$58 407.39
4	$58 407.39	$408.85	$58 816.24	$57 869.07
5	$57 869.07	$405.08	$58 274.15	$57 326.98

4 Jamie borrowed $5000 and agreed to repay it over 3 years. Interest of 8.5% p.a. is calculated annually and added to the loan. Jamie agreed to repay $2000 per year at the end of the first 2 years and pay the balance owing at the end of the third year.

a Find the interest charged at the end of the first year.

b Find the amount owing at the beginning of the second year.

c Find the balance that Jamie must pay at the end of the third year.

5 Lincoln wants to borrow $16 000 and intends to repay the loan in annual instalments over 4 years. He has the option of choosing a flat-rate loan of 7% p.a. interest or a reducing-balance loan where the annual interest is 8.5%.

a How much are the equal annual instalments on the flat-rate loan?

b If Lincoln chooses the reducing-balance loan, the first three instalments will be $4900. By drawing up a table, find the amount of the final instalment.

c With which loan will Lincoln pay back the least and by how much?

6* Allan borrowed $34 000 and agreed to repay it with four yearly instalments of $10 726.00. The reducible-interest rate was 10% p.a. By drawing up a table show that the loan will be repaid after 4 years.

Answers ➲ pp. 127–128

6 Tables and graphs

- **Financial institutions publish tables** that give the repayments required to repay reducing-balance loans.
- **A graph can show the balance owing** over the term of a loan.
- **A graph can also clearly show the benefits,** or otherwise, of changes in repayments.

EXAMPLE 1

A bank supplied the following table showing the amount of each monthly repayment required to repay certain loans at its current interest rate.

Loan amount	Term of loan (years)			
	10	15	20	25
$40 000	$506.70	$405.71	$359.89	$335.68
$70 000	$886.73	$709.99	$629.81	$587.44
$150 000	$1900.14	$1521.40	$1349.59	$1258.79
$180 000	$2280.16	$1825.68	$1619.51	$1510.55

a What is the amount of each monthly instalment if a loan of $70 000 is repaid over 15 years?

b If $150 000 is borrowed over 25 years, find the total amount that will be repaid.

c Find the total amount of interest paid on a $40 000 loan taken over 10 years.

d How much extra will be repaid on a loan of $180 000 taken over 25 years, than on a loan for the same amount over 20 years?

a Loan: $70 000, term: 15 years

Monthly instalment = $709.99

b Loan: $150 000, term: 25 years

Monthly instalment = $1258.79

25 years = 25 × 12 months
= 300 months

Total repaid = 300 × $1258.79
= $377 637

If $150 000 is borrowed over 25 years a total of $377 637 will be repaid.

c Loan: $40 000, term: 10 years

Monthly instalment = $506.70

10 years = 10 × 12 months
= 120 months

Total repaid = 120 × $506.70
= $60 804

Interest = $60 804 − $40 000
= $20 804

The difference between the amount borrowed and the amount repaid is the interest.

If $40 000 is borrowed over 10 years, a total of $20 804 will be paid in interest.

d Loan: $180 000, term: 25 years

Monthly instalment = $1510.55

25 years = 25 × 12 months
= 300 months

Total repaid = 300 × $1510.55
= $453 165

Loan: $180 000, term: 20 years

Monthly instalment = $1619.51

20 years = 20 × 12 months
= 240 months

Total repaid = 240 × $1619.51
= $388 682.40

The table doesn't give the interest rate, although it could be worked out. The values are only accurate for the rate used.

Difference = $453 165 − $388 682.40
= $64 482.60

If a $180 000 loan is taken over 25 years compared to 20 years, an additional $64 482.60 would need to be repaid.

EXAMPLE 2

The table shows the monthly repayment (in dollars) for every $1000 borrowed at different interest rates over different numbers of years.

Monthly repayments per $1000					
Term (years)	Interest rate				
	6%	6.5%	7%	7.5%	8%
5	19.33	19.57	19.80	20.04	20.28
7	14.61	14.85	15.09	15.34	15.59
12	9.76	10.02	10.28	10.55	10.82
15	8.44	8.71	8.99	9.27	9.56

a **Find the amount of each monthly instalment if $25 000 is borrowed over 7 years at 7.5% p.a. interest.**

b **Leanne borrows $87 500 at 6% p.a. interest over 15 years. How much in total must she repay?**

c **How much interest will be paid in total if $60 000 is borrowed at 7% p.a. interest over 12 years?**

d **How much more will each monthly instalment be if $18 000 is borrowed over 5 years rather than over 7 years, if the interest rate is 6.5% p.a.?**

e **Bruce wants to take out a loan of $44 000 over 12 years. How much will he save each month if the interest rate is 7.5% p.a. instead of 8% p.a.?**

a 7.5% p.a., 7 years, $25 000

Monthly repayment = $15.34 per $1000
= 25 × $15.34
= $383.50

The table gives the repayment required for every $1000 of the loan amount. You need to multiply the appropriate value in the table by the number of thousands.

b 6% p.a., 15 years, $87 500

Monthly repayment = 87.5 × $8.44
= $738.50

15 years = 15 × 12 months
= 180 months

Total repaid = 180 × $738.50
= $132 930.00

Leanne must repay a total of $132 930.

c 7% p.a., 12 years, $60 000

Monthly repayment = 60 × $10.28
= $616.80

12 years = 12 × 12 months
= 144 months

Total repaid = 144 × $616.80
= $88 819.20

Interest = $88 819.20 − $60 000
= $28 819.20

The amount of interest paid will be $28 819.20.

d 6.5% p.a., $180 000

Over 5 years:

Monthly repayment = 18 × $19.57
= $352.26

Over 7 years:

Monthly repayment = 18 × $14.85
= $267.30

Difference = $352.26 − $267.30
= $84.96

But by paying more each month a total of $1317.60 would be saved.

The repayments will be $84.96 more each month if repaid over 5 years rather than 7.

e 12 years, $44 000

At 7.5%:

Monthly repayments = 44 × $10.55
= $464.20

At 8%: Monthly repayments = 44 × $10.82
= $476.08

Difference = $476.08 − $464.20
= $11.88

Bruce would save $11.88 every month.

And Bruce would save $1710.72 over the term of the loan.

EXAMPLE 3

This graph shows the balance owing over time on a loan of $300 000, which is repaid over 30 years.

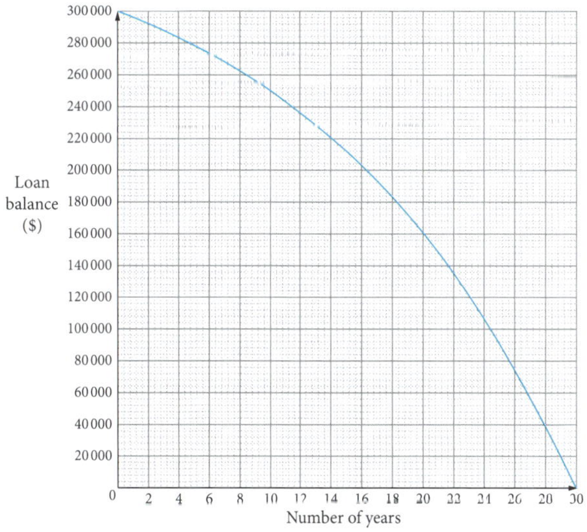

a **Approximately how much is still owing after 2 years?**

b **How much has the loan reduced after 10 years?**

c **After how many years will the loan be half paid off?**

a After 2 years the balance owing is about $292 000.

b After 10 years the balance is about $250 000. So the loan has reduced by about $50 000.

c The loan is half paid off when the balance owing is $150 000. This happens after around 21 years.

The graph doesn't provide accurate answers but is a quick way to find approximate values and gives us a picture of the balance owing at all times over the term of the loan.

EXAMPLE 4

This graph shows the difference in amounts owing on a loan when an additional $500 is paid every month.

a How many years earlier is the loan repaid by paying the extra amount each month?

b How much is still owing if the minimum repayment is made each month, when the loan would be paid off if the extra amount is paid?

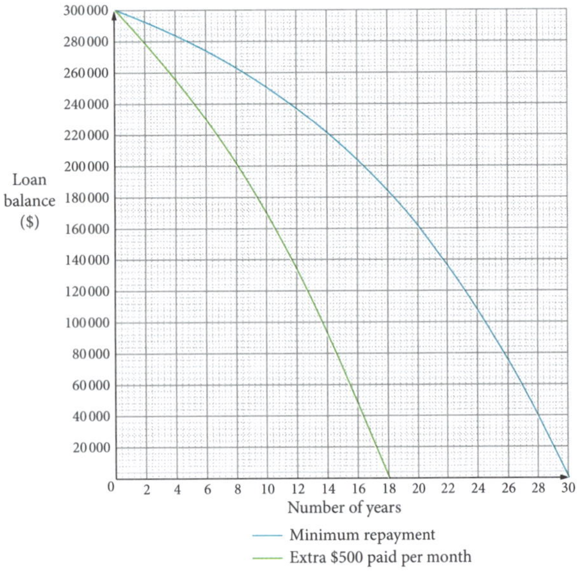

a The loan is repaid after 18 years by making the higher repayments. This is 12 years earlier than with the minimum repayment.

b After 18 years the loan balance, if the minimum repayment is made, is about $180 000.

The graph makes obvious the benefits of paying an extra amount.

→ PRACTICE

1 **The table shows the amount of each monthly repayment required to repay certain loans at a bank's current interest rate.**

Loan Amount	Term of loan (years)			
	15	20	25	30
$50 000	$600.08	$550.54	$526.61	$514.31
$80 000	$960.13	$880.87	$842.58	$822.89
$140 000	$1680.24	$1541.52	$1474.51	$1440.06
$220 000	$2640.37	$2422.39	$2317.09	$2262.95

a What is the amount of each monthly instalment if a loan of $140 000 is repaid over 25 years?

b If $80 000 is borrowed over 15 years, find the total amount that will be repaid.

c Find the total amount of interest paid on a $220 000 loan taken over 30 years.

d How much extra will be repaid on a loan of $50 000 taken over 20 years than on a loan for the same amount over 15 years?

2 **The table shows the monthly repayment (in dollars) for every $1000 borrowed at different interest rates over different numbers of years.**

Repayments per $1000 borrowed					
Term (years)	Interest rate				
	6%	7%	8%	9%	10%
10	11.10	11.61	12.13	12.67	13.22
15	8.44	8.99	9.56	10.14	10.75
20	7.16	7.75	8.36	9.00	9.65
25	6.44	7.07	7.72	8.39	9.09

a Find the amount of each monthly instalment if $35 000 is borrowed at 8% p.a. interest over 10 years.

b Tim borrows $56 400 at 9% p.a. interest over 15 years. How much must he repay?

c How much interest will be paid in total if $135 000 is borrowed at 6% p.a. interest over 25 years?

d How much more will each monthly instalment be if $84 000 is borrowed over 15 years rather than over 20 years, if the interest rate is 10% p.a.?

e Nick wants to take out a loan of $75 000 over 20 years. How much will he save each month if the interest rate is 7% p.a. instead of 8% p.a.?

3 This graph shows the balance owing over time on a loan of $240 000, which is repaid over 25 years.

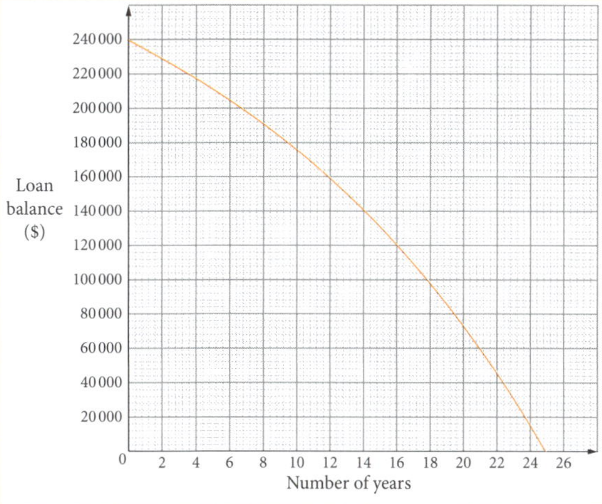

a Approximately how much is still owing after 4 years?

b How much has the loan reduced by after 12 years?

c After how many years will the loan be half paid off?

4 This graph shows the difference in amounts owing when a lump sum is paid off a loan after 5 years.

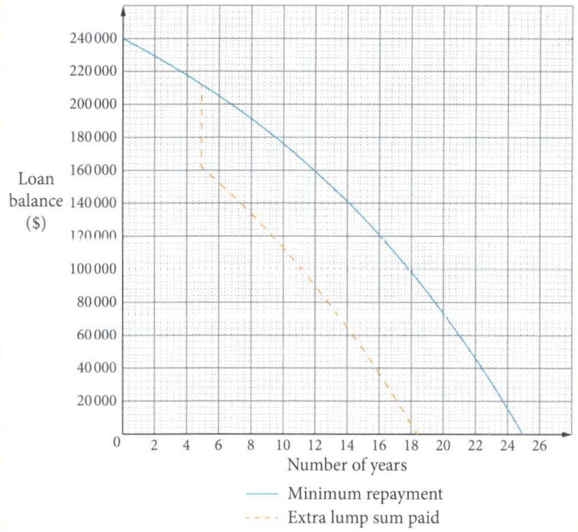

a How many years earlier is the loan repaid by paying the extra amount?

b How much is the lump sum that is paid off?

c How much would still be owing on the loan if the minimum repayments are made at the time when the loan would be paid off if the extra lump sum is paid?

5* The following table shows the monthly instalments required, for each $1000 borrowed, to repay loans over different numbers of years at a particular interest rate.

Repayments per $1000 borrowed	
Number of years	**Repayment**
5	$21.64
7	$17.02
10	$13.66
15	$11.24

a How much is the monthly repayment on a $15 000 loan over 7 years?

b Kay pays $865.60 per month on a loan of $40 000. What is the term of the loan?

c Ian is paying $341.50 per month for 10 years. How much interest will he pay?

Answers ➲ pp. 128–129

7 Credit cards

- **A credit card is a means of borrowing money** to pay for goods and services.
- **Different credit cards have different fees and charges** and ways of charging interest.
- **Some credit cards have an interest-free period** which means that, provided there is no previous outstanding amount, if the amount owing is paid by the due date no interest will be charged. (An outstanding amount is an amount that should have been paid but has not been paid.)
- **A cash advance is an amount of money borrowed as cash** using a credit card.

EXAMPLE 1

Jennifer received this credit card statement from her bank.

STAMELL BANKING

CREDIT ACCOUNT STATEMENT

Customer Enquiries: 13 000
BSB Number: 131-000
Account Number: 485620158
Statement Period: 4/10/2018 to 3/11/2018
Statement No. 63 (page 1 of 2)

Opening balance	$488.00
Purchases, cash advances and other debits	$1425.45
Interest charges	$0.00
Payments and other credits	$488.00
Closing balance	**$1425.45**

Due date	**18/11/18**	Available credit at statement date	$3574.55
Minimum payment	$29.00	Credit limit	$5000.00
		Interest rate purchases:	19.74% p.a.

Transaction details

Date	**Details**	**Amount**	**Balance**
17/10/18	Payment—Thank you	$488.00 CR	$0.00
22/10/18	Surething Insurance	$366.45	$366.45
25/10/18	Pretty Prints	$466.00	$832.45
03/11/18	Woodrow Washers	$593.00	$1425.45

Minimum payment warning: by making only the minimum payment each month you will pay more interest and take longer to pay off your balance.

If you make no additional purchases and each month you pay:	You will pay off the balance shown on this statement in about:	And you will end up paying total interest charges of about:
Only the minimum payment	11 years and 8 months	$2129.12
$72.37	2 years	$311.43, saving around $1817.69

If you are having trouble making credit card repayments please contact us; we may be able to help.

a What is the due date?

b What is the amount that Jennifer should pay by the due date to avoid paying any interest?

c How much did Jennifer spend at Pretty Prints?

d What is the daily interest rate for purchases? Give the answer as a decimal correct to two significant figures.

e If you make only the minimum payment each month, and make no other purchase, about how much does the bank determine you will pay altogether?

a 18/11/18

b $1425.45

the closing balance

c $466.00

d Daily interest rate = 19.74% ÷ 365
= 0.054 082 1917…%
= 0.000 540 821 917…
= 0.000 54 (2 sig. figs.)

e Total paid = $1425.45 + $2129.12
= $3554.57

The minimum payment will not always be $29, so the total paid is not $29 times the number of months in 11 years and 8 months (140).

EXAMPLE 2

Kathryn has a credit card that has up to 55 days interest free and an interest rate of 16.9% p.a., charged daily.

She receives her monthly bill totalling $785 and pays it in full before the due date. How much interest will Kathryn have to pay?

Kathryn pays no interest.

> Because the bill is paid in full by the due date. The interest-free days are from the time of purchase up to the due date on the bill. The amount of interest-free time depends on when a purchase is made.

EXAMPLE 3

Jamie's credit card has an interest-free period and an interest rate of 16.99% p.a., charged daily.

a **What is the daily interest rate correct to five decimal places?**

b **Jamie's bill shows a total owing of $8362. How much interest will Jamie pay if he pays the bill one day late?**

a Daily rate = 16.99% ÷ 365
= 0.046 547 9452…%
= 0.046 55% (5 d.p.)

> The daily rate is found by dividing the yearly rate by 365.

b Interest = 0.046 55% of $8362
= 0.000 4655 × $8362
= $3.89 (nearest cent)

EXAMPLE 4

Maree takes a cash advance of $2500 from her credit card. Cash advances attract immediate interest at the rate of 0.055% per day. If Maree is charged compound interest, how much interest will she pay if she repays the advance in 16 days?

$PV = \$2500$, $r = 0.000\,55$, $n = 16$

$FV = PV(1 + r)^n$
$= \$2500(1.000\,55)^{16}$
$= \$2522.090\,98\ldots$
$= \$2522.09$ (nearest cent)

$I = \$2522.09 - \$2500$
$= \$22.09$

Maree will pay $22.09 interest.

> Take care if the daily interest rate is given as a percentage. It is easy to think that it is already expressed as a decimal rather than a percentage.

→ PRACTICE

1 Jahir received this credit card statement from his bank.

STAMELL BANKING

CREDIT ACCOUNT STATEMENT

Customer Enquiries: 13 000
BSB Number: 131-000
Account Number: 485620158
Statement Period: 1/05/2019 to 31/05/2019
Statement No. 70 (page 1 of 2)

Opening balance	$1753.26
Purchases, cash advances and other debits	$987.00
Interest charges	$0.00
Payments and other credits	$1753.26
Closing balance	**$987.00**

Due date	**19/06/19**	Available credit at statement date	$9013.00
Minimum payment	$25.00	Credit limit	$10 000.00
		Interest rate:	18.93% p.a.

Transaction details

Date	Details	Amount	Balance
10/05/19	Payment—Thank you	$1753.26 CR	$0.00
21/05/19	Dr I M Fine	$278.35	$278.35
29/05/19	Fred's Fuel	$73.15	$351.50
31/05/19	Tablemania	$635.50	$987.00

Minimum payment warning: by making only the minimum payment each month you will pay more interest and take longer to pay off your balance.

If you make no additional purchases and each month you pay:	You will pay off the balance shown on this statement in about:	And you will end up paying total interest charges of about:
Only the minimum payment	5 years and 3 months	$588.00
$49.85	2 years	$209.40, saving around $378.60

If you are having trouble making credit card repayments please contact us; we may be able to help.

a What is the due date?

b What is the amount that Jahir should pay by the due date to avoid paying any interest?

c How much did Jahir spend at Tablemania?

d What is the daily interest rate? Give the answer as a decimal correct to three significant figures.

e If you make only the minimum payment each month, and make no other purchase, about how much does the bank determine you will pay altogether?

2 Harry's credit card has up to 60 days interest free and an interest rate of 15.7% p.a., charged daily. He receives his monthly bill totalling $2175 for goods that he has purchased during the month and pays it in full before the due date. How much interest will Harry have to pay?

3 Scarlett's credit card has an interest-free period and a simple interest rate of 16.06% p.a., charged daily.

a What is the daily interest rate?

b Scarlett's bill shows a total owing of $3475. How much interest will Scarlett pay if she pays the bill one day late?

4 Brooke takes a cash advance of $4000 from her credit card. Cash advances attract immediate interest at the rate of 0.053% per day. If Brooke is charged compound interest, how much interest will she pay if she repays the advance in 28 days?

5* The minimum monthly payment required on Emma's credit card is the larger of \$25 or 2% of the balance owing (provided the balance owing is greater than \$25).

a What is the minimum monthly payment required on a balance of \$2874?

b Emma pays only the minimum amount. She calculates that the minimum payment for the following month, provided she makes no further transactions, will then be \$56.33. Explain why Emma is wrong.

Answers ➲ pp. 129–130

8 Credit card calculations

➔ **Interest calculated on credit cards will be assumed to be compound interest** which is usually charged daily.

➔ **Because different amounts are added to the account at different times**, calculations must be made separately for each.

EXAMPLE 1

Emma's credit card has an interest-free period and an interest rate of 17.5% p.a., charged daily.

a What is the daily interest rate to five decimal places?

b Emma receives her bill for \$3150 and pays \$2000 on the due date. If she makes no further purchases in the next month, how much will she need to pay 30 days later to clear the debt?

a Daily rate = 17.5% ÷ 365
= 0.047 945 205…%
= 0.047 95% (5 d.p.)

b PV = \$3150 − \$2000
= \$1150

$r = 0.000\,4795$, $n = 30$

$FV = PV(1 + r)^n$
= \$1150(1.000 4795)^{30}
= \$1166.658 28…
= \$1166.66 (nearest cent)

Emma would need to pay \$1166.66 to clear the debt.

If she did make further purchases they would attract immediate interest because the interest-free period only applies if there is no outstanding amount.

EXAMPLE 2

Hashim has a credit card that has no interest-free period and a daily interest rate of 0.0435% charged daily from (and including) the date of purchase until (and including) the date of repayment. He uses his credit card to pay for computer equipment on 13 September. If the cost of the equipment was \$3125.00 and he pays the credit card bill on 3 October, find the total he would need to repay to cover the equipment and interest.

From 13 September to 3 October inclusive is 21 days.

September has 30 days. No interest is charged for the first 12 days, so the number of days accruing interest in September is 30 − 12 = 18. Adding another 3 days in October gives 21 days altogether.

PV = \$3125, $r = 0.000\,435$, $n = 21$

$FV = PV(1 + r)^n$
= \$3125(1.000 435)^{21}
= \$3153.671 39…
= \$3153.67 (nearest cent)

Hashim would need to pay \$3153.67.

EXAMPLE 3

Xanthe uses her credit card to get a cash advance of \$1500. A fee is charged. This fee is the greater of \$2 or 3% of the amount advanced.

a How much is the fee Xanthe must pay?

b Cash advances attract immediate (compound) interest of 18.24% p.a. What amount must be paid on the due date, 27 days later, to clear the debt?

c Xanthe only pays the minimum amount. This minimum amount is 2% of the closing balance (rounded up to the nearest dollar) or \$10, whichever is greater. How much will Xanthe need to pay in a further 30 days to clear the debt?

a 3% of \$1500 = 0.03 × \$1500
= \$45

Xanthe must pay a fee of \$45.

b Total owing = \$1500 + \$45
= \$1545

$PV = \$1545$, $r = \dfrac{0.1824}{365}$, $n = 28$

$FV = PV(1 + r)^n$

$= \$1545\left(1+\dfrac{0.1824}{365}\right)^{28}$

= \$1566.764 62…
= \$1566.76 (nearest cent)

Interest is charged on the first day plus a further 27 days making a total of 28 altogether.

The amount to be paid is \$1566.76

c 2% of \$1566.76 = 0.02 × \$1566.76
= \$31.3352

So the minimum payment is \$32.

New total = \$1566.76 − \$32
= \$1534.76

The question said that the amount was rounded **up** to the nearest dollar.

$PV = \$1534.76$, $r = \dfrac{0.1824}{365}$, $n = 31$

$$FV = PV(1 + r)^n$$
$$= \$1534.76\left(1+\frac{0.1824}{365}\right)^{31}$$
$$= \$1558.71482\ldots$$
$$= \$1558.71 \text{ (nearest cent)}$$

Xanthe will need to pay $1558.71 to clear the debt.

EXAMPLE 4

Kelly makes the following purchases using her credit card.

Date	Item	Cost	Total
2 May	Books	$239.00	$239.00
10 May	Coat	$150.00	$389.00
29 May	Shoes	$78.00	$467.00

Kelly pays the total owing plus all interest charges on 14 June. The credit card has no interest-free period and an interest rate of 18.4% p.a., charged daily.

a **Find the daily rate of interest, correct to five decimal places.**

b **Find the amount of interest paid on the books. (Interest is charged from, and including, the date of purchase until, and including, the date of repayment.)**

c **Find the total amount of interest Kelly must pay.**

a Daily rate = 18.4% ÷ 365
= 0.050 410 958...%
= 0.050 41% (5 d.p.)

b From 2 May to 14 June inclusive is 44 days.

$PV = \$239$, $r = 0.000\,5041$, $n = 44$

$$FV = PV(1 + r)^n$$
$$= \$239(1+ 0.000\,5041)^{44}$$
$$= \$244.358\,977\ldots$$
$$= \$244.36 \text{ (nearest cent)}$$

Inclusive means 'including both endpoints'.

Interest = $244.36 – $239.00
= $5.36

Kelly must pay $5.36 interest on the books.

c Coat:

From 10 May to 14 June is 36 days.

$PV = \$150$, $r = 0.000\,5041$, $n = 36$

$$FV = PV(1 + r)^n$$
$$= \$150(1.000\,5041)^{36}$$
$$= \$152.746\,291\ldots$$
$$= \$152.75 \text{ (nearest cent)}$$

Interest = $152.75 – $150.00
= $2.75

Shoes:

From 29 May to 14 June is 17 days.

$PV = \$78$, $r = 0.000\,5041$, $n = 17$

$$FV = PV(1 + r)^n$$
$$= \$78(1.000\,5041)^{17}$$
$$= \$78.671\,1390\ldots$$
$$= \$78.67 \text{ (nearest cent)}$$

Interest = $78.67 – $78.00
= $0.67

Total interest = $5.36 + $2.75 + $0.67
= $8.78

Kelly must pay a total of $8.78 in interest.

EXAMPLE 5

Toby pays 0.048% per day interest on his credit card and there are no interest-free days. From 1 to 12 November his balance is $4275.

a **How much interest will Toby pay for the period from 1 to 12 November?**

b **On 13 November he charges an additional $555 to the account. If Toby wishes to pay the whole amount owing on 30 November, how much will he need to pay?**

a $PV = \$4275$, $r = 0.000\,48$, $n = 12$

$$FV = PV(1 + r)^n$$
$$= \$4275(1.000\,48)^{12}$$
$$= \$4299.69 \text{ (nearest cent)}$$

Interest = $4299.69 – $4275.00
= $24.69

b Balance = $4299.69 + $555
= $4854.69

$PV = \$4854.69$, $r = 0.000\,48$, $n = 18$

$$FV = PV(1 + r)^n$$
$$= \$4854.69(1.000\,48)^{18}$$
$$= \$4896.81 \text{ (nearest cent)}$$

This would include a total of $66.81 in interest.

Toby will need to pay a total of $4896.81.

→ PRACTICE

1 **Mark's credit card has an interest-free period and an interest rate of 16.8% p.a., charged daily.**

a **What is the daily interest rate to five decimal places?**

b **Mark receives his bill for $1750 and pays $1000 on the due date. If he makes no further purchases in the next month, how much will he need to pay 31 days later to clear the debt?**

2 Leila has a credit card that has no interest-free period and a daily interest rate of 0.0495% charged daily from (and including) the date of purchase until (and including) the date of repayment. She uses her credit card to buy furniture costing $5375 on 18 July. If she pays the credit card bill on 6 August, find the total she would need to repay to cover the furniture and interest.

3 Xavier uses his credit card to get a cash advance of $3200. A fee is charged. This fee is the greater of $2 or 1.5% of the amount advanced.

a How much is the fee Xavier must pay?

b Cash advances attract immediate (compound) interest of 20.99% p.a., charged daily. What amount must be paid on the due date, 25 days later, to clear the debt?

c Xavier only pays the minimum amount on the due date. This minimum amount is 2% of the closing balance (rounded up to the nearest dollar), or $10, whichever is greater. How much will Xavier need to pay in a further 29 days to clear the debt?

4 Alison makes the following purchases using her credit card.

Date	Item	Cost	Total
11 March	Clothes	$328.00	$328.00
19 March	Petrol	$75.00	$403.00
26 March	Insurance	$294.00	$697.00

Alison pays the total owing plus all interest charges on 7 April. The credit card has no interest-free period and an interest rate of 17.6% p.a., charged daily from, and including, the date of purchase until, and including, the date of repayment.

a Find the daily rate of interest, correct to three decimal places.

b Find the amount of interest paid on the clothes.

c Find the total amount of interest Alison must pay.

5 Brendan pays 0.056% per day interest on his credit card and there are no interest-free days. From 7 to 23 January his balance is $3680. Interest is calculated from, and including, the day of purchase until, and including, the day of repayment.

a How much interest will Brendan pay for the period from 7 to 23 January?

b On 24 January he charges an additional $1095 to the account. If Brendan wishes to pay the whole amount owing on 31 January, how much will he need to pay?

6* On 24 May, Jonathan takes a $3000 cash advance from his credit card. Cash advances attract immediate interest at the rate of 17.52% p.a. charged daily, from and including the date the advance is taken until and including the date of repayment. Find the amount of interest that Jonathan will pay if he doesn't repay the money until 15 September.

Answers ⮊ pp. 130–131

HSC EXAM-TYPE QUESTIONS

Total marks: 30

PART A

(Suggested time: 12 minutes)

(1 mark each)

1 This is part of the statement Aldo received for his credit card.

STAMELL BANKING

CREDIT ACCOUNT STATEMENT

Opening balance	$673.00
Purchases and other debits	$1642.85
Payments and other credits	$673.00
Closing balance	**$1642.85**
Available credit	$4357.15
Credit limit	$6000.00

What amount must Aldo pay by the due date to avoid paying any interest?

A $673.00 **B** $1642.85 **C** $4357.15 **D** $6000.00

2 An amount of $70 000 is borrowed. The reducible interest rate is 1% per month and monthly repayments are $1004.30. How much interest will be paid in the second month?

A $700 **B** $689.96 **C** $679.91 **D** $696.96

3 An asset originally valued at $22 500 is depreciated using the declining-balance method. The rate of depreciation is 12% p.a. By how much (to the nearest dollar) has the asset depreciated after 5 years?

A $11 874 **B** $10 626 **C** $13 500 **D** $9000

4 The following table was drawn up to show the principal and interest over the first 3 months of a loan.

Principal	Interest	$P + I$	$P + I - R$
$60 000.00	$480.00	$60 480.00	$59 217.00
$59 217.00	$473.74	$59 690.74	$58 427.74
$58 427.74	$467.42	$58 895.16	$57 632.16

What is the monthly rate of reducible interest?

A 0.008% **B** 0.08% **C** 0.8% **D** 8%

5 The monthly repayments on loans over 12 years at a particular rate of interest are $10.91 per $1000 borrowed. How much interest in total (to the nearest dollar) would be paid on a loan of $90 000?

A $13 092 **B** $40 920 **C** $51 394 **D** $98 190

6 Yasmin is borrowing some money to buy a motorbike priced at $8500. She pays 15% deposit and agrees to repay the balance in equal monthly instalments over 3 years. The flat rate of interest charged is 6% p.a. Find the amount of each instalment.

A $236.82 **B** $243.19 **C** $272.24 **D** $320.40

7 Freya has a credit card with up to 50 days interest free and an interest rate of 0.056% per day. How much interest will Freya pay if she has an amount of $4200 overdue for 30 days?

A $0 **B** $70.56 **C** $71.14 **D** $78.40

8 The graph shows the value of a van over the first 10 years.

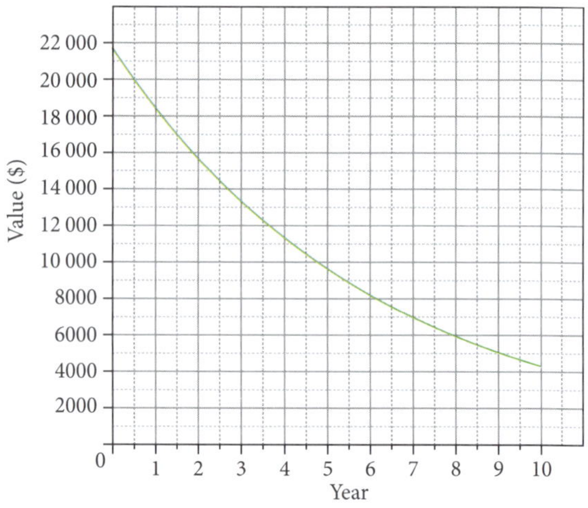

During which year did its value fall by $2000?

A 3rd **B** 4th **C** 5th **D** 6th

PART B

(Suggested time: 33 minutes)

Show all working.

9 Penny borrows $8400 for a holiday. She agrees to repay the loan over 4 years with equal monthly payments of $225.40.

- **a** How much does Penny repay in total? (1 mark)
- **b** How much interest does she pay? (1 mark)
- **c** What is the annual rate of simple interest charged? (2 marks)

10 The following repayment table shows the amounts for the first 3 months of a loan.

Mth	Principal	Interest	$P + I$	$P + I - R$
1	$90 000.00	$675.00	$90 675.00	$89 650.77
2	$89 650.77	$672.38	$90 323.15	$89 298.92
3	$89 298.92	$669.74	$89 968.66	$88 944.43

- **a** What is the amount of each repayment? (1 mark)
- **b** How can we easily tell that the interest charged is a reducible rate and not a flat rate? (1 mark)
- **c** Complete the next row of the table. (3 marks)

11 A machine was bought for $38 000 and is being depreciated by the declining-balance method.

- **a** After one year the value of the machine has reduced to $34 960. What is the rate of depreciation? (1 mark)
- **b** How much did the machine depreciate in the tenth year? (2 marks)

12 Tina's credit card has no interest-free period and an interest rate of 0.06% per day calculated from, and including, the date of purchase until, and including, the date of payment. Her balance was $0 until she used the card to buy some furniture for $2100 on 16 August. Find the amount she would need to pay on 1 September to clear the card. (2 marks)

13 Bailey wants to borrow $85 000 and repay it over 15 years. One option he has is a fixed (reducible) interest loan with monthly repayments of $773.54.

- **a** How much will Bailey repay altogether if he takes this option? (1 mark)
- **b** Another option Bailey is considering is a loan with a variable interest rate. Bailey has calculated that with current interest rates he will save $1717 with this loan. Give a reason why Bailey should not take this loan. (1 mark)

14 The table shows the monthly payment required per $1000 borrowed for different terms and interest rates.

Term (years)	5%	6%	7%	8%
5	$18.87	$19.33	$19.80	$20.28
10	$10.61	$11.10	$11.61	$12.13
15	$7.91	$8.44	$8.99	$9.56

How much interest would be paid on a $12 000 loan over 10 years at 7% p.a.? (2 marks)

15 Kyle receives the following credit card statement. His card offers up to 50 days interest free and an interest rate of 16.9725% is charged daily.

Date	Details	Amount	Total
01/06/19	Opening balance	$0.00	$0.00
09/06/19	Insurance	$936.00	$936.00
23/06/19	Clothes	$419.00	$1355.00
Due date: 08/07/19		Amount due:	$1355.00

- **a** What is the interest rate per day? (1 mark)
- **b** How much interest will be charged if the bill is paid by the due date? (1 mark)
- **c** Find the amount of interest paid if Kyle pays the bill 15 days late. (2 marks)

ANSWERS

PRACTICE

1 Depreciation ➲ p. 106

1 $V_0 = \$26\,400$, $D = \$1850$, $n = 8$

$S = V_0 - Dn$

$= \$26\,400 - \$1850 \times 8$

$= \$11\,600$

After 8 years the value of the forklift will be $11 600.

2 $S = \$15\,700$, $V_0 = \$24\,500$, $n = 8$

$S = V_0 - Dn$

$\$15\,700 = \$24\,500 - D \times 8$

$8D = \$24\,500 - \$15\,700$

$= \$8800$

$D = \$1100$

The amount of depreciation allowed per year is $1100.

3 $S = \$4625$, $D = \$675$, $n = 6$

$S = V_0 - Dn$

$\$4625 = V_0 - \$675 \times 6$

$\$4625 = V_0 - \$4050$

$V_0 = \$4625 + \$4050$

$= \$8675$

The purchase price of the machine was $8675.

4 $S = \$3638$, $r = 0.08$, $n = 6$

$S = V_0(1 - r)^n$

$\$3638 = V_0(1 - 0.08)^6$

$\$3638 = V_0 \times 0.606\,355\,001\ldots$

$V_0 = \$3638 \div 0.606\,355\,001\ldots$

$= \$5999.785\,59\ldots$

$= \$6000$ (nearest dollar)

The purchase price of the tools was $6000.

5 $S = \$26\,400$, $V_0 = \$43\,000$, $n = 3$

$S = V_0(1 - r)^n$

$\$26\,400 = \$43\,000(1 - r)^3$

$(1 - r)^3 = \dfrac{\$26\,400}{\$43\,000}$

$1 - r = \sqrt[3]{\dfrac{\$26\,400}{\$43\,000}}$

$= 0.849\,920\,863\ldots$

$= 0.85$ (2 d.p.)

$r = 1 - 0.85$

$= 0.15$

The machine has been depreciated at 15% p.a.

6 Value at end of 3rd year:

$V_0 = \$56\,000$, $r = 0.09$, $n = 3$

$S = V_0(1 - r)^n$

$= \$56\,000(1 - 0.09)^3$

$= \$42\,199.976$

$= \$42\,200$ (nearest dollar)

Depreciation in 4th year = 9% of $42 200

= $3798

The amount of depreciation allowed during the fourth year is $3798.

2 The declining-balance method of determining depreciation ➲ pp. 107–108

1 $V_0 = \$42\,000$, $r = 0.12$, $n = 7$

$S = V_0(1 - r)^n$

$= \$42\,000(1 - 0.12)^7$

$= \$17\,164.3750\ldots$

$= \$17\,200$ (nearest hundred dollars)

After 7 years the car is worth approximately $17 200.

2 $V_0 = \$1400$, $r = 0.08$, $n = 3$

$S = V_0(1 - r)^n$

$= \$1400(1 - 0.08)^3$

$= \$1090.1632$

$= \$1090$ (nearest dollar)

Depreciation = $1400 – $1090

= $310

The cement mixer has depreciated by $310, to the nearest dollar, in that time.

3 a $V_0 = \$76\,000$, $r = 0.125$, $n = 5$

$S = V_0(1 - r)^n$

$= \$76\,000(1 - 0.125)^5$

$= \$38\,981.079\ldots$

= **$38 981** (nearest dollar)

b Depreciation in sixth year

= 12.5% of $38 981

= $4872.625

= **$4873** (nearest dollar)

3 Comparing methods of depreciation ➲ pp. 109–110

1 a **After 3 years the value is about $15 000.**

b **The value of the asset is $8000 after 7 years.**

c When $t = 0$, $v = \$25\,000$

When $t = 1$, $v = \$21\,250$

Depreciation = $25 000 – $21 250

= $3750

% depreciation $= \dfrac{\$3750}{\$25\,000} \times 100\%$

= 15%

The asset is decreasing in value by 15% per year.

2 a **The purchase price was $4000.**

b **After 4 years depreciation at 20% the machine is worth $1650.**

c **At 10% depreciation the machine halves in value after about 7 years.**

d **At 15% depreciation the machine lost about $500 in the second year.**

e **After 6 years there is about $450 difference in salvage value if depreciated at 15% or 20%.**

3 a $V_0 = \$32\,500$, $D = \$2700$, $n = 5$

$S = V_0 - Dn$
$= \$32\,500 - \$2700 \times 5$
$= \$19\,000$

If depreciated by the straight-line method the van will be worth $19 000 after 5 years.

b $V_0 = \$32\,500$, $r = 0.09$, $n = 5$

$S = V_0(1 - r)^n$
$= \$32\,500(1 - 0.09)^5$
$= \$20\,281.0447\ldots$
$= \mathbf{\$20\,300}$ (nearest $100)

c After a further year under the straight-line method of depreciation, the van will lose a further $2700 in value.

Salvage value = $19 000 − $2700
= $16 300

Under the declining-balance method, the van will lose a further 9% of its value.

Depreciation = 0.09 × $20 300
= $1827

Salvage value = $20 300 − $1827
= $18 473

The van will be worth more after 6 years using the declining-balance method.

4 $V_0 = \$7600$

a $r = 0.1$, $n = 10$

$S = V_0(1 - r)^n$
$= \$7600(1 - 0.1)^{10}$
$= \$2649.956\,14\ldots$
$= \mathbf{\$2650}$ (nearest dollar)

b $S = V_0 - Dn$

$\$2650 = \$7600 - D \times 10$
$10D = \$7600 - \$2650$
$= \$4950$
$D = \mathbf{\$495}$

c

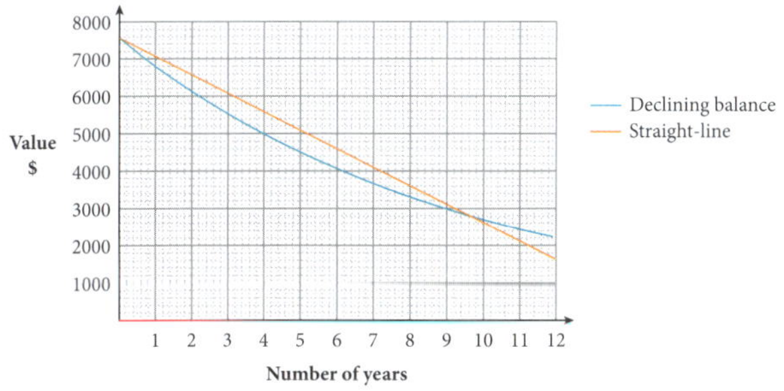

4 Flat-rate loans ➲ p. 111

1 a $P = \$6100$, $r = 0.008$, $n = 60$

$I = Prn$
$= \$6100 \times 0.008 \times 60$
$= \$2928$

Jordan pays $2928.00 in interest.

b Total to be repaid = $6100 + $2928
= $9028

Jordan must repay a total of $9028.00.

2 a Total repaid = $247.50 × 60
= **$14 850**

b Interest = $14 850 − $11 000
= **$3850**

c Total interest = $3850

Interest per year = $3850 ÷ 5
= $770

Interest rate = $\dfrac{\$770}{\$11\,000} \times 100\%$
= 7%

The interest rate is 7% p.a.

3 $P = \$32\,000$, $r = 0.055$, $n = 6$

$I = Prn$
$= \$32\,000 \times 0.055 \times 6$
$= \$10\,560$

Total to be repaid = $32 000 + $10 560
= $42 560

Number of instalments = 6 × 12
= 72

Each instalment = $42 560 ÷ 72
= $591.111 11…
= **$591.11** (nearest cent)

4 a Price = $21 000, deposit = $6000

Principal = $21 000 − $6000
= $15 000

$r = 0.08$, $n = 4$

$I = Prn$
$= \$15\,000 \times 0.08 \times 4$
$= \mathbf{\$4800}$

b Total to be repaid = $15 000 + $4800
= $19 800

Number of instalments = 4 × 12
= 48

Each instalment = $19 800 ÷ 48
= **$412.50**

c Total paid
= total to be repaid + deposit
= $19 800 + $6000
= **$25 800**

5 a Total paid = $7000 + 60 × $800
= **$55 000**

b Amount borrowed = $37 000 − $7000
= $30 000

Amount repaid = $55 000 − $7000
= $48 000

Interest = $48 000 − $30 000
= $18 000

$I = \$18\,000$, $P = \$30\,000$, $n = 5$

$I = Prn$
$\$18\,000 = \$30\,000 \times r \times 5$
$\$18\,000 = \$150\,000r$
$r = 0.12$

The interest rate was 12% p.a.

5 Reducing-balance loans ➲ pp. 113–114

1 a **$24 790.12**

b **$196.63**

c After 3 months the amount owing is $24 365.31

Reduction = $25 000 − $24 365.31
= $634.69

The loan had reduced by $634.69.

d Total repaid = 3 × $409.88
= **$1229.64**

e Difference = $1229.64 – $634.69
= **$594.95**

The difference is the amount of interest paid.

2 Amount borrowed: $22 000

Interest rate: 1% per month

Repayment: $489.38 per month

	Principal	Interest	$P + I$	$P + I - R$
1	$22 000.00	$220.00	$22 220.00	$21 730.62
2	$21 730.62	$217.31	$21 947.93	$21 458.55
3	$21 458.55	$214.59	$21 673.14	$21 183.76
4	$21 183.76	$211.84	$21 395.60	$20 906.22

3 a R = $60 420.00 – $59 472.83
= **$947.17**

b Interest rate = $\frac{\$420}{\$60\,000} \times 100\%$
= 0.7%

The interest rate per month is 0.7%.

c Month 6: P = $57 326.98

Interest = 0.007 × $57 326.98
= $401.29 (nearest cent)

$P + I$ = $57 326.98 + $401.29
= $57 728.27

$P + I - R$ = $57 728.27 – $947.17
= $56 781.10

The principal in month seven will be $56 781.10

4 a Interest 1st year = 0.085 × $5000
= **$425**

b $P + I - R$ = $5000 + $425 – $2000
= **$3425**

c Interest 2nd year
= 0.085 × $3425
= $291.13 (nearest cent)

Amount owing at start of 3rd year
$3425 + $291.13 – $2000
= $1716.13

Interest 3rd year
= 0.085 × $1716.13
= $145.87 (nearest cent)

$P + I$ = $1716.13 + $145.87
= $1862.00

Jamie must pay $1862 to pay out the loan at the end of the third year.

5 a P = $16 000, $r = 0.07$, $n = 4$

$I = Prn$
= $16 000 × 0.07 × 4
= $4480

Total to repay = $16 000 + $4480
= $20 480

Each instalment = $20 480 ÷ 4
= **$5120**

If Lincoln chooses the flat-rate loan, each instalment will be $5120.

b P = $16 000, r = 8.5%, R = $4900

	Principal	Interest	$P + I$	$P + I - R$
1	$16 000.00	$1360.00	$17 360.00	$12 460.00
2	$12 460.00	$1059.10	$13 519.10	$8619.10
3	$8619.10	$732.62	$9351.72	$4451.72
4	$4451.72	$378.40	$4830.12	–

The final instalment will be $4830.12.

c With the flat-rate loan Lincoln will repay $20 480.

With the reducing-balance loan:

Total repaid = 3 × $4900 + $4830.12
= $19 530.12

Difference = $20 480 – $19 530.12
= $949.88

Lincoln will pay $949.88 less with the reducing-balance loan.

6 Amount borrowed: $34 000

Interest rate: 10% p.a.

Repayments: $10 726.00

	Principal	Interest	$P + I$	$P + I - R$
1	$34 000.00	$3400.00	$37 400.00	$26 674.00
2	$26 674.00	$2667.40	$29 341.40	$18 615.40
3	$18 615.40	$1861.54	$20 476.94	$9750.94
4	$9750.94	$975.09	$10 726.03	$0.03

With the payment of an additional 3 cents the loan will be repaid after 4 years.

6 Tables and graphs ➲ pp. 116–117

1 a Loan: $140 000, term: 25 years

Monthly instalment = $1474.51

b Loan: $80 000, term: 15 years

Monthly instalment = $960.13

15 years = 15 × 12 months
= 180 months

Total repaid = 180 × $960.13
= $172 823.40

If $80 000 is borrowed over 15 years a total of $172 823.40 will be repaid.

c Loan: $220 000, term: 30 years

Monthly instalment = $2262.95

30 years = 30 × 12 months
= 360 months

Total repaid = 360 × $2262.95
= $814 662

Interest = $814 662 – $220 000
= $594 662

If $220 000 is borrowed over 30 years, a total of $594 662 will be paid in interest.

d Loan: $50 000, term: 20 years

Monthly instalment = $550.54

20 years = 20 × 12 months
= 240 months

Total repaid = 240 × $550.54
=$132 129.60

Loan: $50 000, term: 15 years

Monthly instalment = $600.08

15 years = 15 × 12 months
= 180 months

Total repaid = 180 × \$600.08
= \$108 014.40

Difference = \$132 129.60 – \$108 014.40
= \$24 115.20

If a \$50 000 loan is taken over 20 years compared to 15 years, an additional \$24 115.20 would need to be repaid.

2 a 8% p.a., 10 years, \$35 000

Monthly repayment
= \$12.13 per \$1000
= 35 × \$12.13
= **\$424.55**

b 9% p.a., 15 years, \$56 400

Repayment = 56.4 × \$10.14
= \$571.90 (nearest cent)

15 years = 15 × 12 months
= 180 months

Total repaid = 180 × \$571.90
= \$102 942

Tim must repay a total of \$102 942.

c 6% p.a., 25 years, \$135 000

Monthly repayment = 135 × \$6.44
= \$869.40

25 years = 25 × 12 months
= 300 months

Total repaid = 300 × \$869.40
= \$260 820

Interest = \$260 820 – \$135 000
= \$125 820

The amount of interest paid will be \$125 820.

d 10% p.a., \$84 000

Over 15 years:

Monthly repayment = 84 × \$10.75
= \$903

Over 20 years:

Monthly repayment = 84 × \$9.65
= \$810.60

Difference = \$903.00 – \$810.60
= \$92.40

The repayments will be \$92.40 more each month if repaid over 15 years rather than 20 years.

e 20 years, \$75 000

At 7%:

Monthly repayments = 75 × \$7.75
= \$581.25

At 8%:

Monthly repayments = 75 × \$8.36
= \$627

Difference = \$627.00 – \$581.25
= \$45.75

Nick would save \$45.75 every month.

3 a **After 4 years the amount owing is about \$218 000.**

b After 12 years the amount owing is about \$160 000. **This means the loan has reduced by \$80 000.**

c The loan will be half paid off when the balance owing is \$120 000. **This happens after 16 years.**

4 a By paying the extra amount the loan is repaid after about $18\frac{1}{2}$ years. **So the loan is paid off about $6\frac{1}{2}$ years earlier.**

b Before the lump sum was paid the balance was \$212 000.

After the lump sum was paid the balance was \$162 000.

Difference = \$212 000 – \$162 000
= \$50 000

The lump sum was \$50 000.

c **At the time the loan with extra lump sum is repaid, the balance of the original loan is about \$94 000.**

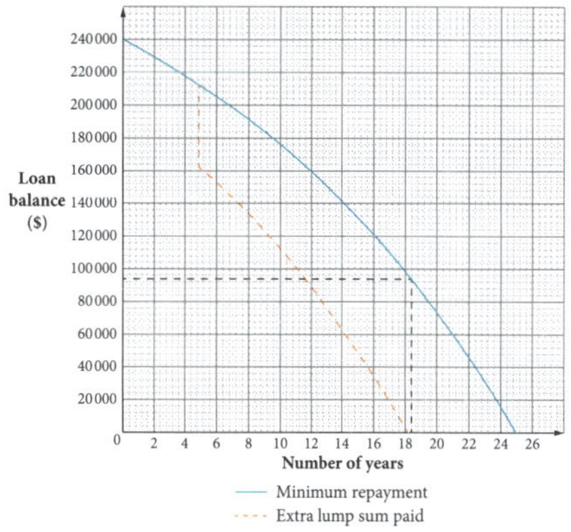

5 a 7 years:

Repayment = \$17.02 per \$1000

\$15 000 loan:

Repayment = \$17.02 × 15
= **\$255.30**

b Repayment = \$865.60 on \$40 000 loan

Repayment per \$1000 = \$865.60 ÷ 40
= \$21.64

The term of the loan is 5 years.

c 10 years:

Repayment = \$13.66 per \$1000

Amount of loan
= (\$341.50 ÷ \$13.66) × \$1000
= \$25 000

Total repayments = \$341.50 × 10 × 12
=\$40 980

Interest = \$40 980 – \$25 000
= \$15 980

Ian will pay a total of \$15 980 in interest.

7 Credit cards ➲ pp. 120–121

1 a **19/06/19**

b **Jahir should pay \$987.00.**

c **At Tablemania Jahir spent \$635.50.**

d Interest rate = 18.93% p.a.

Daily rate = 18.93% ÷ 365
= 0.051 863 013…%
= 0.000 518 630 13…
= **0.000 519** (3 sig.figs)

e Extra paid = \$588.00

Total paid altogether = \$987.00 + \$588.00
= **\$1575.00**

2 **Harry will pay no interest because the bill is paid in full by the due date.**

3 a Daily rate = 16.06% ÷ 365
= **0.044%**

b Interest = 0.044% of $3475
$= 0.000\,44 \times \$3475$
$= \mathbf{\$1.53}$ (nearest cent)

4 $PV = 4000,\ r = 0.000\,53,\ n = 28$
$FV = PV(1 + r)^n$
$= \$4000(1.000\,53)^{28}$
$= \$4059.786\,67\ldots$
$= \$4059.79$ (nearest cent)
$I = \$4059.79 - \$4000$
$= \$59.79$
Brooke will pay $59.79 interest.

5 a 2% of $2874 = 0.02 × $2874
= $57.48
So the minimum payment is $57.48.
b Now $2874 – $57.48 = $2816.52
2% of $2816.52 = $56.33 (nearest cent)
So Emma would be right if she didn't have to pay interest, but interest will be added to the $2816.52 that is outstanding so the balance owing will be higher and **the minimum payment will therefore be more than $56.33.**

8 Credit card calculations ➲ pp. 122–123

1 a Daily rate = 16.8% ÷ 365
$= 0.046\,027\,397\ldots\%$
$= \mathbf{0.046\,03\%}$ (5 d.p.)
b $P = \$1750 - \$1000$
$= \$750$
$r = 0.000\,4603,\ n = 31$
$FV = PV(1 + r)^n$
$= \$750(1.000\,4603)^{31}$
$= \$760.78$ (nearest cent)
Mark would need to pay $760.78 to clear the debt.

2 From 18 July to 6 August inclusive is 20 days.
$PV = \$5375,\ r = 0.000\,495,\ n = 20$
$FV = PV(1 + r)^n$
$= \$5375 \times (1.000\,495)^{20}$
$= \$5428.463\,47\ldots$
$= \$5428.46$ (nearest cent)
Leila would need to pay $5428.46.

3 a 1.5% of $3200 = $48
So the fee is $48.
b Total owing = $3200 + $48
= $3248
$PV = \$3248,\ r = \dfrac{0.2099}{365},\ n = 26$
$FV - PV(1 + r)^n$
$= \$3248 \times \left(1+\dfrac{0.2099}{365}\right)^{26}$
$= \$3296.914\,08\ldots$
$= \$3296.91$ (nearest cent)
Xavier must pay $3296.91 to clear the debt.
c 2% of $3296.91 = $65.9382
= $66 (nearest dollar)
So the minimum payment is $66.
Balance = $3296.91 – $66.00
= $3230.91
$PV = \$3230.91,\ r = \dfrac{0.2099}{365},\ n = 30$
$FV = PV(1 + r)^n$
$= \$3230.91 \times \left(1+\dfrac{0.2099}{365}\right)^{30}$
$= \$3287.117\,12\ldots$
$= \$3287.12$ (nearest cent)
Xavier must pay $3287.12 to clear the debt.

4 a Daily rate = 17.6% ÷ 365
$= 0.048\,219\,178\ldots\%$
$= \mathbf{0.048\%}$ (3 d.p.)
b From 11 March to 7 April inclusive is 28 days.
$PV = \$328,\ r = 0.000\,48,\ n = 28$
$FV = PV(1 + r)^n$
$= \$328 \times (1.000\,48)^{28}$
$= \$332.437\,005\ldots$
$= \$332.44$ (nearest cent)
Interest = $332.44 – $328.00
= $4.44
Alison must pay $4.44 interest on the clothes.
c Petrol: From 19 March to 7 April is 20 days.
$PV = \$75,\ r = 0.000\,48,\ n = 20$
$FV = PV(1 + r)^n$
$= \$75 \times (1.000\,48)^{20}$
$= \$75.723\,2926\ldots$
$= \$75.72$ (nearest cent)
Interest = $75.72 – $75.00
= $0.72
Insurance: From 26 March to 7 April is 13 days.
$PV = \$294,\ r = 0.000\ 48,\ n = 13$
$FV = PV(1 + r)^n$
$= \$294 \times (1.000\,48)^{13}$
$= \$295.839\,852\ldots$
$= \$295.84$ (nearest cent)
Interest = $295.84 – $294.00
= $1.84
Total interest = $4.44 + $0.72 + $1.84
= $7.00
Alison must pay a total of $7 in interest.

5 a From 7 to 23 January is 17 days.
$PV = \$3680,\ r = 0.000\,56,\ n = 17$
$FV = PV(1 + r)^n$
$= \$3680 \times (1.000\,56)^{17}$
$= \$3715.19$ (nearest cent)
Interest = $3715.19 – $3680.00
= **$35.19**
b Balance = $3715.19 + $1095
= $4810.19
$PV = \$4810.19,\ r = 0.000\,56,\ n = 8$
$FV = PV(1 + r)^n$
$= \$4810.19 \times (1.000\,56)^{8}$
$= \$4831.781\,93\ldots$
$= \$4831.78$ (nearest cent)
Brendan will need to pay a total of $4831.78.

6 Daily rate = 17.52% ÷ 365
= 0.048%

Number of days = (31 − 23) + 30 + 31 + 31 + 15
= 115

$FV = \$3000 \times (1.000\,48)^{115}$
= $3170.213 84...
= $3170.21 (nearest cent)

I = $3170.21 − $3000
= $170.21

Jonathan will pay $170.21 in interest.

HSC EXAM-TYPE QUESTIONS

Part A

1 **B.** The closing balance is the amount that must be paid.
So $1642.85 is the amount to be paid.

2 **D.** 1st month: P = $70 000
I = 0.01 × $70 000 = $700
$P + I - R$ = $70 000 + $700 − $1004.30
= $69 695.70
2nd month: P = $69 695.70
I = 0.01 × $69 695.70
= $696.96 (nearest cent)

3 **B.** $V_0 = \$22\,500$, $r = 0.12$, $n = 5$
$S = V_0(1 - r)^n$
$= \$22\,500(1 - 0.12)^5$
= $11 874 (nearest dollar)
Total depreciation = $22 500 − $11 874
= $10 626

4 **C.** Interest $= \dfrac{480}{60\,000} \times 100\%$
= 0.8%

5 **C.** Repayments = 90 × $10.91 = $981.90
Total repaid = $981.90 × 12 × 12
= $141 393.60
Interest = $141 393.60 − $90 000
= $51 393.60
= $51 394 (nearest dollar)

6 **A.** Deposit = 0.15 × $8500
= $1275
Balance = $8500 − $1275
= $7225
Interest = $7225 × 0.06 × 3
= $1300.50
Total to repay = $7225 + $1300.50
= $8525.50
Repayment = $8525.50 ÷ 36
= $236.82 (nearest cent)

7 **C.** The amount is overdue so interest must be paid.
PV = $4200, r = 0.000 56, n = 30
$FV = PV(1 + r)^n$
$= \$4200 \times (1.000\,56)^{30}$
= $4271.14 (nearest cent)
I = $4271.14 − $4200
= $71.14

8 **B.** At the end of 3 years the value is approximately $13 200.
At the end of 4 years the value is approximately $11 200.
So the decrease in value of $2000 was in the 4th year.

Part B

9 a Total repaid = $225.40 × 12 × 4
= **$10 819.20** ✓

b Interest = $10 819.20 − $8400
= **$2419.20** ✓

c Interest per year = $2419.20 ÷ 4
= $604.80 ✓

Interest rate $= \dfrac{604.8}{8400} \times 100\%$
= **7.2%** ✓

10 a Repayment = $90 675 − $89 650.77
= **$1024.23** ✓

b **The interest is not the same each month. It is reducing.** ✓

c Interest rate $= \dfrac{675}{90\,000} \times 100\%$
= 0.75% ✓

4th month: P = $88 944.43
I = 0.0075 × $88 944.43
= $667.08 (nearest cent) ✓
$P + I$ = $88 944.43 + $667.08
= $89 611.51
$P + I - R$ = $89 611.51 − $1024.23
= $88 587.28 ✓

Month	Principal	Interest	$P+I$	$P+I-R$
1	$90 000.00	$675.00	$90 675.00	$89 650.77
2	$89 650.77	$672.38	$90 323.15	$89 298.92
3	$89 298.92	$669.74	$89 968.66	$88 944.43
4	**$88 944.43**	**$667.08**	**$89 611.51**	**$88 587.28**

11 a Amount of depreciation = $38 000 − $34 960
= $3040

Rate of depreciation $= \dfrac{\$3040}{\$38\,000} \times 100\%$
= 8%

The rate of depreciation is 8% p.a. ✓

b $V_0 = \$38\,000$, $r = 0.08$
$S = V_0(1 - r)^n$
When $n = 9$,
$S = \$38\,000(1 - 0.08)^9$
= $17 942.131...
= $17 942 (nearest dollar) ✓
When $n = 10$,
$S = \$38\,000(1 - 0.08)^{10}$
= $16 506.7612...
= $16 507 (nearest dollar)
Amount of depreciation in tenth year
= $17 942 − $16 507
= **$1435** ✓

12 PV = $2100, r = 0.0006, n = 17
$FV = \$2100(1 + 0.0006)^{17}$ ✓
= $2121.523 12...
= $2121.52 (nearest cent)

Tina would need to pay $2121.52 to clear the debt. ✓

13 a Total repaid = $\$773.54 \times 12 \times 15$
$= \mathbf{\$139\,237.20}$ ✓

b **If interest rates rise, Bailey could pay a lot more.** ✓

14 Monthly payment = $\$11.61 \times 12$
$= \$139.32$ ✓

Total repaid = $\$139.32 \times 12 \times 10$
$= \$16\,718.40$

Interest = $\$16\,718.40 - \$12\,000$
$= \mathbf{\$4718.40}$ ✓

15 a Daily rate = $16.9725\% \div 365$
$= \mathbf{0.0465\%}$ ✓

b **If the bill is paid by the due date no interest will be charged.** ✓

c $PV = \$1355$, $r = 0.000\,465$, $n = 15$

$FV = PV(1 + r)^n$
$= \$1355(1.000\,465)^{15}$
$= \$1364.48$ (nearest cent) ✓

$I = \$1364.48 - \$1355$
$= \$9.48$

Kyle will pay \$9.48 in interest. ✓

CHAPTER 7

FURTHER STATISTICAL ANALYSIS

TOPIC: STATISTICAL ANALYSIS

The statistical investigation process for a survey

1 The statistical investigation process

- **To properly undertake a statistical investigation**, a certain process should be followed.
- **An issue needs to be identified** and a statistical question posed.
- **A decision must be made on a way to gather information** and then the data collected.
- **The data needs to be organised, summarised and represented**.
- **The data must be analysed**, and the results interpreted and communicated.

EXAMPLE 1

Nigel listed the following steps he intended to take when statistically investigating an issue:

- **posing a question**
- **collecting data**
- **displaying data**
- **analysing**
- **concluding.**

Nigel missed an important step. What is it?

After collection, the data needs to be organised so that it can be displayed.

It is possible to move back and forth in the process. For example, after organising the data Nigel might decide to collect some new data or refine his question.

EXAMPLE 2

A café owner was considering adding a book library to her business. She was considering what type of books she should have. In a survey of her customers, people were asked what their favourite type of book was. Nineteen customers preferred adventure, 23 mystery, 14 romance, 11 science fiction and 5 replied 'other'.

a **How many people replied to the survey?**

b **Show the results in a sector graph.**

c **What conclusion, if any, can be made?**

a Total replies = 19 + 23 + 14 + 11 + 5
= 72

b Angle for adventure = $\frac{19}{72} \times 360°$
= 95°

Mystery: $\frac{23}{72} \times 360° = 115°$

Romance: $\frac{14}{72} \times 360° = 70°$

Science fiction: $\frac{11}{72} \times 360° = 55°$

Other: $\frac{5}{72} \times 360° = 25°$

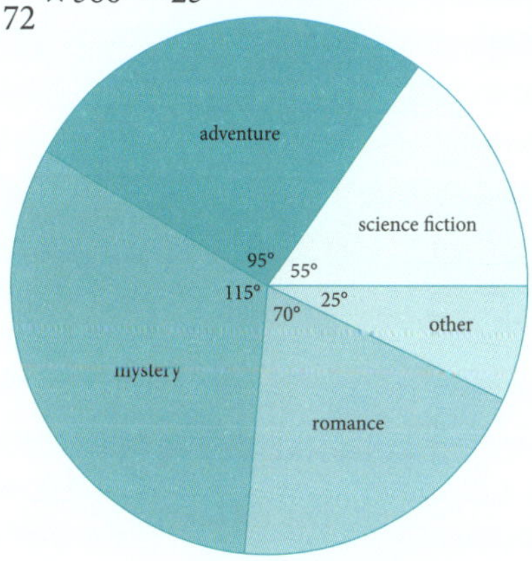

The best type of display to use depends on the type of data. Here it is categorical data so a sector graph is an appropriate means of display.

c The results clearly show that mysteries were the most popular, but that the other genres mentioned all received a reasonable number of votes and there were not many votes for 'other'. Because the survey was of the customers it would seem reasonable that it was representative and that the owner could confidently provide plenty of mystery and adventure books, with perhaps slightly fewer numbers of romance and science fiction.

EXAMPLE 3

A survey was conducted to see the number of dogs that people owned. The results were:

1, 0, 2, 1, 1, 4, 2, 0, 1, 0, 1, 2, 2, 1, 3.

a Draw a frequency distribution table showing x, f, fx and the cumulative frequency.

b Draw a frequency histogram to illustrate the results.

c What is the mode?

d How many dogs were owned altogether?

e What is the mean?

f What is the median?

g In how many houses did the people have less than three dogs?

h In what percentage of houses were there two or more dogs?

i What conclusions, if any, can be made?

a

Score (x)	Frequency (f)	fx	Cum. freq.
0	3	0	3
1	6	6	9
2	4	8	13
3	1	3	14
4	1	4	15
Total	15	21	

b

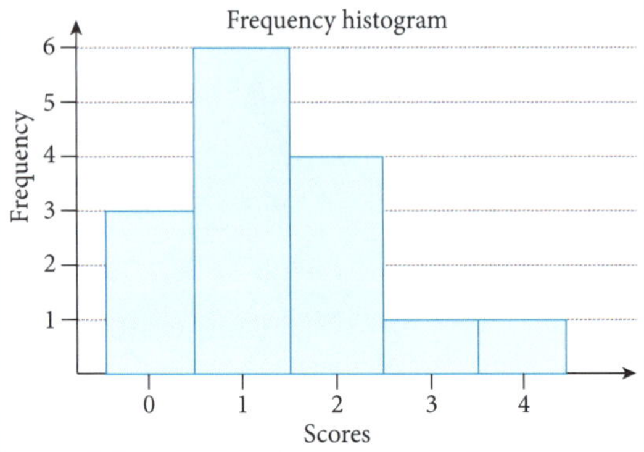

c Mode = 1

The highest frequency is 6. The score with the highest frequency is 1.

d 15

e $\bar{x} = \frac{21}{15}$

$= 1.4$

There are 15 scores altogether. So the middle score is the 8th. From the cumulative frequency we can see that the 8th score (as well as the 4th, 5th, 6th, 7th and 9th) is 1.

f Median = 1

g In 13 houses the people had fewer than three dogs.

h Number of homes with two or more dogs is 15 – 9 or 6.

$$\text{Percentage} = \frac{6}{15} \times 100\%$$
$$= 40\%$$

i There were only 15 houses surveyed so it is a very small sample and might not be representative of the general population. The most likely outcome for a house is that there is one dog. The histogram is positively skewed and the spread is not large.

EXAMPLE 4

250 people were surveyed prior to an election to determine which issues they thought were most important. The table shows the results.

Issue	Number	Issue	Number
Accountability	12	Equality	10
Cost of living	15	Health	30
Economy	60	Immigration	17
Education	25	Security	8
Employment	20	Other	5
Environment	48		

a Draw a table showing the issues arranged in order from highest to lowest and show the cumulative total and cumulative percentage.

b Draw a Pareto chart to illustrate this data.

c What conclusions can be made?

a

Issue	Number	Cumulative total	Cumulative percentage
Economy	60	60	24%
Environment	48	108	43.2%
Health	30	138	55.2%
Education	25	163	65.2%
Employment	20	183	73.2%
Immigration	17	200	80%
Cost of living	15	215	86%
Accountability	12	227	90.8%
Equality	10	237	94.8%
Security	8	245	98%
Other	5	250	100%

The data had been collected and is now organised.

b

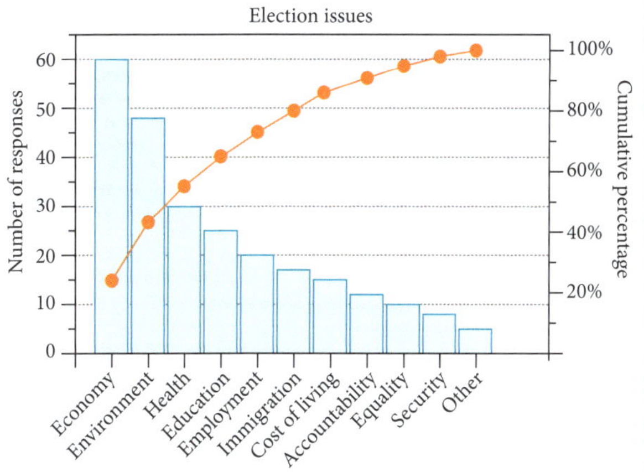

c From the Pareto chart it is immediately obvious that the economy is the most important issue, followed by the environment. These figures can be read from the table but the relative importance of each of the issues can be more clearly seen from the chart.

EXAMPLE 5

A teacher wanted to know if there was a difference between the results for boys and for girls in a particular subject.

Forty students sat for an exam and the results (out of 100) for boys and for girls are listed below.

Boys:

61 78 65 45 46 53 59 64 68 73 82
62 66 65 58 49 54 64 71 57

Girls:

56 68 65 62 59 75 70 66 74 45 78
82 69 61 71 68 77 72 63 74

a Draw up a back-to-back stem-and-leaf plot.

b Does there appear to be a difference?

a

Exam results

Boys		Girls
9 6 5	4	5
9 8 7 4 3	5	6 9
8 6 5 5 4 4 2 1	6	1 2 3 5 6 8 8 9
8 3 1	7	0 1 2 4 4 5 7 8
2	8	2

b On the whole the girls do appear to have performed slightly better on the test than the boys. The range of scores is the same for both groups but the median mark for the girls is 68.5, 5.5 marks higher than the median mark for the boys (63). The girls' marks are negatively skewed whereas the boys' marks are more symmetrical. This conclusion can only be made from these results and might not always be the same for the subject.

Different interpretations of the data can be given as long as they are supported by the data and valid arguments.

→ PRACTICE

1 **Nigella listed the following steps she intended to take when statistically investigating an issue:**

- **posing a question**
- **collecting data**
- **organising data**
- **analysing**
- **concluding.**

Nigella missed an important step. What is it?

2 **A car dealer was considering adding some more demonstration vehicles and wanted to have a good variety of colours from that model. In a survey, people were asked what their favourite car colour was. Twenty-seven people nominated red, 18 blue, 33 silver, 9 black and 42 white, while 6 voted for another colour.**

a How many people replied to the survey?

b Show the results in a sector graph.

c What conclusion, if any, can be made?

3 **These are the number of runs scored per over in the first few overs of a cricket match:**

1, 3, 0, 1, 0, 2, 3, 0, 0, 1, 5, 1,
3, 0, 2, 0.

a Draw a frequency distribution table showing x, f, fx and the cumulative frequency ($c.f.$).

b Draw a frequency histogram.

c What is the mode?

d How many runs were scored altogether?

e What is the mean?

f What is the median?

g In how many overs did the batsmen score less than two runs?

h What conclusions, if any, can be made?

4 **A number of people were surveyed about the reason for their complaints to a bus company. The results are shown in the table.**

Reason	Number of complaints
Cleanliness	8
Driver attitude	9
Overcrowding	12
Running late	19
Running early	2
Length of the trip	6
Route taken	3
Other	1

a Draw a table showing the reasons arranged in order from highest to lowest number of complaints and show the cumulative total and cumulative percentage.

b Draw a Pareto chart to illustrate this data.

c What conclusions can be made?

5 A teacher wanted to know if there was a difference between the results for boys and for girls in a particular subject.

Forty students sat for an exam and the results (out of 100) for boys and for girls are listed below.

Boys: 65 73 79 64 82 93 81 77 69
71 80 76 76 68 87 72 79 66
88 80

Girls: 74 82 76 62 77 85 91 73 75
71 90 84 69 73 78 86 77 70
77 76

a Draw a back-to-back stem-and-leaf plot of the results.

b Does there appear to be a difference?

6* A teacher wanted to know which of the sports offered by the school was most dangerous.

He collected some data from 240 students and formed these graphs.

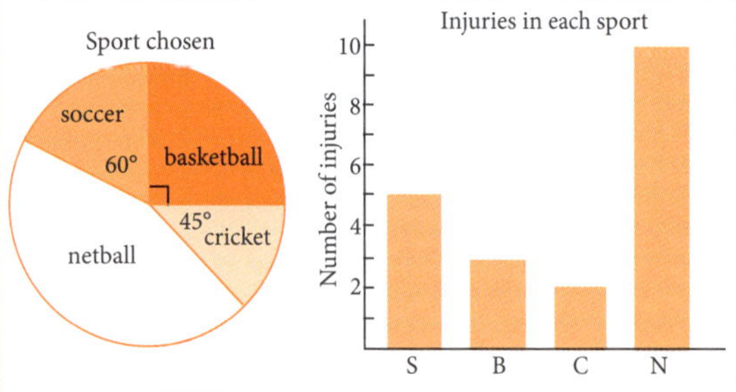

Based on this information, which sport is the most dangerous?

Answers ➲ pp. 151–152

2 Designing questionnaires

- **When designing a questionnaire** it is important to be clear about what you are trying to find out and how you will achieve that.
- **You need to identify and find the target population.**
- **Questions should use simple language** and not be ambiguous or biased.
- **It is important to bear in mind issues such as privacy and ethnic and cultural sensitivities.**

EXAMPLE 1

Tony made this questionnaire when he wanted to find from his workmates the mode of transport they preferred to use to travel to work.

Which is your favourite means of transport?
☐ car ☐ train
☐ bus ☐ bicycle
Please place a tick in one of the boxes.

Briefly comment on some things that Tony should improve.

Tony asked for their favourite means of transport. This is not the same thing as the preferred means of travelling to work. Tony only gave four options, which doesn't cover all the possibilities. Someone might walk to work, for example. It also doesn't allow for those workers who use a combination of different means of transport. It might be helpful to give respondents an opportunity to explain why they choose the method of travel they do; there might not be a train, for example.

Careful consideration should be given to every question in a questionnaire to not only make sure that it follows all the conditions of a good questionnaire but that it will also give answers to the information that is required.

EXAMPLE 2

Lisa believed that the majority of people would agree that Australians eligible to receive a pension should be able to receive that pension even if they are living overseas. To confirm this belief she made up a questionnaire that included the question:

'Do you think the Australian government should pay a pension to overseas people?'

Lisa was surprised to find that the responses were totally different to what she expected. Why might that have been? Comment in relation to the question that was posed.

The question is ambiguous. It doesn't specify that the pension might be paid to Australians who are overseas. People responding to the question might have taken it to mean anyone who is overseas.

A good idea is to 'pilot' a questionnaire before actually implementing it. By trying out the questionnaire on just a few people any ambiguities should be found.

a The company expected to receive many more replies than they received. Give a reason why they might not have had many replies.

b Because they had not received many replies the company offered prizes of a selection of products to be won. What results might this have? Briefly comment.

Answers ➲ p. 152

➔ PRACTICE

1 Meghan didn't like the colours of the school uniform and decided to distribute a questionnaire to students.

> The school colours need to be changed. Which combination do you prefer?
> ☐ blue/yellow
> ☐ green/yellow
> ☐ blue/green

Briefly comment on Meghan's questionnaire and some things that Meghan should improve.

2 Les believed that most people would prefer to buy fresh fish from a market rather than cooked fish. He distributed a questionnaire about the market and one question asked:

'Do you prefer to have the fish cooked?'

Les was very surprised to find that nearly everyone replied 'yes' to this question. Why might that have been? Comment in relation to the question that was posed.

3* A marketing company wanted to conduct an online survey. The questionnaire included some questions that were compulsory and these were marked with an asterisk on the form.

> *Name: ____________________
> *Age: ____________________
> *Email: ____________________
> Occupation: ____________________
> Which of our products do you use regularly?
> ____________________
> Which is your favourite? ____________________
> Which is your least favourite and why?
> ____________________
> What other of these types of products would you like to be able to buy? ____________________
> ____________________

Exploring and describing data arising from two quantitative variables

1 Scatterplots

➔ **Bivariate data is data involving two sets of variables.**

➔ **A scatterplot can be used to show the relationship** between two separate sets of numerical data. One set of data is shown on the horizontal axis and the other on the vertical axis.

➔ **Statistical software or spreadsheet programs can be used to produce scatterplots.**

EXAMPLE 1

A scatterplot has been drawn showing the age and current value of certain cars.

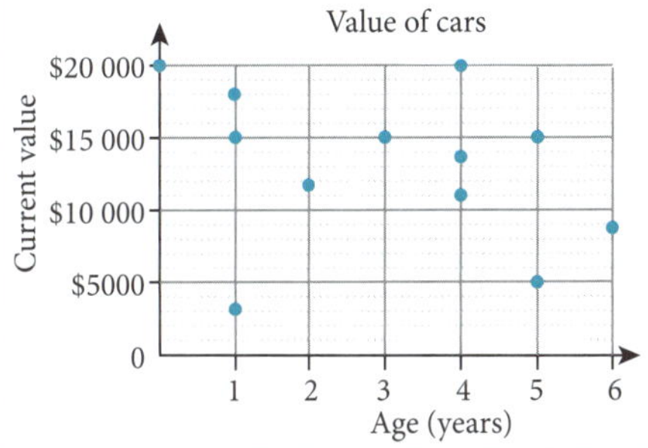

a **How many vehicles are shown on the scatterplot?**

b **What is the current value of the car that is 2 years old?**

c **What ages are the cars that are valued at $15 000?**

d **What is the difference in value between the two cars that are 5 years old?**

e **One vehicle has been extensively damaged in an accident. Which vehicle do you think this might be? Justify your answer.**

a There are 12 vehicles shown.

There are 12 dots on the scatterplot, including one on the vertical axis.

b The 2-year-old car is valued at $12 000.

c There are three cars valued at $15 000. One is 1 year old, one is 3 years old and the other is 5 years old.

d Of the 5-year-old cars, one is valued at \$15 000 and one at \$5000.

Difference = \$15 000 – \$5000
= \$10 000

e The 1-year-old car valued at \$3000 is worth a lot less than you would expect for a car that is almost new and so would be the car that has been extensively damaged.

EXAMPLE 2

A scatterplot has been drawn to show the height and weight of some female students.

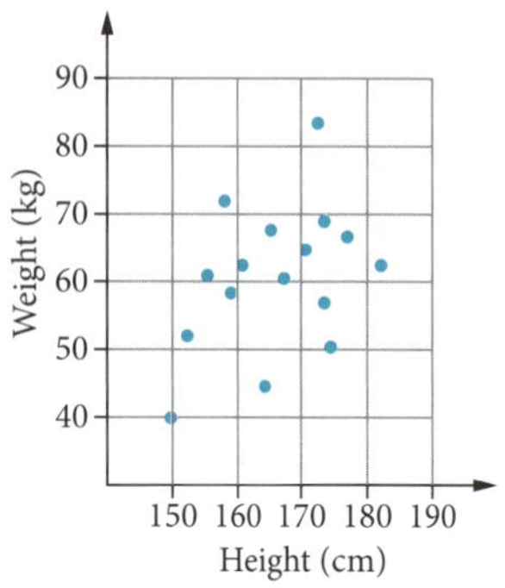

a **The data from how many students is shown on the scatterplot?**

b **What is the approximate weight of the tallest girl?**

c **How tall is the girl who weighs 72 kg?**

d **Ideally anyone over 160 cm tall should weigh at least 50 kg. Give an example of a girl who does not satisfy this ideal.**

a 16

b 62 kg

The tallest girl is the only one over 180 cm.

c 157 cm

d One girl is about 165 cm tall and weighs about 45 kg.

EXAMPLE 3

The table shows the length and width (both in mm) of the hands of 20 children.

Length	Width	Length	Width
101	51	94	46
88	42	79	48
85	43	104	50
96	47	92	44
103	49	76	39
81	40	105	53
78	46	83	42
75	35	94	52
108	53	90	44
78	38	86	40

Show the results on a scatterplot, placing the length on the horizontal axis and the width on the vertical axis.

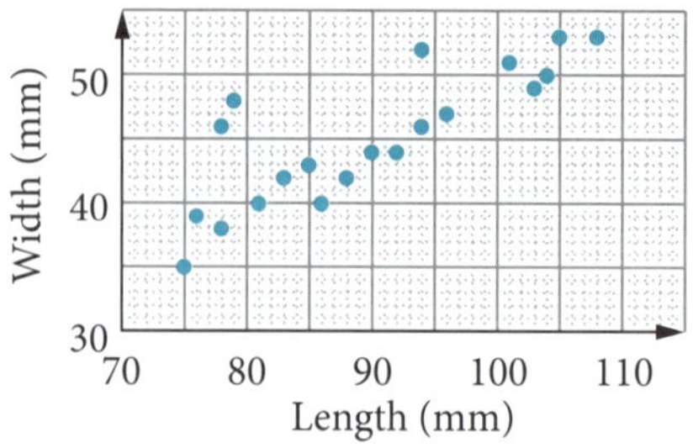

When drawing a scatterplot make sure that both axes are clearly labelled and that the scale is shown correctly. (All gaps between values on the axes must be the same.)

→ PRACTICE

1 **A scatterplot has been drawn showing the age and number of hospital stays of a certain group of children with a particular disease.**

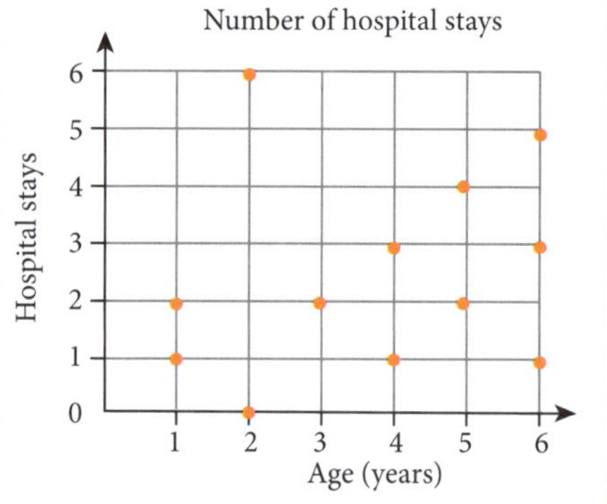

a **How many children are shown on the scatterplot?**

b **How many hospital stays had the child who was 3 years old?**

c **What ages are the children who have had three hospital stays?**

d **What is the difference in the number of hospital stays between the two children who are 5 years old?**

e **One child has been very ill a number of times for her age. Which child do you think this might be? Justify your answer.**

2 A scatterplot has been drawn to show the height and weight of some male students.

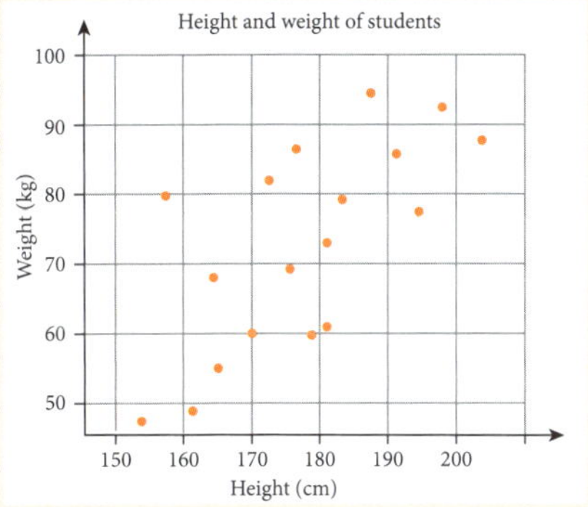

a The data from how many students is shown on the scatterplot?

b What is the approximate weight of the tallest boy?

c How tall is the boy who weighs 55 kg?

d Ideally anyone who weighs more than 70 kg should be at least 167 cm tall. Give an example of a boy who does not satisfy this ideal.

3 The table shows the results of a survey of 20 high school students who were asked how many hours of television they watched each night.

Year	Hours	Year	Hours
7	3	8	$4\frac{1}{2}$
10	1	9	3
8	$2\frac{1}{2}$	7	$2\frac{1}{2}$
12	2	7	5
11	4	12	$2\frac{1}{2}$
12	3	9	$3\frac{1}{2}$
11	$3\frac{1}{2}$	10	4
9	4	12	0
10	2	8	3
7	$4\frac{1}{2}$	9	$1\frac{1}{2}$

Show the results on a scatterplot, placing the year on the horizontal axis and the hours on the vertical axis.

4* A scatterplot has been drawn to show the marks gained (out of 10) by a group of students in both History and Geography.

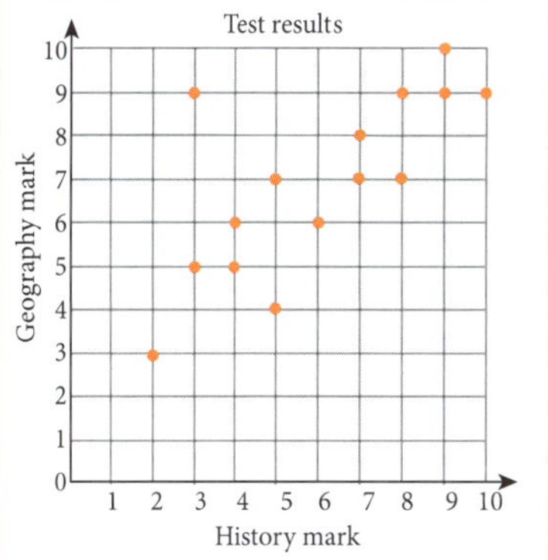

a How many students are in the group?

b What percentage of the students performed better in History than in Geography?

Answers ⮌ p. 153

2 Patterns and features of scatterplots

➔ **From a scatterplot we can easily see whether there appears to be any sort of pattern.** This allows us to draw conclusions about any relationship between the two data sets.

➔ **A pattern is said to be linear if it forms a straight line.** (By this we don't mean that every point falls on a straight line as if it had been drawn with a ruler, but if we can roughly see a line formed by the points.)

➔ **If a pattern is linear, the direction might be positive** (leaning to the right, /) or negative (leaning to the left, \).

EXAMPLE 1

A scatterplot has been drawn showing the value of certain bottles of wine as they age.

Describe any pattern.

The dots follow a pattern. The value increases slowly with age at first, then much more quickly.
The pattern resembles an exponential curve.

It is an example of exponential growth.

EXAMPLE 2

For each scatterplot, determine whether there is a pattern and, if so, whether it is linear. If it is linear, state the direction.

a
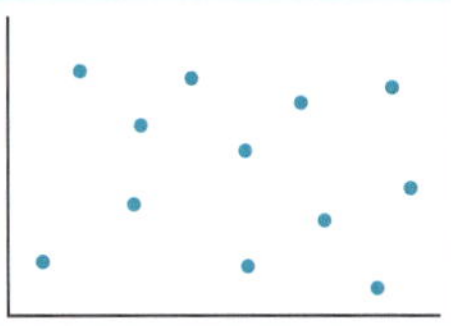

b
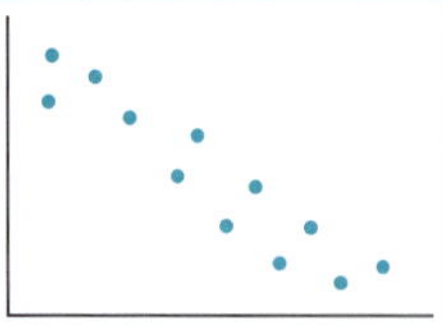

c
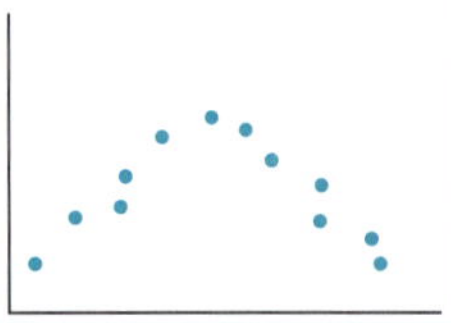

d
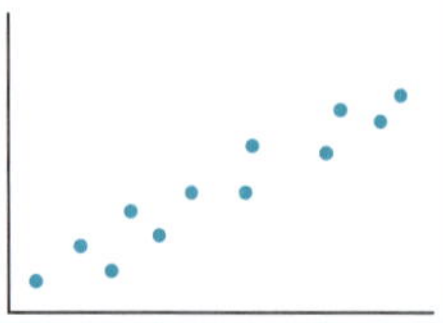

a No pattern

b Linear pattern, negative

As one variable increases, the other decreases.

c Non-linear pattern

It appears to be parabolic.

d Linear pattern, positive

Both variables increase.

EXAMPLE 3

Twelve students had to each write two stories, one on cats and one on dogs. Each story was marked out of 10 and the results are shown in the table.

Cats	5	7	8	6	7	9	4	9	6	8	10	8
Dogs	6	7	10	5	8	9	5	7	4	9	7	7

a Draw a scatterplot, placing the score for cats on the horizontal axis.

b Does there appear to be any pattern?

a
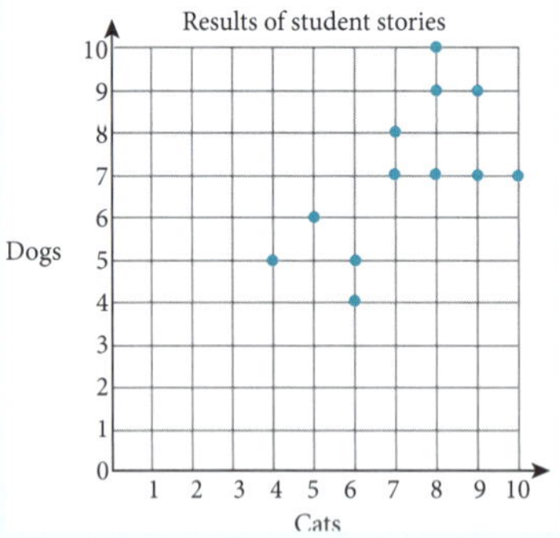

b There does appear to be a slight linear pattern. The direction is positive.

By drawing a scatterplot it is much easier to see a pattern or relationship between the scores.

→ PRACTICE

1 A scatterplot has been drawn showing the value of certain cars as they age.

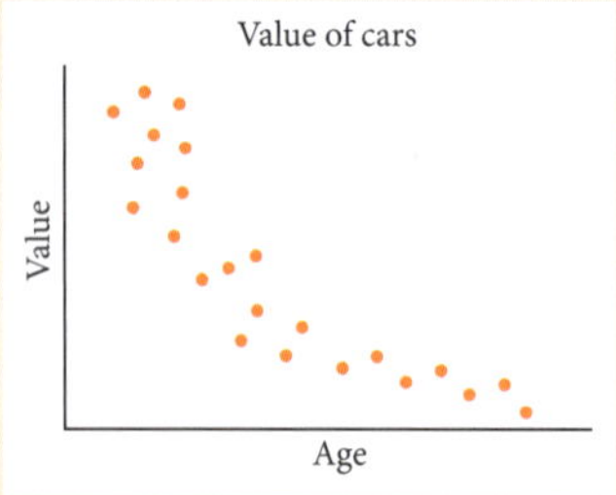

Describe any pattern.

2 For each scatterplot, determine whether there is a pattern and, if so, whether it is linear. If it is linear, state the direction.

a
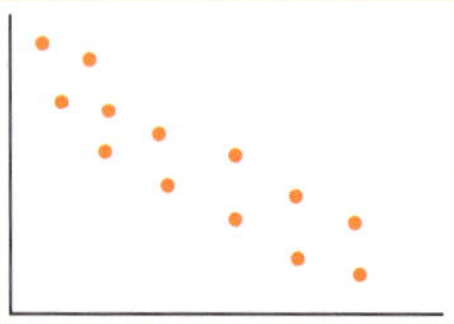

b

c
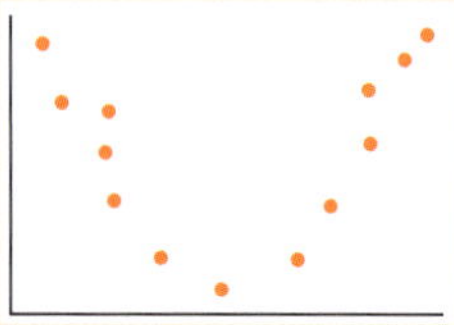

d
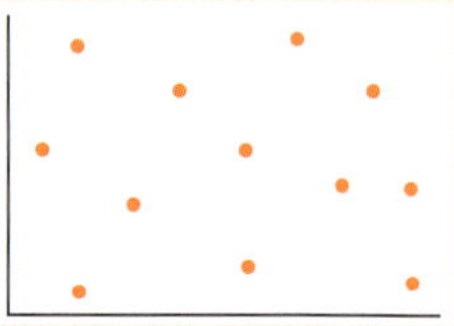

3 Twelve students sat for tests in both English and Maths and their results, given as deciles, are shown in the table below.

English	Maths	English	Maths
1	3	5	3
4	2	2	3
9	8	4	5
2	1	6	4
7	2	3	1
5	4	9	4

a Draw a scatterplot, placing the score for English on the horizontal axis.

b Does there appear to be any pattern?

4* A scatterplot has been drawn to show the star ratings given by two reviewers of the same movies.

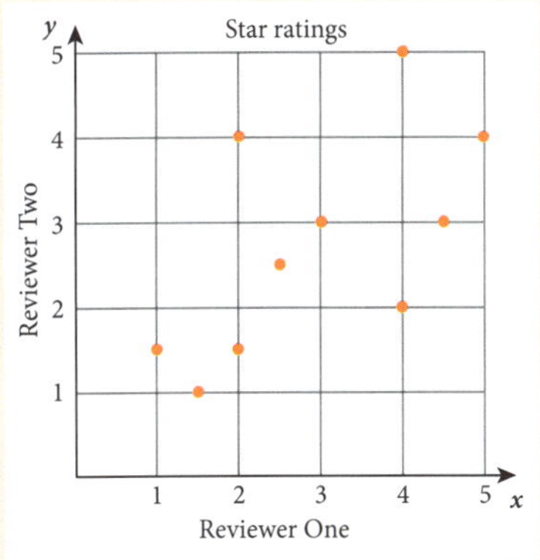

a How many movies were reviewed?

b How many movies received a 5-star rating?

c What was the rating by reviewer two for the movie that reviewer one gave just one star?

d What type of pattern (if any) does the scatterplot show?

Answers ➲ p. 153

3 Line of best fit

→ **A line of best fit is a straight line that can be drawn to approximately show where most points lie**. For example, in the scatterplot at right, the line shown is a line of fit.

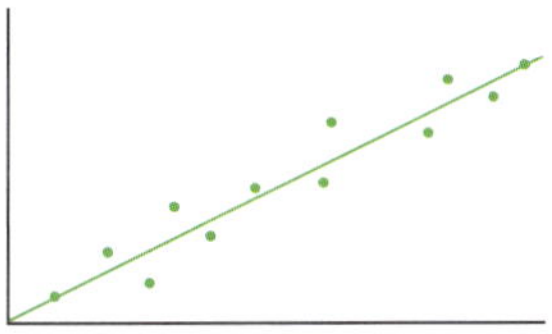

→ **A line of fit can be drawn by eye** which means you can simply look at the scatterplot and try to draw the line so that it goes roughly through the middle of the set of points. There is no exact answer for such a line of fit.

→ **The least-squares regression line is a special line of fit.** The sum of all the squared vertical distances from the line to the actual data points is minimised which means that it is the closest possible line to the data points.

→ **The trendline feature in a spreadsheet program is an example of a line of best fit**.

EXAMPLE 1

A scatterplot has been drawn showing the heights and ages of certain children.

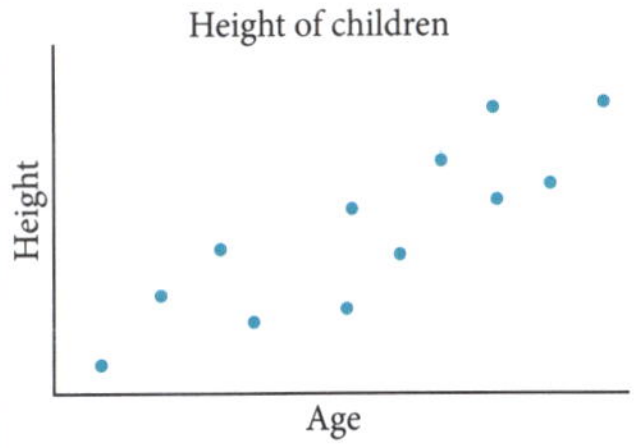

Draw in a line of fit.

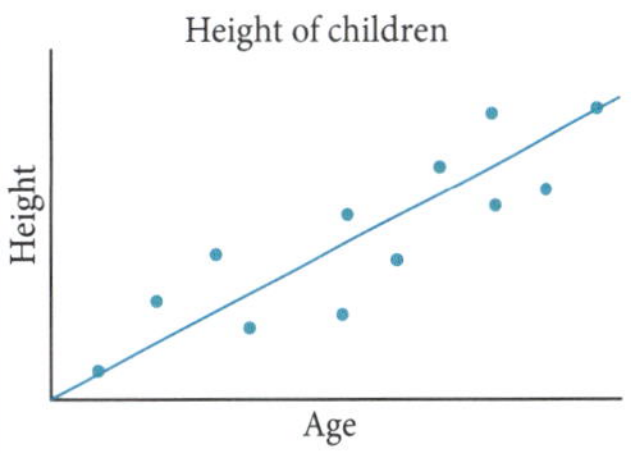

Try to have roughly the same number of data points either side of the line.

EXAMPLE 2

As an exercise in proving the value of study, a teacher gave 10 students a sheet of strange facts. She gave them each a different number of minutes, from one to 10, to study the facts. She then gave a quick quiz. The table below shows the results.

Minutes studied	1	2	3	4	5	6	7	8	9	10
Test mark	0	3	2	5	6	7	6	8	7	9

Draw a scatterplot of the results and draw in a line of fit.

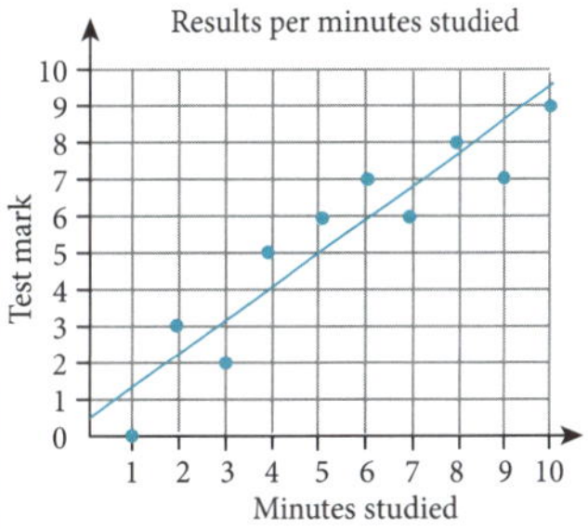

Slightly different lines would also be correct.

→ PRACTICE

1 A scatterplot has been drawn showing the numbers of questions asked at a lecture over time.

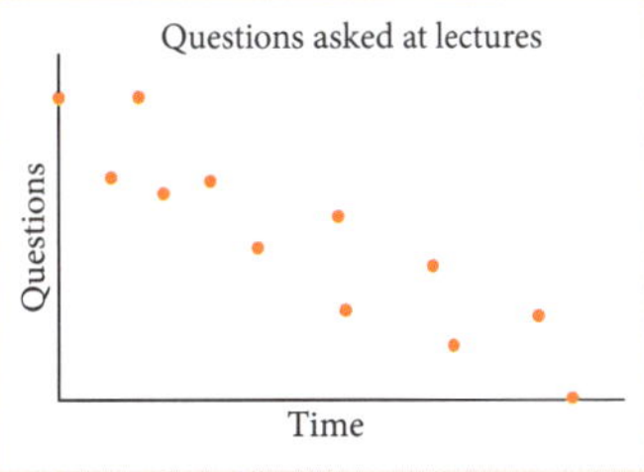

Draw in a line of fit.

2 As an experiment, a group of people were given different amounts of time to memorise a set of objects. The people were then tested to see how many of the objects they could recall. The table below shows the results.

Seconds studied	5	10	15	20	25	30	35	40	45	50	55	60
Objects remembered	3	7	5	8	6	10	9	12	17	15	13	20

Draw a scatterplot of the results and draw in a line of fit.

3* Pupils at a school received marks out of 10 for conduct and attendance. The results for one class of 12 pupils are given in the table.

Conduct	Attendance	Conduct	Attendance
1	6	3	3
7	4	7	5
5	7	6	4
4	4	10	7
9	8	2	3
2	5	5	3

Show the results on a scatterplot. Place conduct on the horizontal axis and attendance on the vertical axis. Draw in a line of fit.

Answers ➲ pp. 153–154

4 The equation of a line of fit

- **The equation of a line of fit is of the form $y = mx + c$,** where m is the gradient and c the vertical intercept.
- **The gradient of a line is its slope.** The gradient is positive if the line leans to the right (/) and is negative if the line leans to the left (\). The gradient is found by dividing the vertical change in position by the horizontal change in position.
- **The vertical intercept is where a line meets the vertical axis of a graph**. In the equation $y = mx + c$ it is the value of y when $x = 0$.
- **The gradient and vertical intercept can be interpreted in terms of the data.**

EXAMPLE 1

A line of fit has been drawn for the points shown on the scatterplot as shown in the diagram.

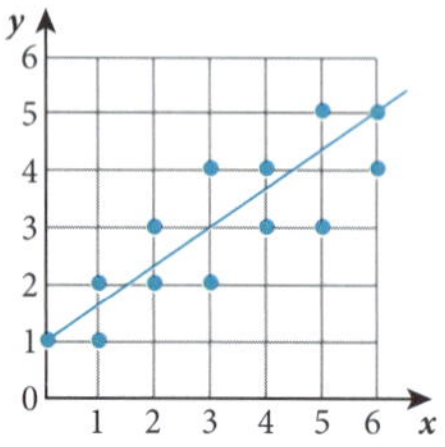

a What is the gradient?

b What is the vertical intercept?

c What is the equation of the line of fit?

a $m = \dfrac{\text{vertical change in position}}{\text{horizontal change in position}}$

$= \dfrac{4}{6}$

$= \dfrac{2}{3}$

Using the two points that the line passes through ((0, 1) and (6, 5)) rise up 4 units and run across 6 units to move from the first to the second point.

b $c = 1$

c $y = mx + c$

$y = \dfrac{2}{3}x + 1$

EXAMPLE 2

What is the equation of the line of fit shown on the scatterplot below?

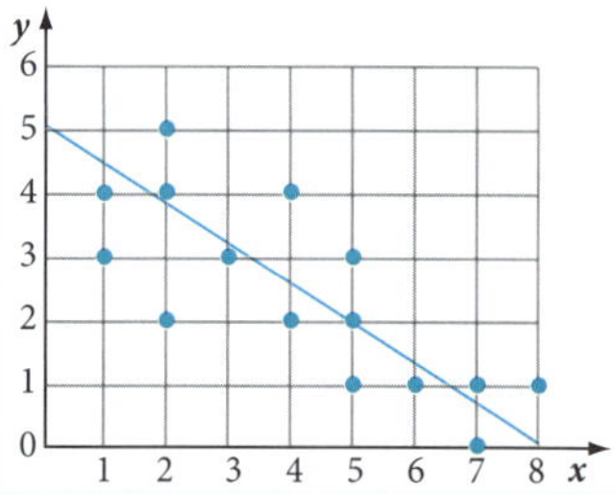

$m = \dfrac{\text{vertical change in position}}{\text{horizontal change in position}}$

$= \dfrac{-5}{8}$

$= -\dfrac{5}{8}$

The vertical change in position is a fall of 5 units so the vertical change is negative, meaning that the gradient will be negative.

$c = 5$

$y = mx + c$

$y = -\dfrac{5}{8}x + 5$

EXAMPLE 3

Here is some data on the weights of some babies.

x Age (months)	6	11	4	3	9	8	10	7	5	8	12	9
y Weight (kg)	7.2	11.5	5.8	6.1	9.4	8.9	9.0	8.4	7.5	9.6	11.1	8.7

a Show the information on a scatterplot.

b The equation of the least-squares regression line is $y = \dfrac{3}{5}x + 4$. Show this line on the scatterplot.

c What does the vertical intercept represent?

d What does the gradient represent?

a, b

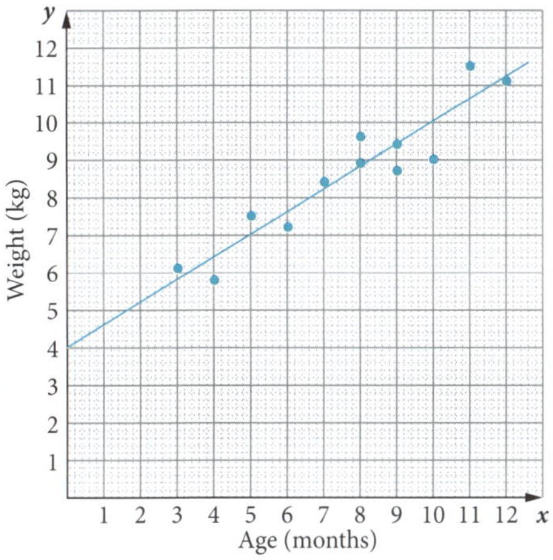

c The vertical intercept is 4. It represents the weight of a newborn baby.

For this set of data the weight at 0 months of age is about 4 kg.

d The gradient is $\frac{3}{5}$. It means that every 5 months the weight of a baby tends to increase by 3 kg.

EXAMPLE 4

Some information was collected on the estimated cholesterol intake of a group of adults and the results are shown plotted against their ages. A line of fit has been drawn on the scatterplot.

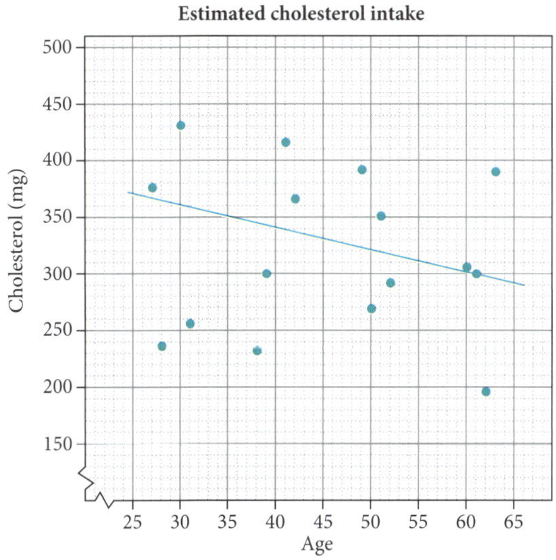

a **Using the line of fit, estimate the cholesterol intake of a person who is 45 years old.**

b **Using the line of fit, at what age would a person have a cholesterol intake of 300 mg?**

c **What is the gradient of the line of fit?**

d **What is the equation of the line of fit?**

a For an age of 45 the line shows a cholesterol estimate of 330 mg.

b 60 years old

c gradient $= \dfrac{\text{vertical change in position}}{\text{horizontal change in position}}$

$= \dfrac{-30}{15}$

$= -2$

Using the two points from parts *a* and *b*.

d $y = mx + c$

$y = -2x + c$

When $x = 60$, $y = 300$

$300 = -2 \times 60 + c$

$= -120 + c$

$420 = c$

So $y = -2x + 420$.

c cannot be read as the *y*-intercept because of the gaps in the axes.

→ PRACTICE

1 A line of fit has been drawn for the points shown on the scatterplot as shown in the diagram.

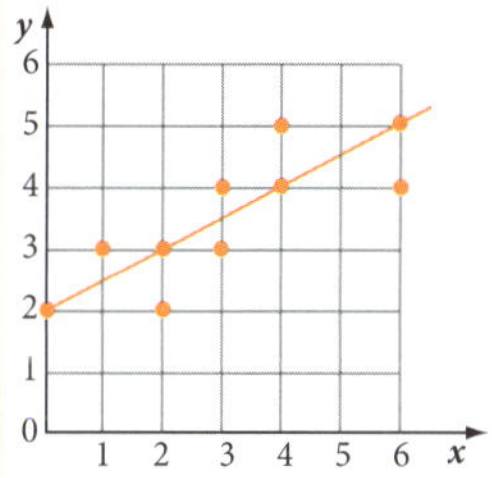

a What is the gradient?

b What is the vertical intercept?

c What is the equation of the line of fit?

2 What is the equation of the line of fit shown on the scatterplot?

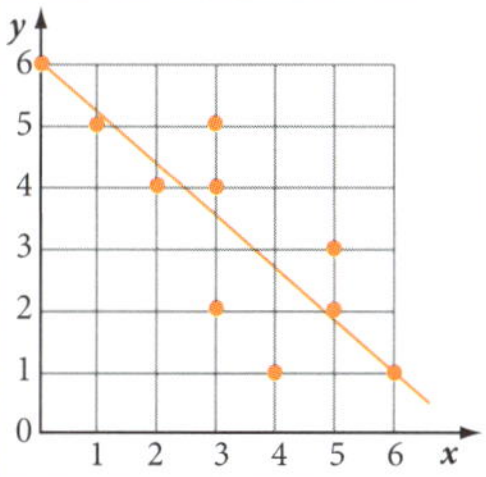

3 Some information was collected about the head circumference of a group of 12 babies. The table shows their ages and head circumferences.

Age (months)	Head circumference (cm)
8	42
14	46
22	48
13	46
5	40
13	45
7	45
16	46
4	40
24	49
11	43
19	47

a Show the information on a scatterplot.

b The equation of the least-squares regression line is $y = \frac{2}{5}x + 39$. Show this line on the scatterplot.

c What does the vertical intercept represent?

d What does the gradient represent?

4 Some information was collected on the BMI (Body Mass Index) of a group of adults and the results are shown plotted against their ages. A line of fit has been drawn on the scatterplot.

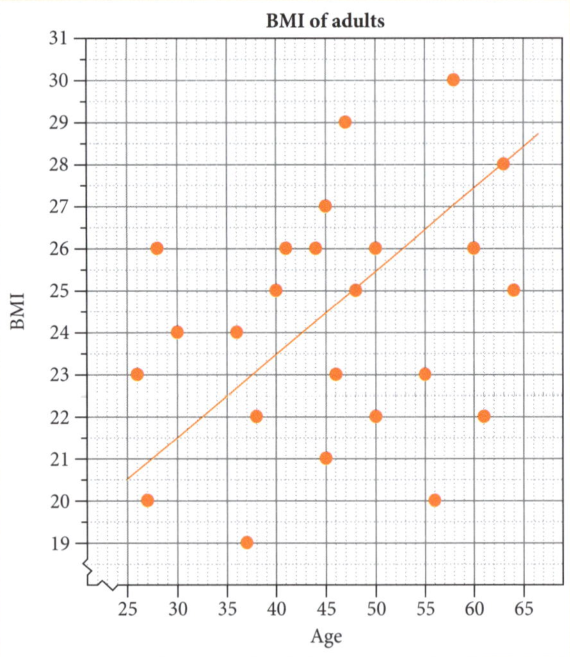

a Using the line of fit, estimate the BMI of a person who is 48 years old.

b Using the line of fit, at what age would a person have a BMI of 28?

c What is the gradient of the line of fit? (Give the answer as a decimal.)

d What is the equation of the line of fit?

5* Joseph drew this scatterplot from a set of data.

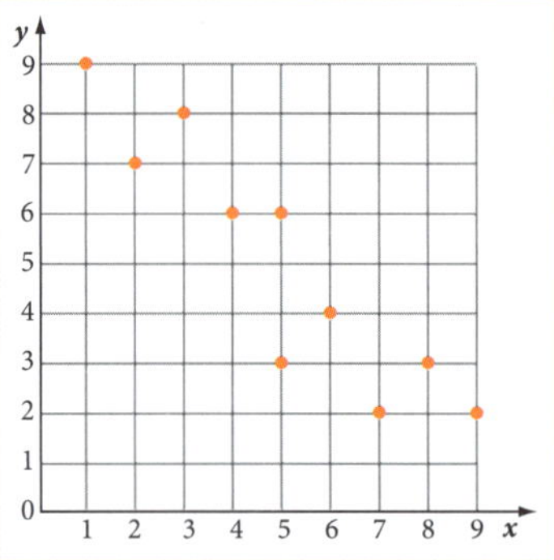

He found that the equation of the least-squares regression line was $y = -0.88x + 9.4$.

a Show that the regression line will pass through the point (5, 5).

b Copy the scatterplot and show the least-squares regression line.

c What is the gradient of the line of fit? What feature of the data is highlighted by the line?

Answers ➲ p. 154

5 Interpolation and extrapolation

➔ **The equation of a line of fit can be used** to make predictions about the data. This is done by substituting values into the equation.

➔ **Interpolation is when predictions are made** concerning values that lie within the range of the original data and extrapolation is when predictions are made concerning values outside the range of the original data. Extreme care must be taken when extrapolating because predictions made for values outside the data range may not be accurate or even sensible.

➔ **The more data points that are used**, the more accurate the predictions will be and they will be most accurate when the relationship between the two sets of variables is strongest.

EXAMPLE 1

A scatterplot was created showing the milk production of various cows, in litres per day, that have been milking for a different number of days. A line of fit was drawn and its equation was found to be

$y = -\frac{1}{12}x + 35$ where y is the daily milk production and x is the number of days since the cow calved. Use this equation to answer the following questions.

a How many litres of milk would you expect a cow to produce if she has been milking for 180 days?

b Buttercup produced 25 L per day. For how long would you expect that she has been milking?

c How much more milk would you expect a cow to produce that has been milking for 60 days compared to one that has been milking for 300 days?

d How many litres does the equation give for a cow that has been milking for 444 days? Do you think this is an accurate prediction? Explain.

a $y = -\frac{1}{12}x + 35$

Once the equation of the line of fit has been determined, it can be used to give estimates for any value of *x*.

When $x = 180$,

$y = -\frac{1}{12} \times 180 + 35$

$= 20$

She would be expected to produce 20 L.

b $y = -\frac{1}{12}x + 35$

When $y = 25$,

$25 = -\frac{1}{12}x + 35$

$-10 = -\frac{1}{12}x$

$x = 120$

Buttercup is likely to have been milking for 120 days.

c $y = -\frac{1}{12}x + 35$

When $x = 60$,

$y = -\frac{1}{12} \times 60 + 35$

$= 30$

When $x = 300$,

$y = -\frac{1}{12} \times 300 + 35$

$= 10$

Extra production $= 30 - 10$

$= 20$

You would expect the cow to produce 20 more litres.

d $y = -\frac{1}{12}x + 35$

When $x = 444$,

$y = -\frac{1}{12} \times 444 + 35$

$= -2$

In fact, cows generally only milk for around 305 days in each lactation, have 60 days dry and calve again one year after the previous calving.

A cow cannot produce a negative amount of milk so this prediction cannot be valid. The equation can only apply to scores in the range of the original data and this must be outside the range.

EXAMPLE 2

A survey was undertaken about the number of safety incidents in which children of different ages had been involved over a certain time period. The results are shown in the scatterplot below.

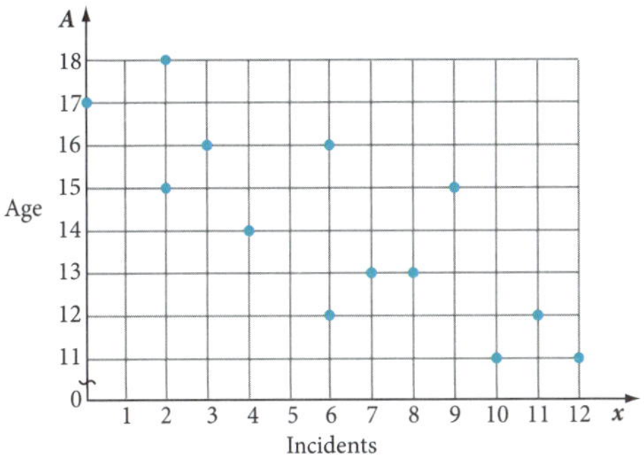

a **Draw in a line of fit and find its equation.**

b **If a child was involved in 16 incidents, what is the likely age?**

c **If a child who is five was surveyed, how many incidents would you predict that they would be involved in?**

a

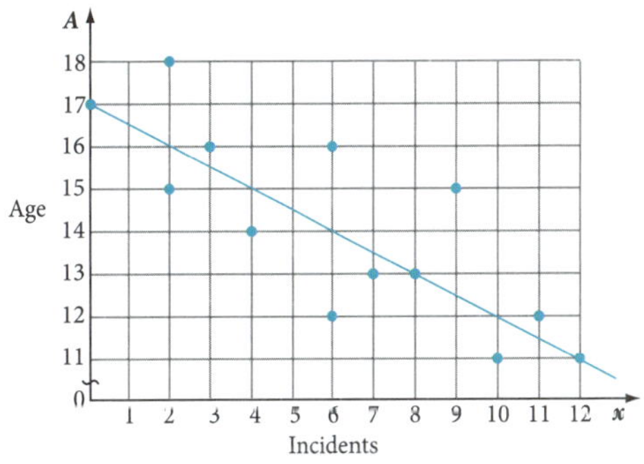

The equation for a line of fit might involve different variables instead of always *x* and *y*.

$m = \frac{\text{vertical change in position}}{\text{horizontal change in position}}$

$= \frac{-6}{12}$

$= -\frac{1}{2}$

The line of fit is not unique. A slightly different line will have a different equation.

Vertical intercept = 17

The equation is $A = -\frac{1}{2}x + 17$.

b When $x = 16$,

$A = -\frac{1}{2} \times 16 + 17$

$= 9$

The child is likely to be 9 years old.

c When $A = 5$,

$$5 = -\frac{1}{2}x + 17$$

$$-12 = -\frac{1}{2}x$$

$$24 = x$$

There are likely to be 24 incidents.

➔ PRACTICE

1 A scatterplot is drawn showing the profit made by a worker for different amounts of output. A line of fit is drawn and its equation found to be $y = \frac{1}{20}x + 2.5$ where y is the profit in dollars and x is the number of units produced. Use this equation to answer the following questions.

a How much profit would you expect would be made if 100 units are produced?

b One day a profit of \$30.50 was made. How many units would you expect were produced?

c How much more profit would you expect from the production of 2500 units than 1000 units?

d How much profit does the equation give for a worker producing 10 million units? Do you think this is an accurate prediction? Explain.

2 A record was kept of the assignment marks and final results achieved by students during a course. These are shown in the scatterplot below.

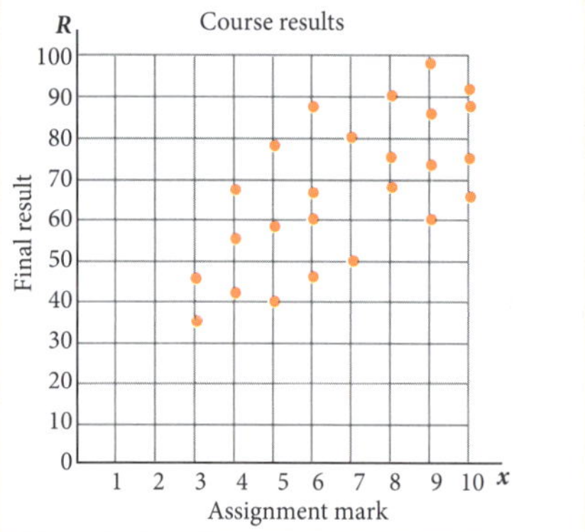

a Draw in a line of fit and find its equation.

b If a student scored 7 on their assignment, what would you expect their final result to be?

c If a student scored 76% as their final result, what was their likely assignment mark?

3* This scatterplot was drawn to show the weight loss of a group of people on a particular diet over time. A least-squares line of best fit is shown on the scatterplot. Use this line of fit to answer the following questions.

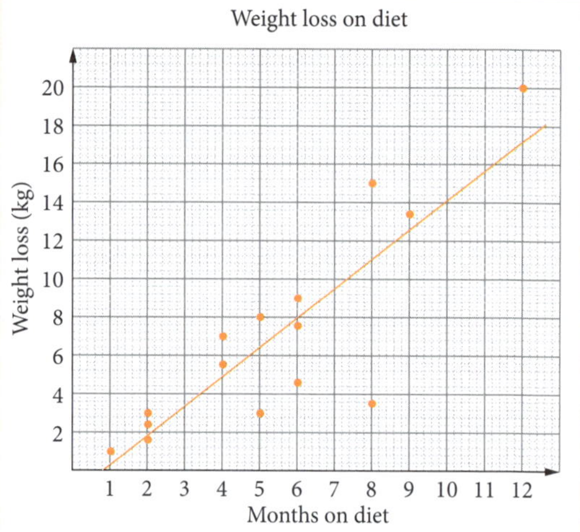

a About how many kilograms would you expect someone who has been on the diet for 3 months to have lost?

b Susan has been on the diet for 7 months and has lost 10 kg. Is this more or less than you would expect?

c Alicia wants to lose 14 kg. About how long would you predict that she needs to spend on the diet before achieving her goal?

d Mara has been on the diet for 18 months. Can you predict how much weight she has lost? Justify your answer.

e Serena uses a different diet. She has been on her diet for 11 months. About how much would you expect her to have lost? Justify your answer.

Answers ➲ pp. 154–155

6 Analysing and describing associations

➔ **Any of the processes of statistical investigation** can be applied to analyse bivariate data.

➔ **Biometric data is data relating to the measurements** of the human body.

➔ **It is important to consider issues** such as the reliability of the data and problems of bias, and ethical considerations such as privacy and cultural sensitivity.

➔ **If there is a relationship between two variables**, it does not mean that one causes the other.

EXAMPLE 1

A study was done of children aged from 2 to 18. This table shows median height at different ages. Use the table to answer the following questions.

Median height (cm)		
Age	Boys	Girls
2	86	85
4	102	101
6	115	115
8	128	128
10	139	138
12	149	151
14	164	160
16	173	162
18	176	163

a **What is the difference in the median height of boys and the median height of girls at age 16?**

b **At what ages are girls taller than boys?**

c **How much do girls grow in height between the ages of 14 and 18?**

d **In what two-year period do girls grow the most? Is this the same two-year period for boys?**

e **Comment on any other trends that can be seen from the data.**

f **Comment on how reliable you believe the data to be.**

a At 16 the median height for boys is 173 cm and the median height for girls is 162 cm.

Difference = (173 – 162) cm
= 11 cm

b The median height for girls is greater than the median height for boys at age 12.

c The difference in the median height for girls aged 14 and the median height for girls aged 18 is (163 – 160) cm or 3 cm.

d Between the ages of two and four the median height for girls increases by (101 – 85) cm or 16 cm. This is the largest increase, so girls grow the most between the ages of two and four. It is also the age when boys grow the most.

e The median heights of the boys and girls are very similar up to age 12. After that age the difference between the heights continues to increase, the largest difference being at age 18. Girls grow very little after age 14 while boys continue to grow.

f There is no information about the size of the sample, or when or where the information was collected. Without this information it is impossible to make judgements about the reliability of the data.

Always consider the source of data and the size of the sample when making judgements.

EXAMPLE 2

This scatterplot was drawn to show the life expectancy at birth of males and females over the years.

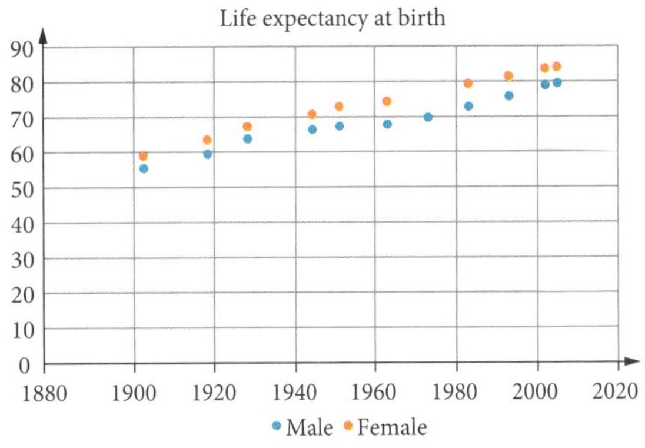

Based on Australian Bureau of Statistics data

a **The life expectancy for males appears for the year 1976 but that for females does not. What do you predict that the life expectancy of a female born in 1976 would be?**

b **James worked out the equation for the least-squares line of best fit for males. He claimed that it was $y = 0.2x - 324$. Find the life expectancy of a boy born in the year 1600 using this equation. Does this suggest that James has the wrong equation? Comment.**

c **James has read media reports that suggest life expectancy may actually decline in coming years because of obesity-related illnesses. He believes that the value given by his equation for a baby boy born in 2020 supports that belief. Do you agree? Comment.**

a The life expectancy for males in 1976 was about 70. The life expectancy for females would have been about 77.

b $y = 0.2x - 324$

When $x = 1600$,

$y = 0.2 \times 1600 - 324$
$= -4$

It is impossible for a person to have a negative life expectancy so the equation cannot be used for a person born in 1600. But this does not mean that James's equation was wrong. The line of fit should not be used for values outside the data points for which it is fitted.

c $y = 0.2x - 324$

When $x = 2020$,

$y = 0.2 \times 2020 - 324$
$= 80$

Although this value is only slightly more than the last value shown on the graph, it does not show that life expectancy is decreasing or even slowing its increase. The equation is that of a straight line. The gradient is 0.2, which means that the life expectancy is increasing by 0.2 years every year. Because the points on the scatterplot do not form an exact linear pattern, some of the points on the scatterplot will fall above or below the line. So, for example, the last point shown on the scatterplot may have been above the line. If it had been on the line, the increase in life expectancy in 2020 might more clearly be seen. However, as 2020 is again outside the data points, predictions shouldn't be made using the line.

→ PRACTICE

1 A study was made of children aged from 2 to 18 months. This table shows the median weight. Use the table to answer the following questions.

Median weight (kg)		
Age (months)	Boys	Girls
2	5.5	4.7
4	7.0	6.4
6	8.0	7.3
8	8.6	8.0
10	9.2	8.6
12	9.7	9.0
14	10.1	9.4
16	10.5	9.9
18	11.0	10.2

a What is the difference in the median weight of boys and the median weight of girls at age 16 months?

b At what age(s) is the biggest difference between the median weight of girls and boys?

c How much weight do girls put on between the ages of 12 and 18 months?

d In what two-month period do girls put on the most weight? Is this the same two-month period for boys?

e Comment on any other trends that can be seen from the data.

f How reliable do you believe the data to be? Comment.

2 This scatterplot was drawn to show the extra years of life after the age of 60 that can be expected by men and women in 10 different countries of the world.

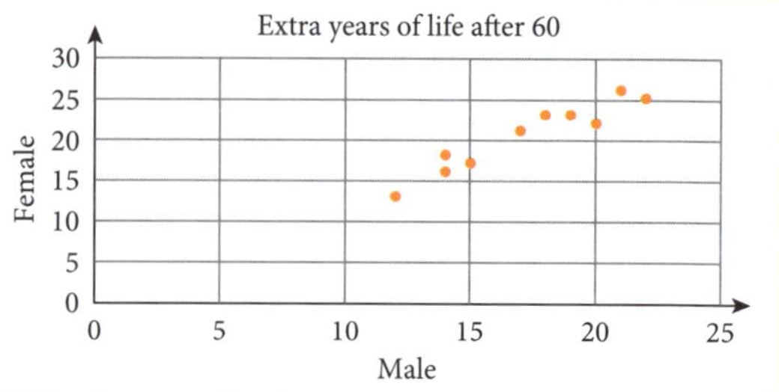

a A 60-year-old woman in one country can expect to live to 86. To what age can a 60-year-old man in that country expect to live?

b Jane worked out the equation for the least-squares line of best fit. She found that it was $y = 1.2x - 0.24$. Using this equation, find the life expectancy of a woman when the life expectancy of a man is 61. Does this suggest that Jane has the wrong equation? Comment.

c Jane has read media reports that suggest life expectancy may actually decline in coming years because of obesity-related illnesses. Does this equation support that belief? Comment.

3* The scatterplot shows the results of two different maths tests taken by students in year 12.

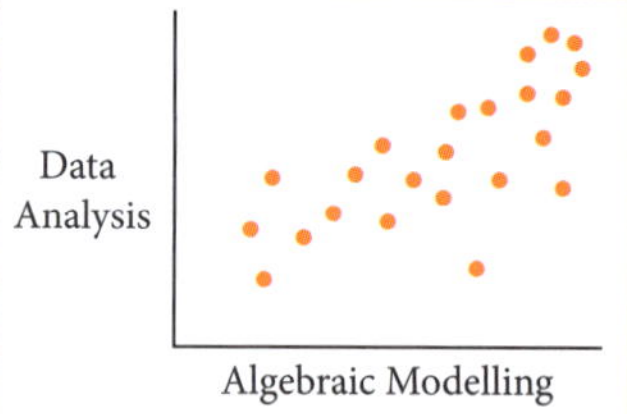

David commented that to do well in Data Analysis you must be good at Algebraic Modelling. Is this correct? Justify your answer.

Answers ➲ p. 155

HSC EXAM-TYPE QUESTIONS

Total marks: 30

PART A

(Suggested time: 12 minutes)

(1 mark each)

1 **In which scatterplot does the data not form a linear pattern?**

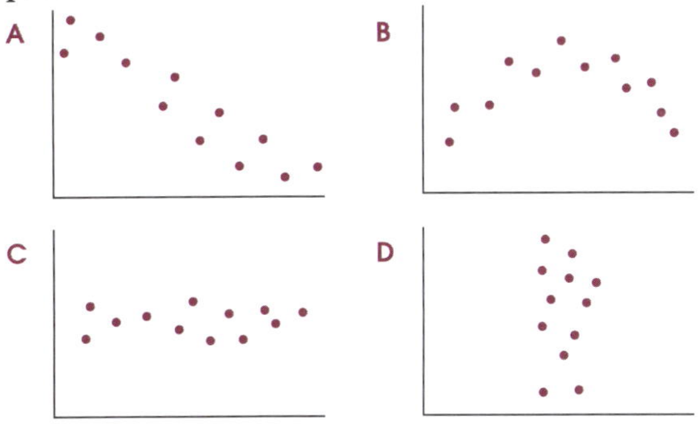

2 **A scatterplot has been drawn to show the age of vehicles traded in at a car dealer's and the price paid for those vehicles. What is the difference in age between the two vehicles that received $13 000 as a trade-in?**

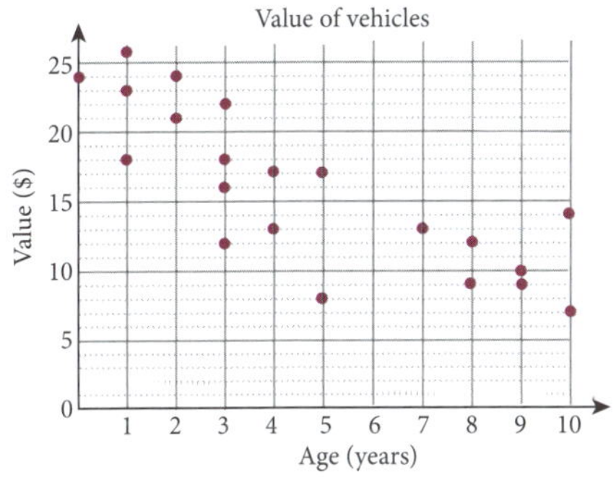

A 2 years B 3 years C 4 years D 5 years

3 **What is the equation of the line of fit shown in this scatterplot?**

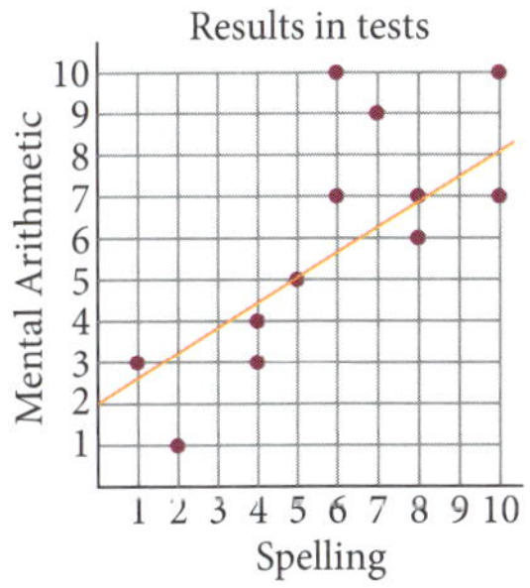

A $y = \frac{3}{5}x + 2$ B $y = \frac{5}{3}x + 2$

C $y = 2x + 3$ D $y = \frac{2}{3}x + 5$

4 **Which of these is not an important quality for questions in a questionnaire?**

A unambiguous B uncomplicated
C unbiased D unethical

5 **Which would be most likely to show a positive linear pattern when plotted on a scatterplot?**

A the price of bread against the number of loaves sold
B the number of cars garaged at a house against the age of the occupants
C the age of babies against the size of their clothes
D the height of a person against the height of their partner

6 **The equation of a line of fit is $y = 0.6x + 4$ when x is the age in months and y the weight in kilograms of some babies. Using the line of fit, what is the best estimate for the age in months of a baby who weighs 7 kg?**

A 5 B 6 C 8 D 9

7 **Samara wanted to conduct a survey about native birds. Once her data has been collected what are her next steps in order?**

A display, organise, analyse, conclude
B analyse, organise, display, conclude
C organise, analyse, display, conclude
D organise, display, analyse, conclude

8 **Twenty people were questioned about the number of minutes they spend exercising each day and the answers are shown on the graph, plotted against their age.**

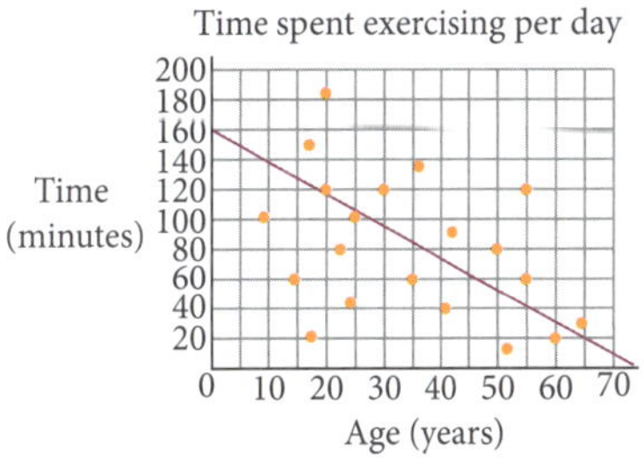

What percentage of those questioned said they exercised for less than an hour each day?

A 45% B 35% C 30% D 25%

PART B

(Suggested time: 33 minutes)

Show all working.

9 A teacher has prepared a scatterplot, and drawn a line of fit, of the scores by students in an essay and in an exam.

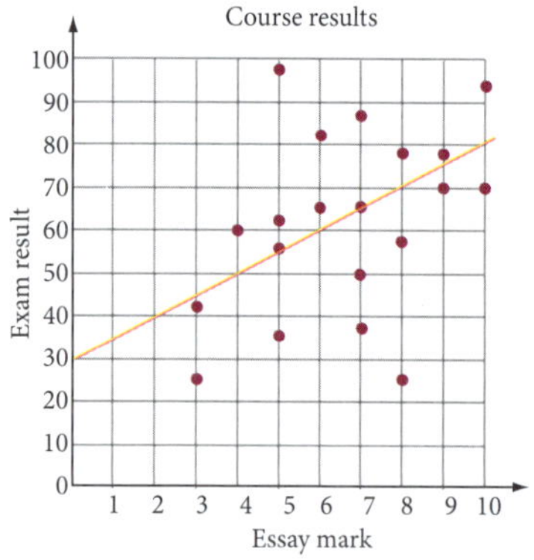

a What did the student who received an essay mark of 4 score in the exam? (1 mark)

b Nathan said: 'Everyone who did well in the essay also did well in the exam.' Give an example to show that Nathan was wrong. (1 mark)

c Using x for the essay mark and y for the exam result, find the equation of the line of fit. (2 marks)

d Natalie received 7 for her essay but was away for the exam. The teacher used the line of fit to give her an estimate. What was Natalie's estimated mark? (1 mark)

10 A survey was taken of the length of time (in months) employees were employed and the time (in minutes) each took to complete a particular task. The results are shown in the table.

Time employed (months)	3	10	18	7	5	12	9	15	4	14
Time to complete task (minutes)	37	29	22	32	35	30	30	25	35	27

a Draw a scatterplot and plot the information. (2 marks)

b Draw in a line of fit. (1 mark)

c Find the equation of the line of fit. (2 marks)

d Use the equation of the line of fit to predict the length of time it would take someone who has been employed for 2 years to complete the task. (1 mark)

e If an employee takes 20 minutes to complete the task, for how long would you expect they have been employed? (1 mark)

f Briefly explain why the equation of the line of fit could not be used to predict the amount of time needed to complete the task by an employee of 4 years standing. (1 mark)

11 This table shows the height and average stride lengths of some male athletes. (All measurements are in centimetres.)

Height	Stride length	Height	Stride length
186	212	180	210
176	200	164	213
178	214	187	219
165	188	174	209
175	201	180	211
188	244	181	217

a Show the points on a scatterplot. (2 marks)

b Draw in a line of fit. (1 mark)

c Another athlete has a height of 170 cm. What would you predict his stride length to be? (1 mark)

12 This scatterplot shows the expected additional years of life of women at age 45 and at 65 over time.

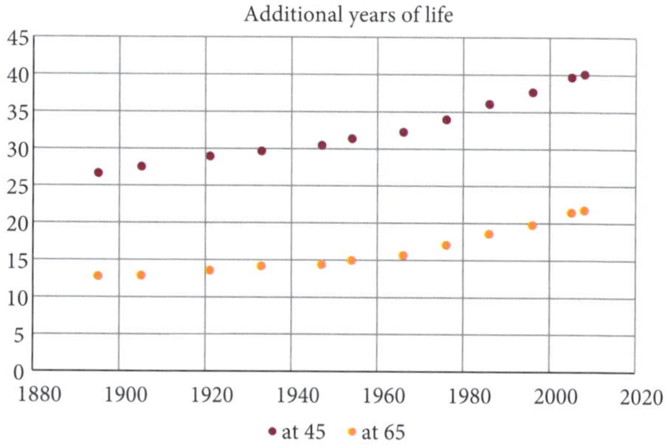

Based on Australian Bureau of Statistics data

a How many additional years of life were expected for a woman who was 65 in 1921? (1 mark)

b What was the life expectancy of a 45-year-old woman in 1905? (1 mark)

c What had happened to the life expectancy of a woman who was 45 in 1976 by the time she was 65? (1 mark)

d Who had the greater life expectancy in 2008: a woman of 45 or a woman of 65? Comment on the reasons why this is so. (2 marks)

PRACTICE

THE STATISTICAL INVESTIGATION PROCESS FOR A SURVEY

1 The statistical investigation process ➲ pp. 135–136

1 Displaying data

2 a Total replies = 27 + 18 + 33 + 9 + 42 + 6
= **135**

b Angle for red = $\frac{27}{135} \times 360°$
= 72°

Blue: $\frac{18}{135} \times 360° = 48°$

Silver: $\frac{33}{135} \times 360° = 88°$

Black: $\frac{9}{135} \times 360° = 24°$

White: $\frac{42}{135} \times 360° = 112°$

Other: $\frac{6}{135} \times 360° = 16°$

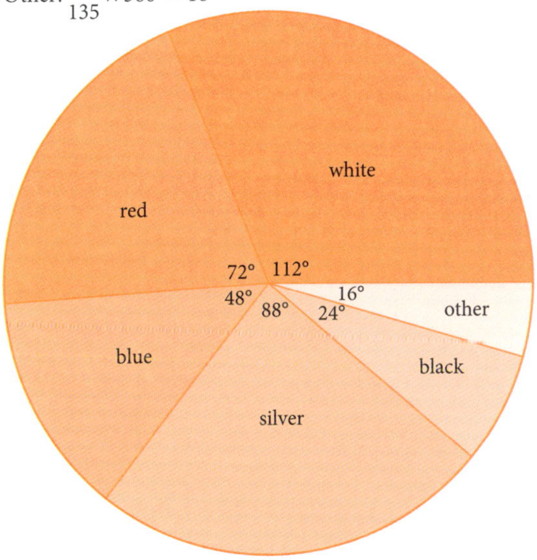

c The results clearly show that white was the most popular, closely followed by silver and then red. Of the suggested colours, black was the least popular. There is no mention of how the people surveyed were chosen, so it is not possible to say how representative the results are for car buyers. **Using the results it would seem that the dealer would be best to have cars that were white, silver or red.**

3 a

x	*f*	*fx*	*c.f.*
0	6	0	6
1	4	4	10
2	2	4	12
3	3	9	15
4	0	0	15
5	1	5	16
Total	16	22	

b

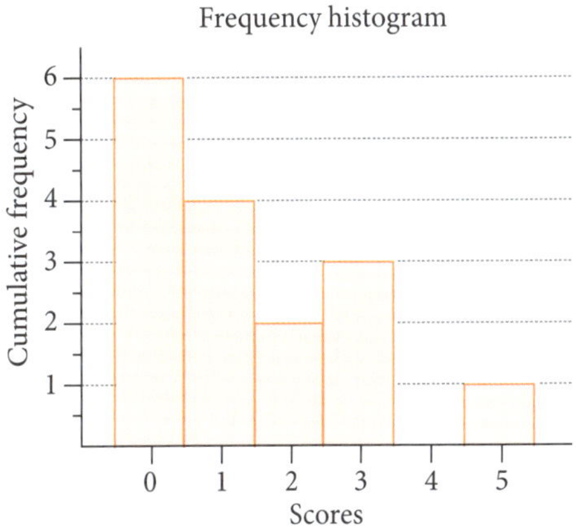

c **Mode = 0**

d **16**

e Mean = 22 ÷ 16
= **1.375**

f **Median = 1**

g **10**

h There were only 16 overs surveyed so it is a small sample and as they were from the beginning of the match they might not be representative of the whole match. The mode is 0 runs per over but that is likely to improve over time. **The histogram is positively skewed and the spread is not large.**

4 a

Reason	Number of complaints	Total complaints	Cumulative percentage
Running late	19	19	31.7%
Overcrowding	12	31	51.7%
Driver attitude	9	40	66.7%
Cleanliness	8	48	80%
Length of the trip	6	54	90%
Route taken	3	57	95%
Running early	2	59	98.3%
Other	1	60	100%

b

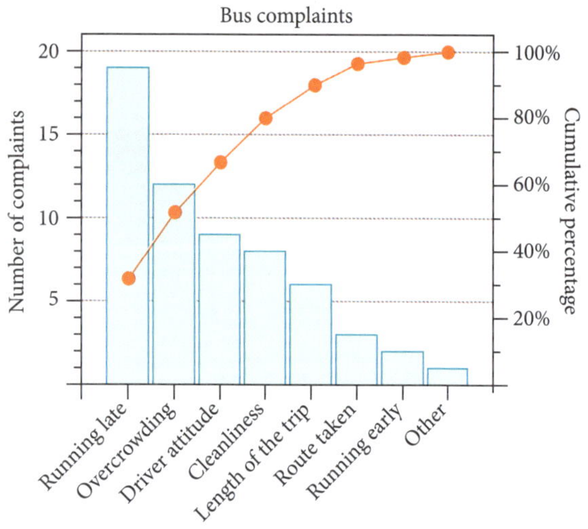

c From the graph it is easy to see that about one-third of the complaints are due to the bus running late. **Over 50% of complaints relate to just two issues.**

5 a

Exam results

Boys		Girls
9 8 6 5 4	6	2 9
9 9 7 6 6 3 2 1	7	0 1 3 3 4 5 6 6 7 7 7 8
8 7 2 1 0 0	8	2 4 5 6
3	9	0 1

b **No, there doesn't appear to be much difference.** The range of scores is the same for both groups (29) and the median mark is the same (76.5). Both sets of marks are positively skewed.

6

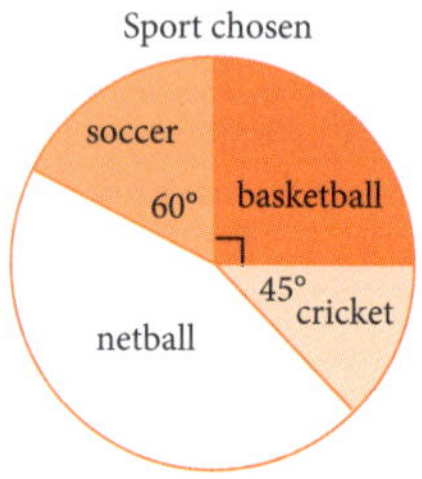

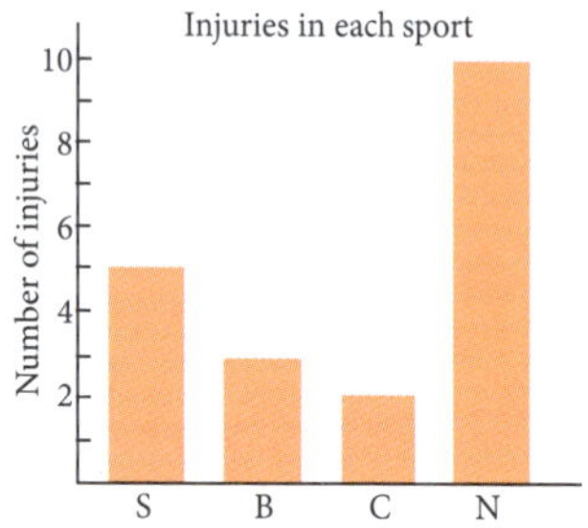

3 basketball players were injured.

Basketball players $= \frac{90}{360} \times 240$

$= 60$

Percentage injured $= \frac{3}{60} \times 100\%$

$= 5\%$

2 cricket players were injured.

Cricket players $= \frac{45}{360} \times 240$

$= 30$

Percentage injured $= \frac{2}{30} \times 100\%$

$= 6.66666\ldots\ \%$

5 students were injured playing soccer.

Soccer players $= \frac{60}{360} \times 240$

$= 40$

Percentage injured $= \frac{5}{40} \times 100\%$

$= 12.5\%$

Angle for netball

$= 360° - (90 + 45 + 60)°$

$= 165°$

Netball players $= \frac{165}{360} \times 240$

$= 110$

110 students played netball.

Percentage injured $= \frac{10}{110} \times 100\%$

$= 9.090909\ldots\ \%$

Based on these figures, soccer is the most dangerous sport.

2 Designing questionnaires ➲ p. 137

1 **Meghan begins with a statement that might not be correct. She could have begun her questionnaire by asking whether people thought that the school colours should be changed. She then only gave three options. These might only be ones that Meghan would like. She should at least have included an 'other' option and allowed students to suggest a different combination.**

2 **The question is not clear. If Les had put 'buy' instead of 'have' the answers might have been different. Depending on the type of fish he was referring to, most people would want to have the fish cooked at some stage.**

3 **a** **People might be concerned about their privacy and not want to give their name or email address to the company. This would mean they couldn't reply to the survey as those fields are compulsory.**

b **Offering prizes might encourage more people to reply to the survey, but it might also skew the results. People might be less inclined to be critical if they wish to win a prize and give the answers they think the company wants rather than their honest opinions.**

EXPLORING AND DESCRIBING DATA ARISING FROM TWO QUANTITATIVE VARIABLES

1 Scatterplots ➲ pp. 138–139

1 a There are 12 children shown.
  b **The 3-year-old child has had two hospital stays.**
  c There are two children who have had three hospital stays. **One is 4 years old and the other is 6 years old.**
  d Of the 5-year-old children, one has had two stays and the other four stays.
  Difference = two stays
  e **The 2-year-old child who has had six hospital stays (The number of hospital stays is more than any other child and yet she is younger than most.)**

2 a **18**
  b **88 kg**
  c **165 cm**
  d **There is a boy who weighs almost 80 kg who is only about 157 cm tall.**

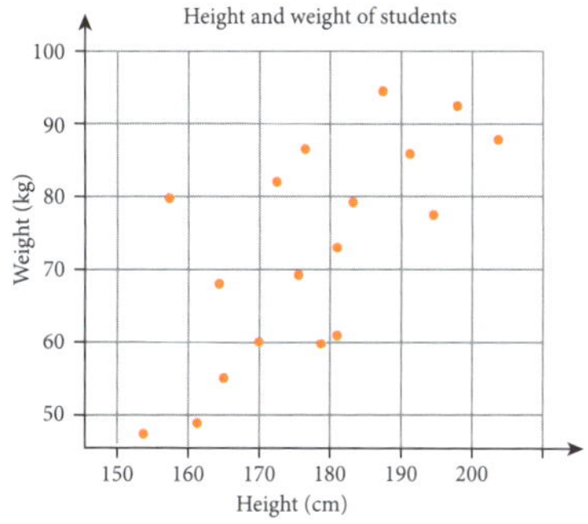

3

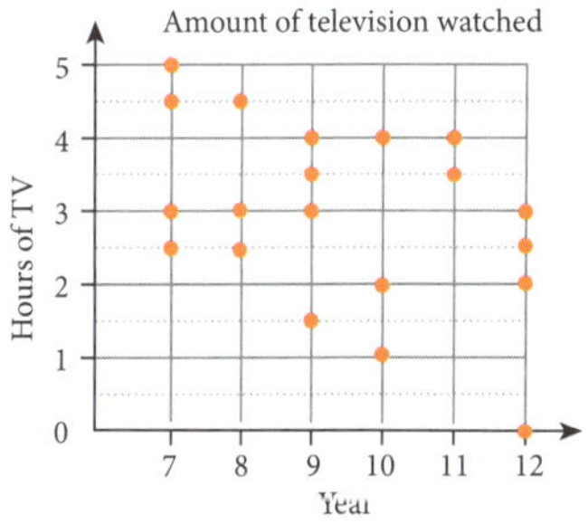

4 a **15**
  b Three students performed better in History than in Geography.

Percentage who performed better $= \frac{3}{15} \times 100\%$

$= \mathbf{20\%}$

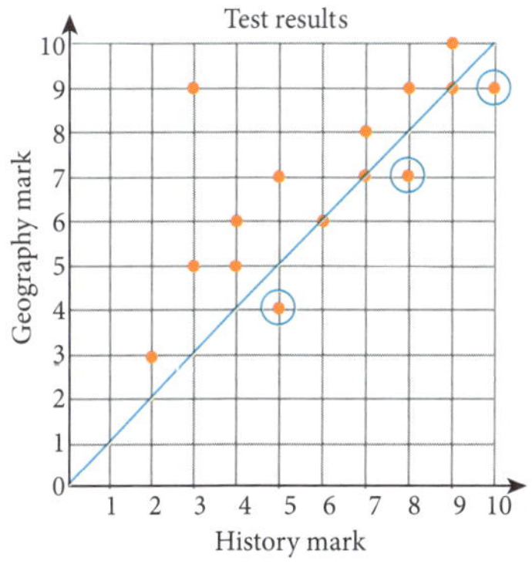

2 Patterns and features of scatterplots ➲ pp. 140–141

1 The dots follow a pattern. At first the value decreases quickly with age, then much more slowly. **The pattern resembles an example of exponential decay.**

2 a **Linear pattern, negative**
  b **Linear pattern, positive**
  c **Pattern, but not linear**
  d **No pattern**

3 a

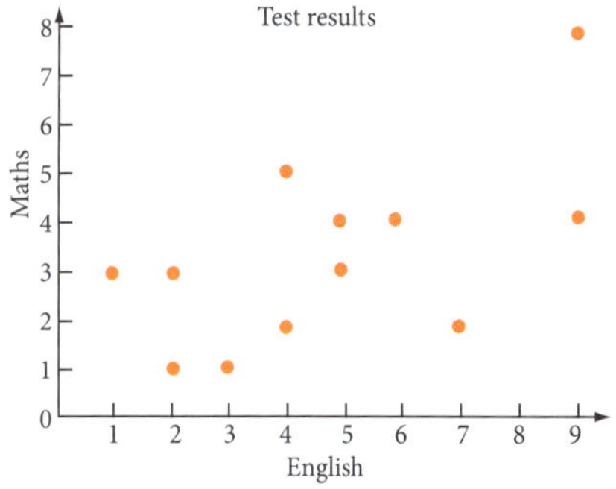

  b **There is a slight linear pattern in the positive direction.**

4 a **10**
  b **2**
  c $1\frac{1}{2}$
  d **There is a slight positive linear pattern.**

3 Line of best fit ➲ pp. 141–142

1

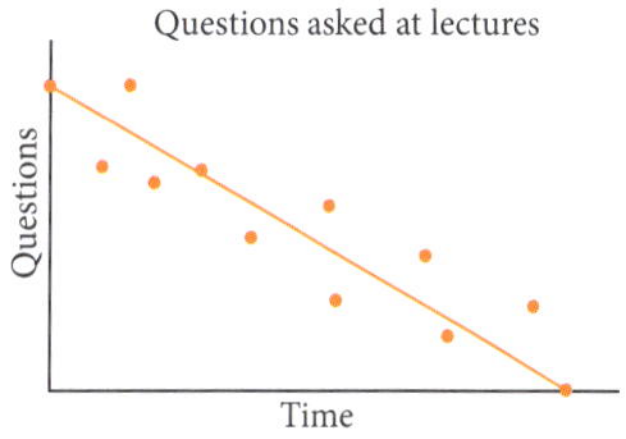

[There is no exact answer. Slight variations are correct.]

2

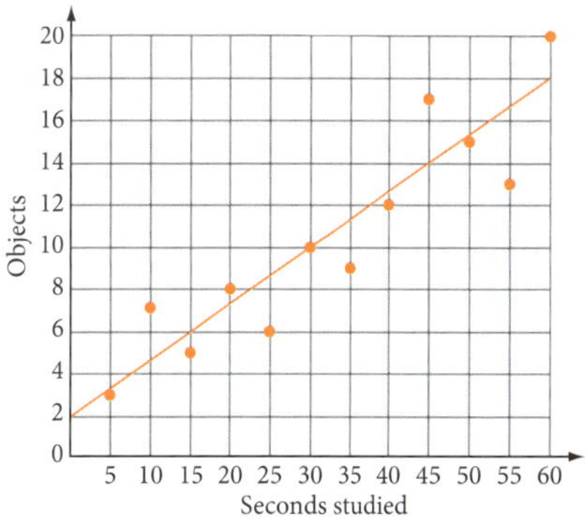

3

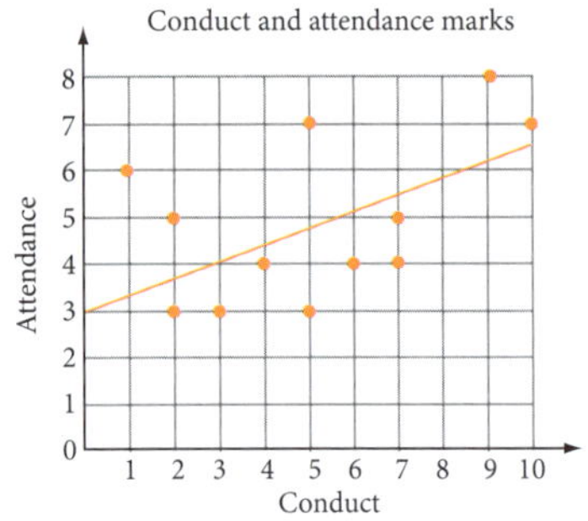

4 The equation of a line of fit ➲ pp. 143–144

1 a $m = \dfrac{\text{vertical change in position}}{\text{horizontal change in position}}$

$= \dfrac{2}{4}$

$= \mathbf{\dfrac{1}{2}}$

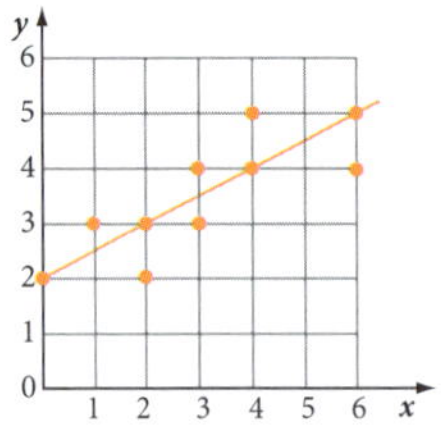

b $c = \mathbf{2}$

c $y = mx + c$

$\mathbf{y = \dfrac{1}{2}x + 2}$

2 $m = \dfrac{\text{vertical change in position}}{\text{horizontal change in position}}$

$= \dfrac{-5}{6}$

$= -\dfrac{5}{6}$

$c = 6$

$y = mx + c$

$\mathbf{y = -\dfrac{5}{6}x + 6}$

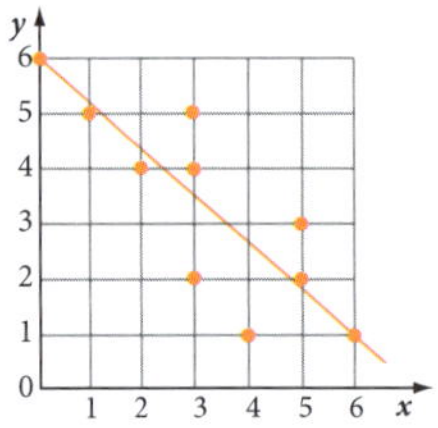

3 a, b

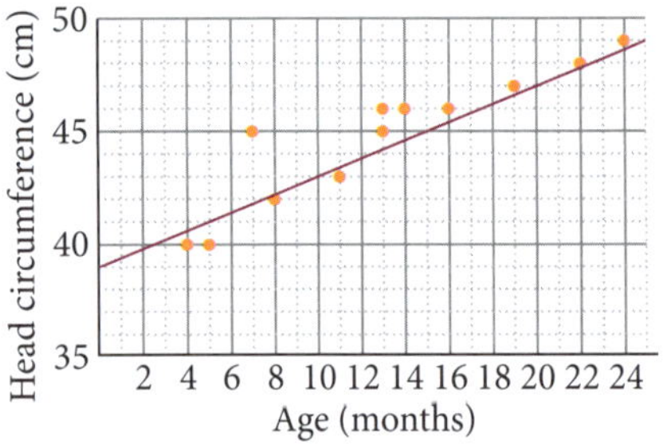

c **The head circumference of a newborn baby**

d **The amount by which a baby's head circumference increases per month**

4 a **For an age of 48 the line shows a BMI of 25.**

b **63 years old**

c gradient $= \dfrac{\text{vertical change in position}}{\text{horizontal change in position}}$

$= \dfrac{3}{15}$

$= \mathbf{0.2}$

d $y = mx + c$

$y = 0.2x + c$

When $x = 63$, $y = 28$

$28 = 0.2 \times 63 + c$

$= 12.6 + c$

$15.4 = c$

So $y = 0.2x + 15.4$.

5 a $y = -0.88x + 9.4$

When $x = 5$,

$y = -0.88 \times 5 + 9.4$

$= 5$

∴ the line will pass through the point (5, 5).

b

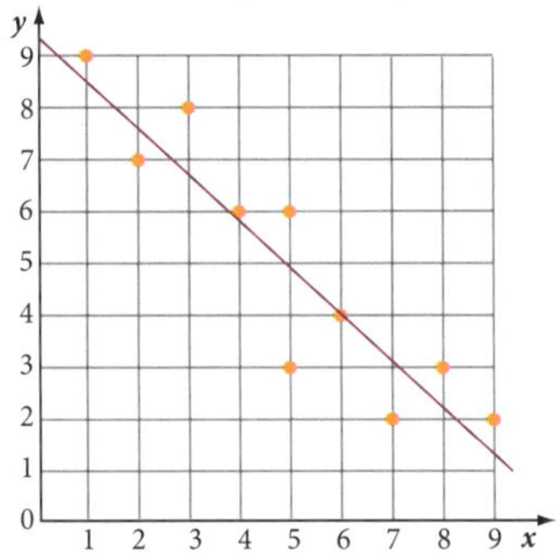

c **Gradient = −0.88. The scatterplot has a negative linear pattern, clearly illustrated by the line of fit with its negative gradient.**

5 Interpolation and extrapolation ➲ p. 146

1 a $y = \dfrac{1}{20}x + 2.5$

When $x = 100$,

$y = \dfrac{1}{20} \times 100 + 2.5$

$= 7.5$

The profit would be expected to be $7.50.

b $y = \dfrac{1}{20}x + 2.5$

When $y = 30.5$,

$30.5 = \dfrac{1}{20}x + 2.5$

$28 = \dfrac{1}{20}x$

$x = 560$

560 units are likely to have been produced.

c $y = \dfrac{1}{20}x + 2.5$

When $x = 1000$,

$y = \dfrac{1}{20} \times 1000 + 2.5$

$= 52.5$

When $x = 2500$,

$y = \dfrac{1}{20} \times 2500 + 2.5$

$= 127.5$

Extra profit = $127.50 − $52.50

= $75

You would expect the profit to be $75 more.

d $y = \frac{1}{20}x + 2.5$

When $x = 10\,000\,000$,

$y = \frac{1}{20} \times 10\,000\,000 + 2.5$

$= 500\,002.5$

The equation gives the profit as $500 002.50. **This is extremely unlikely to be an accurate prediction**. The value of x is much greater than the values used previously and the equation will only be accurate for a limited number of values.

2 a

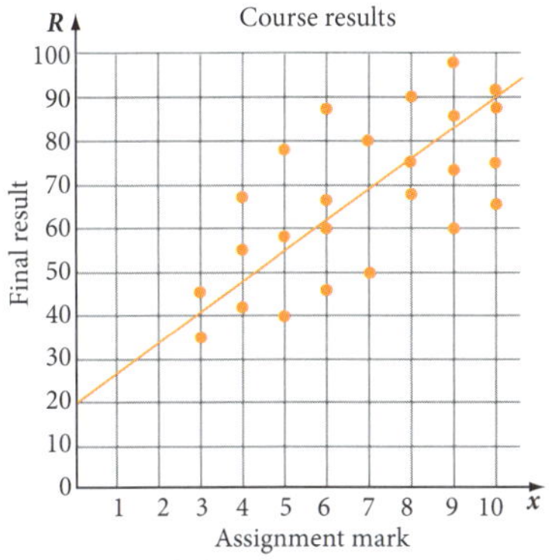

$m = \frac{\text{vertical change in position}}{\text{horizontal change in position}}$

$= \frac{70}{10}$

$= 7$

Vertical intercept = 20

The equation is $R = 7x + 20$.

[The line of fit is not unique. A slightly different line will have a different equation and produce slightly different answers to parts b and c.]

b When $x = 7$,

$R = 7 \times 7 + 20$

$= 69$

The final mark is likely to be 69.

c When $R = 76$,

$76 = 7x + 20$

$56 = 7x$

$8 = x$

The assignment mark was likely to have been 8.

3 a **About 3.5 kg**

b **This is just a little more than you would expect.**

c Alicia would need to be on the diet for **about 10 months**.

d **No, it is not possible to predict the amount of weight Mara has lost. We cannot make predictions outside the data points that have been used to create the line of best fit.**

e **It is impossible to predict how much weight Serena might have lost. We cannot make predictions about a different set of data.**

6 Analysing and describing associations ➲ p. 148

1 a Difference = 10.5 − 9.9

= **0.6 kg**

b **At ages 2 months and 18 months the difference is 0.8 kg.**

c Extra weight = (10.2 − 9) kg

= **1.2 kg**

d **Between 2 and 4 months** is the two-month period when girls put on the most weight: 1.7 kg. Boys also put on the most weight between 2 and 4 months, but slightly less at 1.5 kg.

e The median weight for girls is always less than that for boys, by about roughly the same amount. Both boys and girls tend to put on more weight in the first two-month period and this gradually decreases as they get older.

f Without any explanation of where the data comes from, how it was gathered or how big a sample from which it was taken **we can make no judgements on the reliability of the data.**

2 a If a 60-year-old woman can expect to live to 86 she has an extra life expectancy of 26 years. From the scatterplot an extra 26 for a female matches an extra 21 years for a male. **He can expect to live to 81.**

b A male life expectancy of 61 years means he can expect one extra year after 60.

$y = 1.2x - 0.24$

When $x = 1$,

$y = 1.2 \times 1 - 0.24$

$= 0.96$

So the equation suggests a woman would live to just under one extra year. This does not suggest that Jane has the wrong equation, the value is outside the other values from which the equation was determined so it is not likely to be an accurate prediction.

c **The equation does not support that belief.** The equation gives the extra years for a woman given those for a male so doesn't have anything to do with what happens over time.

3 It is not necessarily correct. All those who did well in Data Analysis also did well in Algebraic Modelling but there is no proof that being good at Algebra causes someone to do well at Data Analysis.

HSC EXAM-TYPE QUESTIONS

Part A

1 **B.** B is not linear.

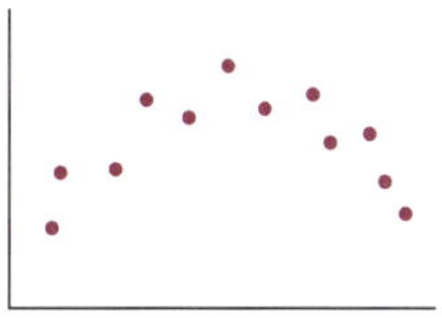

2 **B.** The two vehicles that received $13 000 as a trade-in are 4 years old and 7 years old.

Difference = 3 years

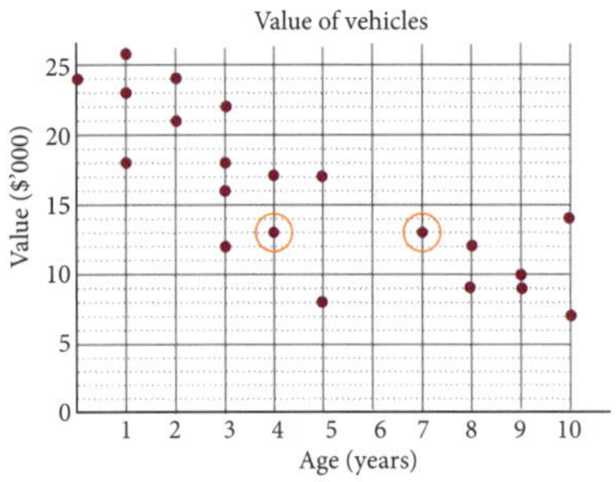

3 **A.** $m = \dfrac{\text{vertical change in position}}{\text{horizontal change in position}}$

$= \dfrac{3}{5}$

$c = 2$

$y = mx + c$

$\mathbf{y = \dfrac{3}{5}x + 2}$

4 **D.** Questions should be ethical (and unambiguous, unbiased and uncomplicated).

5 **C.** As the price of bread rises the number of loaves sold will fall, so this would be a negative pattern. There need not be any relationship between the number of cars garaged at a house and the age of the occupants or the height of two partners.

As the age of babies increases so will the size of their clothes so this will be a positive pattern.

6 **A.** $y = 0.6x + 4$

When $y = 7$,

$7 = 0.6x + 4$

$3 = 0.6x$

$x = \dfrac{3}{0.6}$

$= 5$

7 **D.** Samara would need to organise her data once it has been collected. She would then display and analyse it before drawing conclusions.

8 **C.** From the scatterplot, 6 people said they exercised for less than 60 minutes.

Percentage who exercised for less than an hour

$= \dfrac{6}{20} \times 100\%$

$- 30\%$

Part B

9 **a** **60** ✓

b **One student scored 8 in the essay but only about 25 in the exam.** ✓

c $m = \dfrac{\text{vertical change in position}}{\text{horizontal change in position}}$

$= \dfrac{20}{4}$

$= 5$ ✓

Vertical intercept = 30

The equation is $y = 5x + 30$. ✓

d $y = 5x + 30$

When $x = 7$,

$y = 5 \times 7 + 30$

$= 65$

Natalie's estimated mark was 65. ✓

10 a, b

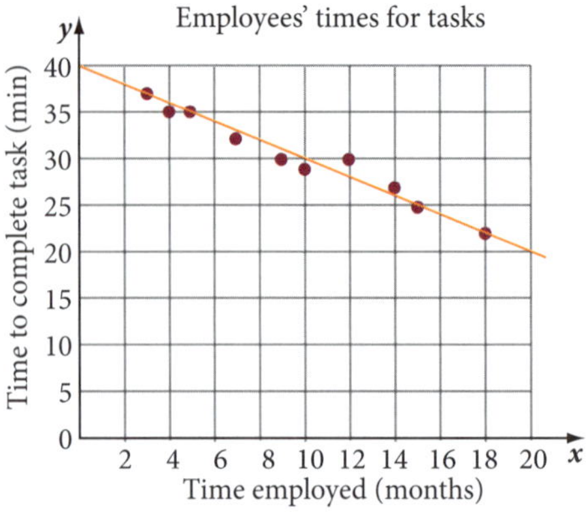

✓✓✓

c $m = \dfrac{\text{vertical change in position}}{\text{horizontal change in position}}$

$= \dfrac{-3}{3}$

$= -1$ ✓

$c = 40$

$y = mx + c$

$\mathbf{y = -x + 40}$ ✓

[A different line of fit will have a different equation.]

d When $x = 24$,

$y = -24 + 40$

$= 16$

It would take someone who has been employed for 2 years about **16 minutes** to complete the task. ✓

e When $y = 20$,

$20 = -x + 40$

$-20 = -x$

$x = 20$

If an employee takes 20 minutes to complete the task, they would probably have been employed for **20 months**. ✓

f When $x = 48$,

$y = -48 + 40$

$= -8$

Because the amount of time cannot be negative, the equation makes no sense for an employee of 4 years standing. ✓

11 a, b

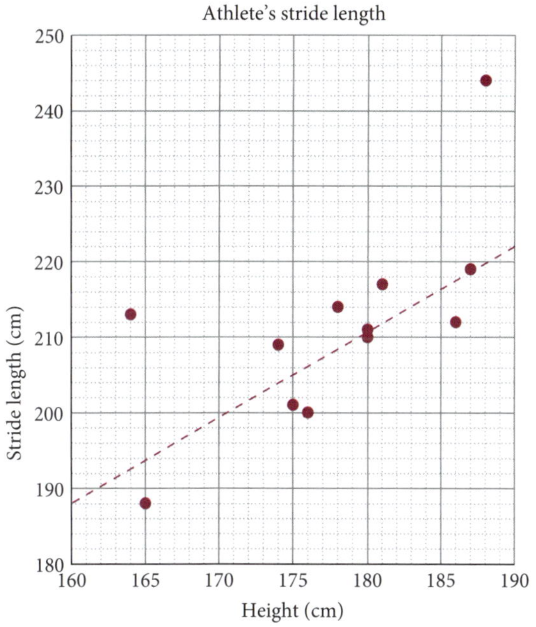

✓✓✓

c With a height of 170 cm, the line of fit gives a stride length of **about 200 cm**. ✓

12 a A woman who was 65 in 1921 could expect an additional **13.5 years of life**. ✓

b In 1905, a 45-year-old woman could expect an additional 28 years of life.

Life expectancy = 45 + 28

= **73 years** ✓

c In 1976, the life expectancy of a 45 year old woman was (45 + 34) years or 79 years.

That woman would be 65 in 1996.

The life expectancy of a 65 year old woman in 1996 was (65 + 19) years or 84 years.

So the life expectancy of that woman had increased by 5 years in that time. ✓

d In 2008, the life expectancy of a 45 year old woman was (45 + 40) years or 85 years.

In 2008, the life expectancy of a 65 year old woman was (65 + 22) years or 87 years.

So **the 65-year-old had the greater life expectancy** by 2 years. ✓

Life expectancy increases as you get older. The 65 year old has not died between the ages of 45 and 65 whereas that possibility still exists for the 45 year old. ✓

CHAPTER 8 NETWORKS AND PATHS

TOPIC: NETWORKS

Networks

1 Network definitions

- **A network is a group of objects** that have some sort of connection.
- **A network is made up of vertices,** representing the objects, and connections which are called edges. Edges do not have to be straight.

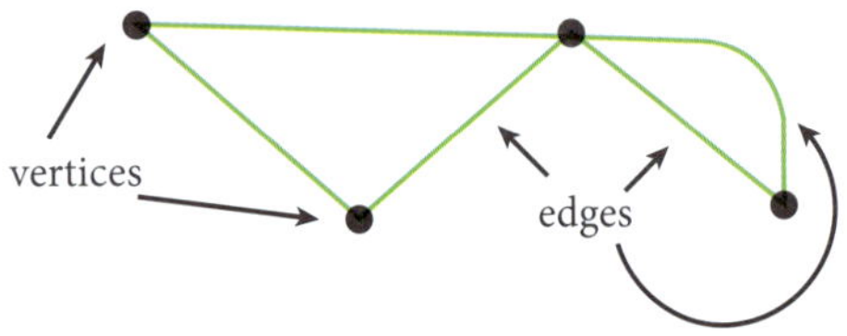

- **A network diagram is a representation of a network.** Diagrams drawn for the same network might differ in appearance but still represent the same connections.
- **A network is sometimes called a graph.**
- **A path is a way through a network,** or part of a network, passing through vertices and travelling along edges at most once.

EXAMPLE 1

Consider this network.

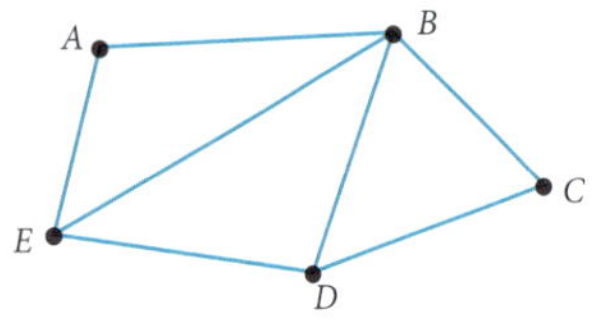

a How many vertices are there?

b How many edges?

c How many different paths are there from A to D? List them.

a There are five vertices.

Vertices might be shown by letters or by names of whatever they represent. Another name for vertices is nodes.

b There are seven edges.

Edges are sometimes called arcs.

c There are six different paths: ABD, $ABCD$, $ABED$, AED, $AEBD$ and $AEBCD$.

A is connected to B and E so, from A, a path must first proceed through either B or E.

EXAMPLE 2

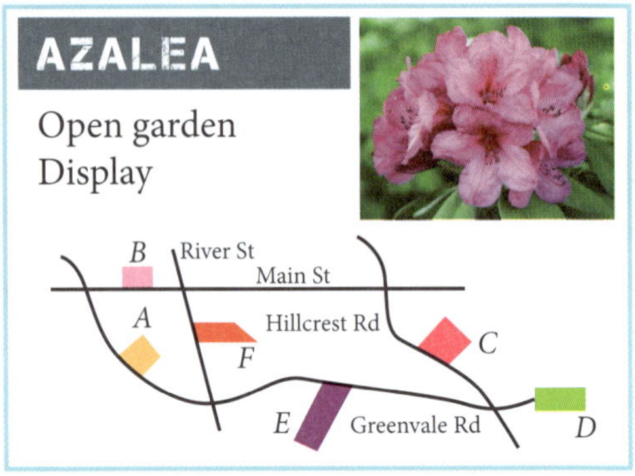

The town of Azalea is holding an open garden display. The map shows the position of the gardens that are available to view.

Draw a network diagram to show the connections between the gardens.

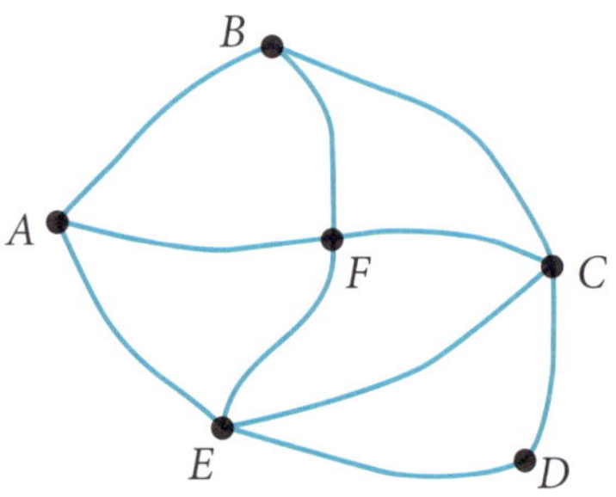

Remember that the edges are not the roads that connect the points; rather they show that there are connections. So, for example, C is connected to D and E is connected to D so these edges are shown. The fact that the roads connecting both C and E to D have a part in common is not relevant.

→ PRACTICE

1 Consider this network:

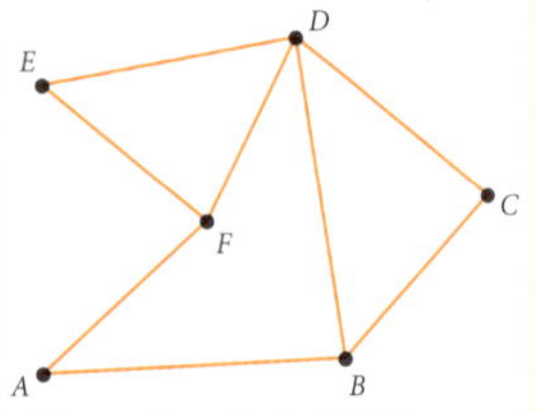

a How many vertices are there?

b How many edges?

c How many different paths are there from A to D? List them.

2 A real estate agent in Rosewood has several properties for sale. The map shows the position of the properties.

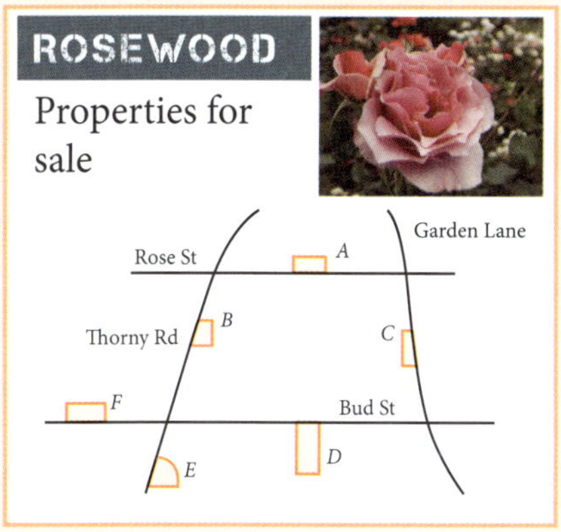

Draw a network diagram to show the connections between the properties.

3* Draw a network diagram to show the connections in this floor plan.

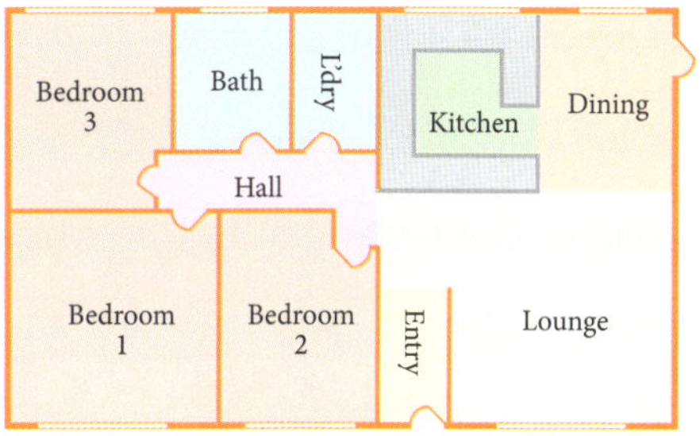

Answers ➲ p. 173

2 Terms and further definitions

- **A connected network is one where there is a path** between any two vertices.
- **A walk is a set of edges in succession**. The end vertex of one edge must be the start vertex of the following edge.
- **A cycle is a closed walk**. The first and last vertices must be the same and no other vertices are repeated.

EXAMPLE 1

Joe drew this network drawing of part of the Sydney rail system.

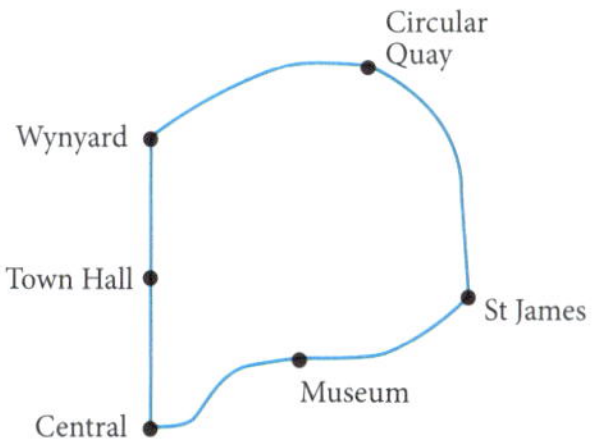

a **Is this an example of a cycle? Justify your answer.**

b **Joe knows there is just one station between Wynyard and St James on the path taken by his usual train. What is that station?**

a Yes, it is a cycle. From any station (vertex) a walk leads to each of the other stations in succession and returns to the original station.

b Circular Quay

Travelling in one direction there is a single station between Wynyard and St James. It is also possible to travel in the opposite direction but you would then pass through three other stations before reaching St James station.

EXAMPLE 2

This network shows the connections between different rooms of a house.

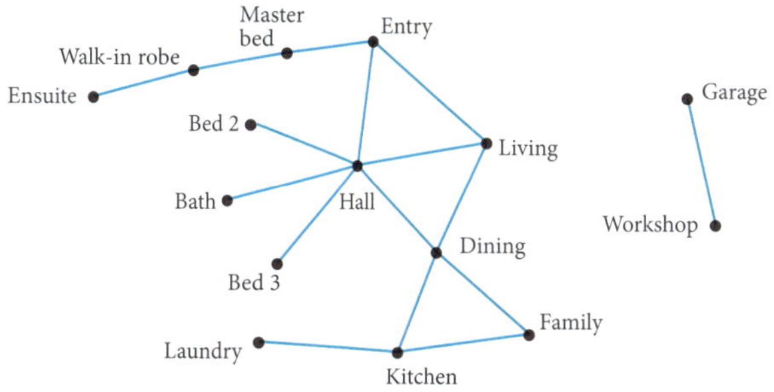

a **Briefly explain why this is not an example of a connected network.**

b **Describe the walk from the entry to the ensuite.**

c **Give an example of three different cycles beginning from the dining room.**

a The garage and workshop are not connected to the other rooms of the house.

b The walk begins at the entry, passes through first the master bedroom and then through the walk-in robe to reach the ensuite.

c One is the dining, hall, living and return to the dining room; another is the dining, living, entry, hall and dining; a third is the dining, family and kitchen then returning to the dining room.

A cycle is sometimes called a circuit.

➔ PRACTICE

1 **Joanne drew this network drawing of a planned trip, starting and finishing in Sydney, showing the towns where stops will be made.**

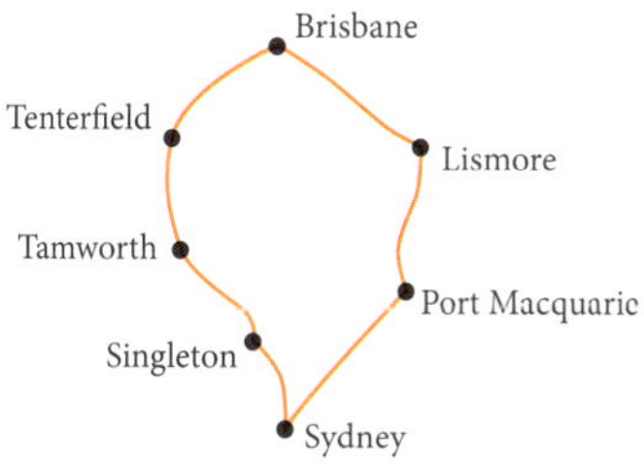

a **Is this an example of a cycle? Justify your answer.**

b Joanne knows there are two stops between Lismore and Tamworth on the route taken on the trip. What are those towns?

2 This network shows the connections between different rooms of a house.

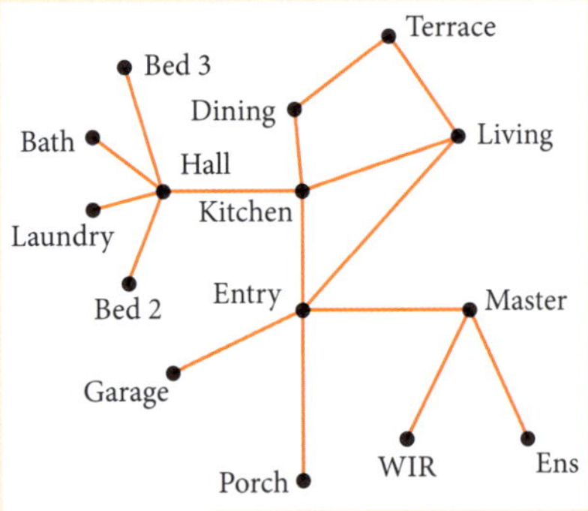

a Briefly explain why this is an example of a connected network.

b Describe a walk from the porch to the bathroom.

c Give an example of three different cycles beginning from the kitchen.

3* a Draw a network diagram showing the connections between the states and territories of Australia.

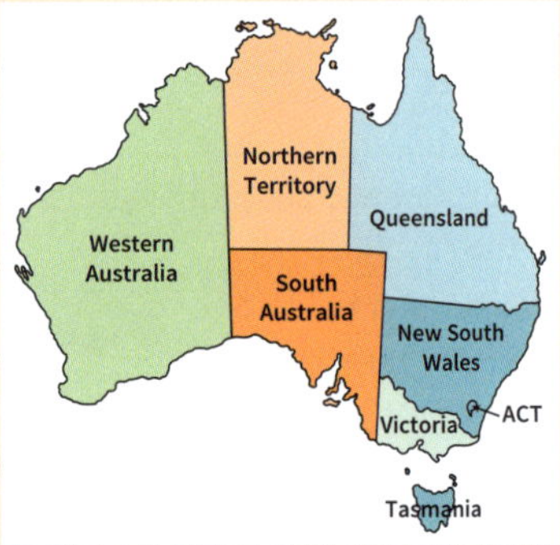

b Is this a connected network? Justify your answer.

c Describe a walk from the ACT to the Northern Territory.

d Give three different cycles beginning and ending in Queensland.

Answers ➲ p. 173

3 Further terms

➔ **The degree of a vertex is the number of edges** that connect to that vertex.

➔ **A directed network is a network** where the edges must be travelled in a certain direction as indicated by arrows.

➔ **Some networks have weighted edges.** The weight of an edge might be a distance, a cost, a time or some other measurement appropriate to the network.

EXAMPLE 1

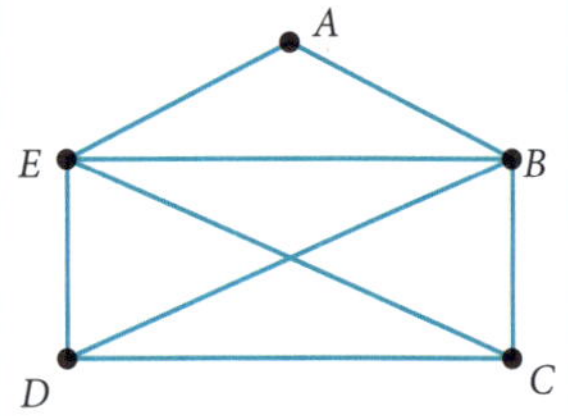

a What is the degree of vertex A?

b How many vertices have degree 3?

c What is the sum of the degrees of all the vertices?

d Redraw the network so that no edges overlap.

a 2

'Vertices' is the plural of 'vertex'. You have one vertex and two or more vertices.

b 2

Vertices C and D have degree 3.

c $\text{Sum} = 2 + 4 + 3 + 3 + 4$
$= 16$

The sum of all the degrees must be twice the number of edges.

d One possibility is

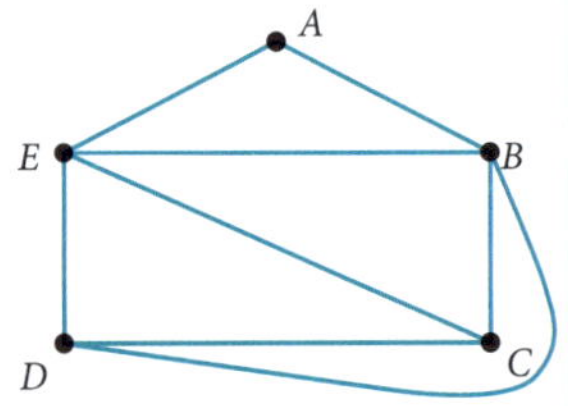

Remember: An edge does not have to be straight.

EXAMPLE 2

This network shows the bus routes in a certain town.

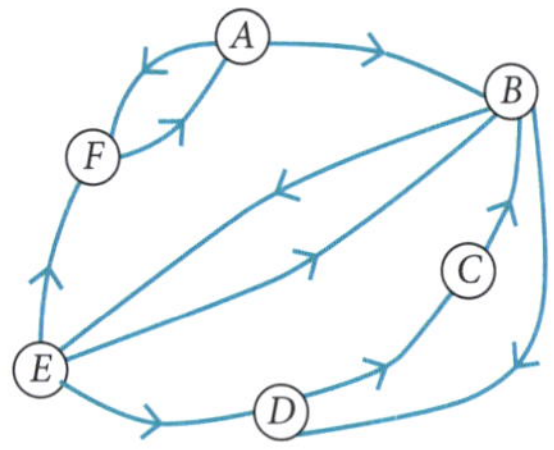

a To travel by bus from A to E, through what other point must you pass?

b Describe the path from C to F.

c In how many different ways is it possible to travel from B to D? Explain.

a B

b From C to B to E to F

Although F is between A and E, the directions on the edges show that it is not possible to travel directly from F to E.

c 2; directly from B to D, or passing through E

A directed network can be thought of as a series of one-way streets.

EXAMPLE 3

This network diagram shows some road distances, in kilometres, between certain NSW towns.

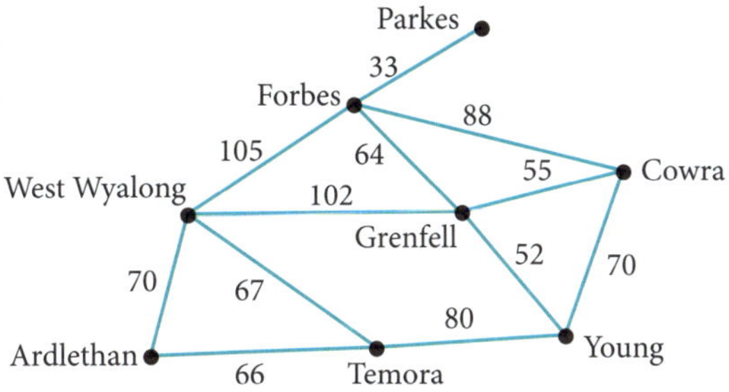

a **Which two distances are the same?**

b **How far is it from Young to Forbes, via Grenfell?**

c **How much further is it to West Wyalong from Parkes than from Ardlethan?**

a From Cowra to Young and from West Wyalong to Ardlethan are both 70 km.

b Distance = (52 + 64) km
= 116 km

> A network diagram is not a map and is not usually drawn to scale.

It is 116 km from Young to Forbes, via Grenfell.

c From Parkes to West Wyalong = (33 + 105) km
= 138 km

Extra distance = (138 − 70) km
= 68 km

It is 68 km further from West Wyalong to Parkes than to Ardlethan.

EXAMPLE 4

Judy drew this diagram to show the links between her house and various other places in her neighbourhood and the time, in minutes, it takes to drive between them.

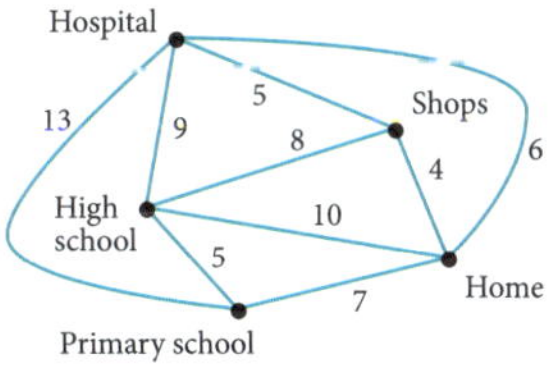

Judy leaves home and travels, by the shortest possible route, first to the high school, then the shops, hospital, primary school and back home.

a **Copy the diagram, showing Judy's journey.**

b **How long should her journey take?**

c **If Judy instead travelled first to the hospital, then the shops, both schools and straight home, how much shorter would her journey be?**

d **Give a reason why Judy might prefer the longer journey.**

a

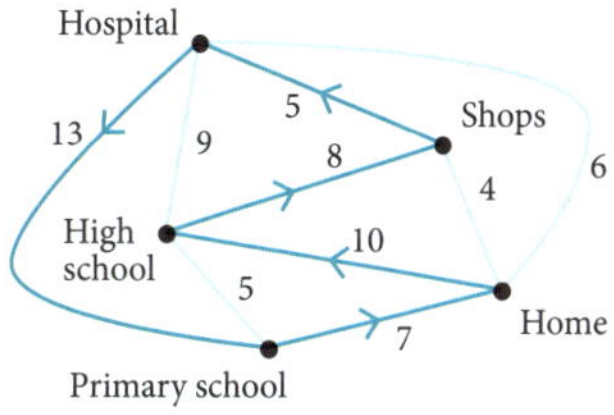

b Time taken = (10 + 8 + 5 + 13 + 7) min
= 43 min

c

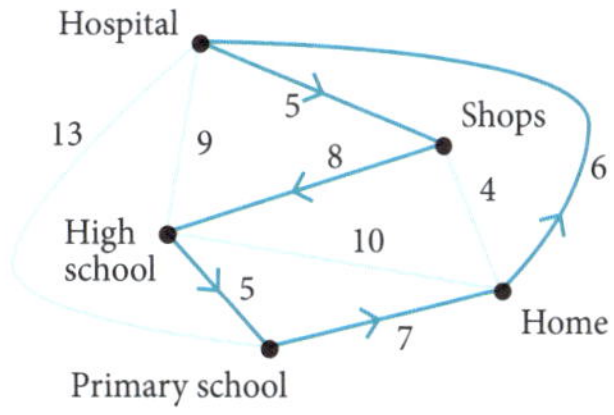

New time = (6 + 5 + 8 + 5 + 7) min
= 31 min

Difference = 43 − 31
= 12 min

Judy's journey could be 12 minutes shorter.

d Possible answers: Judy might have needed to be at the high school at an earlier time and not needed to be at the primary school until later.

The route Judy chose might take longer but be less stressful with less traffic.

> There is no exact answer for this type of question. Think about what is happening and try to come up with an answer that makes sense.

→ PRACTICE

1 a **What is the degree of vertex A?**

b **How many vertices have degree 2?**

c **What is the sum of the degrees of all the vertices?**

d **Redraw the network so that no edges overlap.**

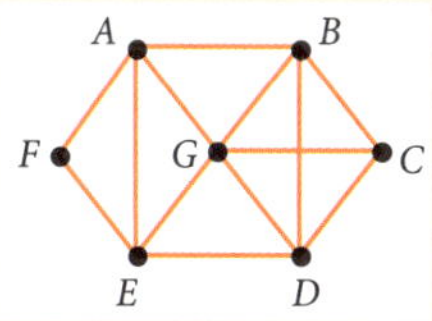

2 **This network shows the bus routes linking certain towns.**

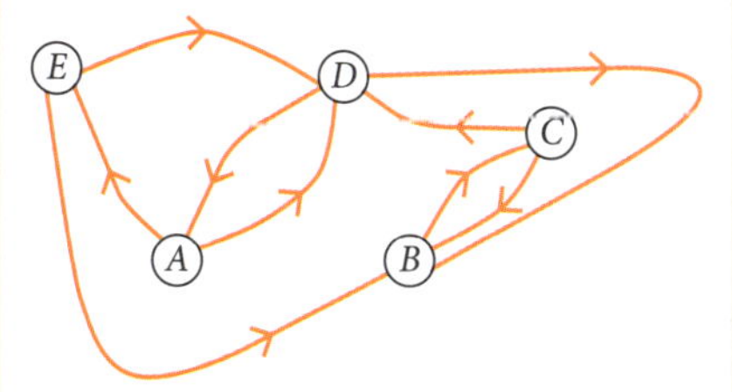

a To travel by bus from D to E, through what other point must you pass?
b Describe the path from B to E.
c In how many different ways is it possible to travel from A to C? Explain.

3 This network diagram shows some road distances, in kilometres, between certain NSW towns.

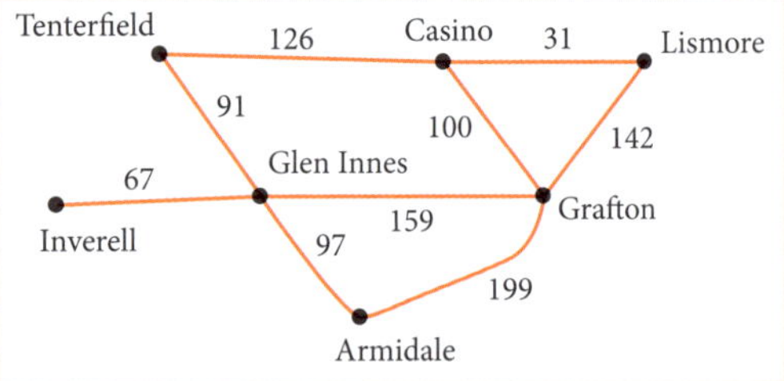

a What is the longest distance between two directly connected towns?
b How far is it from Armidale to Inverell via Glen Innes?
c How much further is it from Tenterfield to Grafton via Glen Innes than via Casino?

4 Ken drew this diagram to show the links between his home and various other places in his neighbourhood and the time, in minutes, it takes to drive between them.

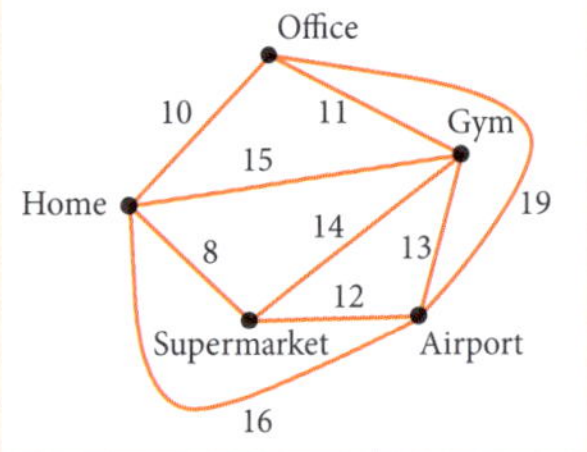

Ken leaves home and travels, by the shortest route, first to the gym, then his office, the airport, supermarket and back home.
a Copy the diagram, showing Ken's journey.
b How long should his journey take?
c If Ken instead travelled first to the office, then the gym, the supermarket, airport and back home, how much shorter would his journey be?
d Give a reason why Ken might prefer the longer journey.

5* The diagram shows the links and times in minutes for a bus to travel between certain towns.

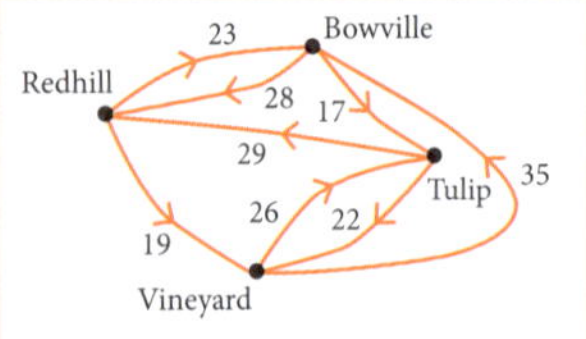

a Describe the path from Redhill to Tulip that doesn't pass through Vineyard.
b How much quicker is it to travel directly from Vineyard to Bowville than via Tulip and Redhill?

Answers ➲ pp. 173–174

4 Applications

- **Networks can be used in problem-solving**; for example, in minimising the cost of connections of cables or pipes or planning efficient routes.
- **A network is said to be traversable** if it is possible to move from one vertex to every other vertex moving along each edge once, and only once. (Vertices are called odd or even according to their degree.)
  - **A network with no odd vertices** is traversable by beginning at any vertex and ending at the same vertex.
  - **A network with exactly two odd vertices** is traversable, by beginning at one of the odd vertices and ending at the other.
  - **A network with more than two odd vertices** is not traversable.

EXAMPLE 1

This is a network diagram of the main roads linking the towns in a shire.

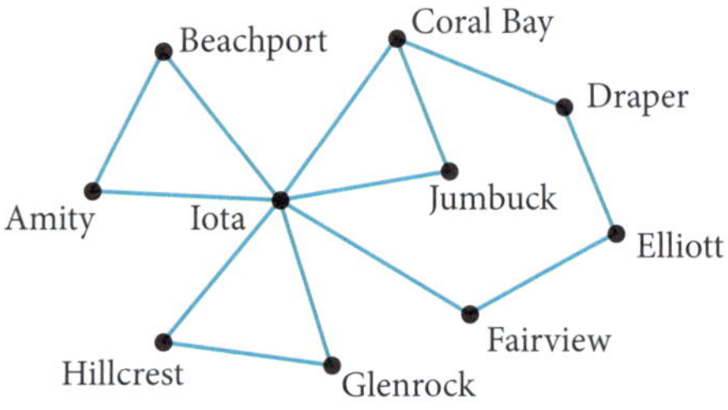

Suggest a route so that workers can fill potholes on all the roads, without travelling on any road more than once.

Begin at Iota, travel to Amity, then Beachport and back to Iota. Then travel to Glenrock and Hillcrest and back to Iota. Next travel to Jumbuck, Coral Bay, Draper, Elliott, Fairview and back to Iota and finally to Coral Bay.

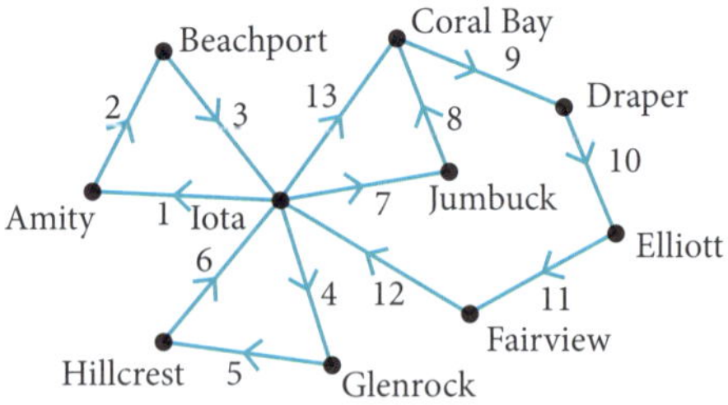

There are several different possible routes but all must start at either Iota or Coral Bay and finish at the other of those towns. Note it is only the roads that are travelled on once. The towns can be passed through more than once.

EXAMPLE 2

By drawing a network diagram, determine whether it is possible to take a walk and cross each of the seven bridges over the Pregel River once.

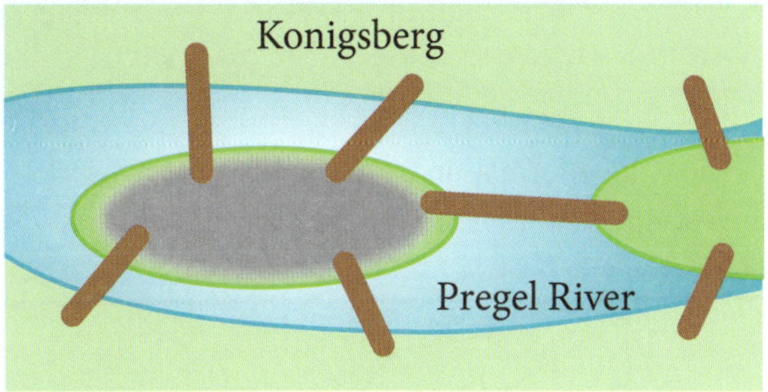

It is not possible to cross each of the bridges once only.
The network diagram has more than two odd vertices, so it is not traversable.

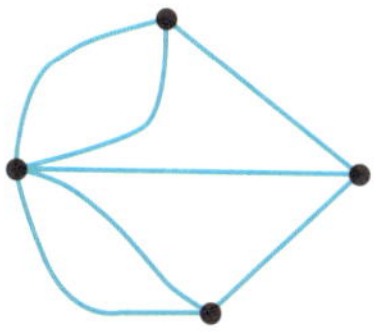

This is a famous problem called the Konigsberg bridge problem. The edges of the diagram represent the bridges and the vertices are the land.

→ PRACTICE

1 **This is a network diagram of the trails linking features of a national park.**

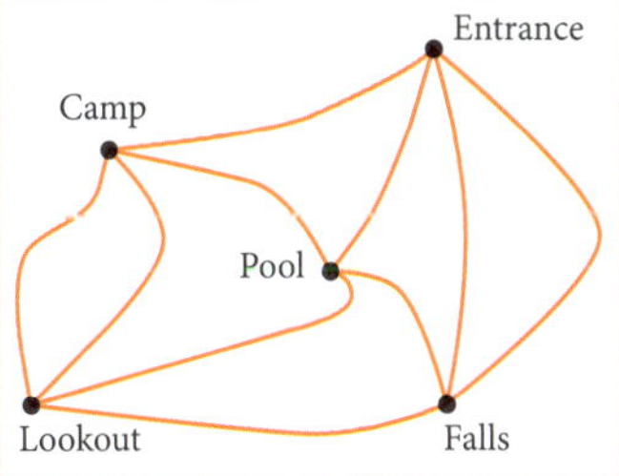

Suggest a route so rangers can check all the trails, without travelling on any trail more than once, beginning and finishing at the entrance.

2 **This picture shows the position of five paddocks and a lane, with the gates between them marked. By drawing a network diagram, determine whether it is possible to take a walk and go through each gate once, but only once.**

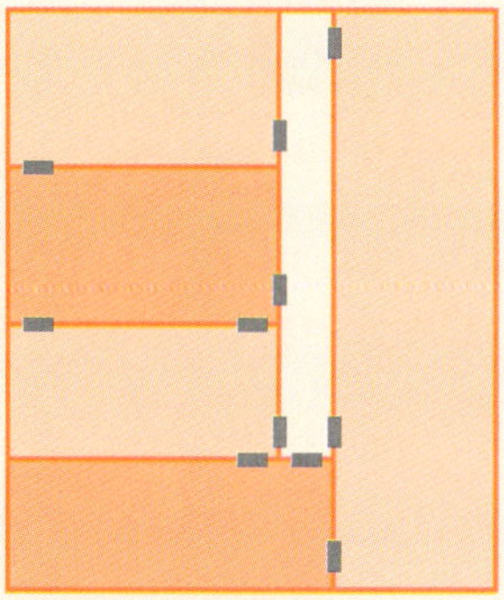

3* **This network diagram shows the cycle paths linking some towns.**

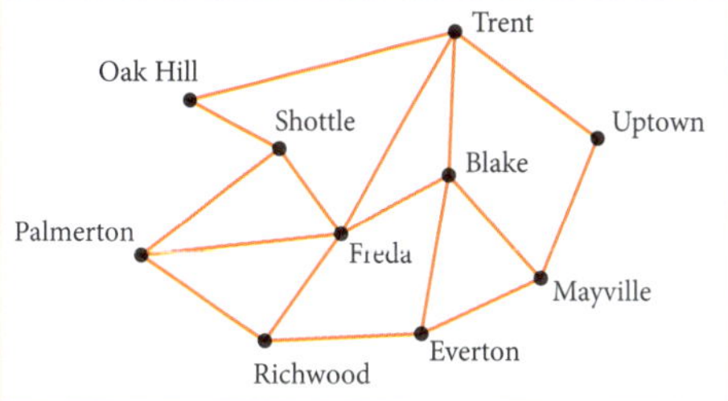

A group of cyclists want to go on a ride from Trent and visit each of the other towns before returning to Trent. Describe a possible route the cyclists could take.

Answers ➲ p. 174

5 Tables and networks

- → **A table can show information about the connections** and weights in a network.
- → **Using a table allows computers to process the information** in a network.
- → **Constructing a network diagram from a table** might give diagrams that differ in appearance but are still basically the same.

EXAMPLE 1

Draw a table to show the information in this network diagram.

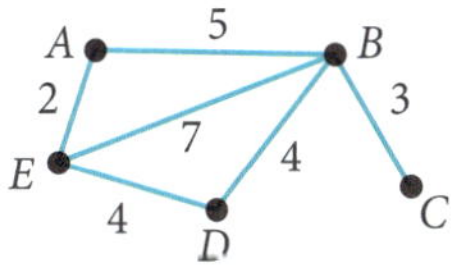

	A	B	C	D	E
A	–	5	–	–	2
B	5	–	3	4	7
C	–	3	–	–	–
D	–	4	–	–	4
E	2	7	–	4	–

The dashes show there is no connection. (Zeroes are sometimes used for the connection between a vertex and itself.)

EXAMPLE 2

Show the information from this directed network diagram in a table.

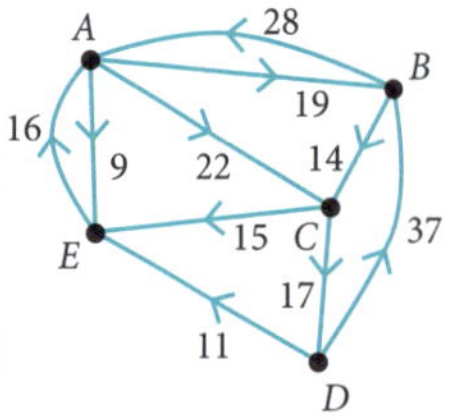

		To				
		A	B	C	D	E
From	A	0	19	22	–	9
	B	28	0	14	–	–
	C	–	–	0	17	15
	D	–	37	–	0	11
	E	16	–	–	–	0

In an undirected network the table will be symmetrical. Because this network is directed, the table is not symmetrical.

EXAMPLE 3

Construct a network diagram from this table.

		To					
		A	B	C	D	E	F
From	A	–	10	–	12	11	–
	B	–	–	7	13	–	–
	C	–	6	–	–	–	14
	D	10	–	9	–	8	–
	E	–	–	12	–	–	5
	F	9	–	–	–	7	–

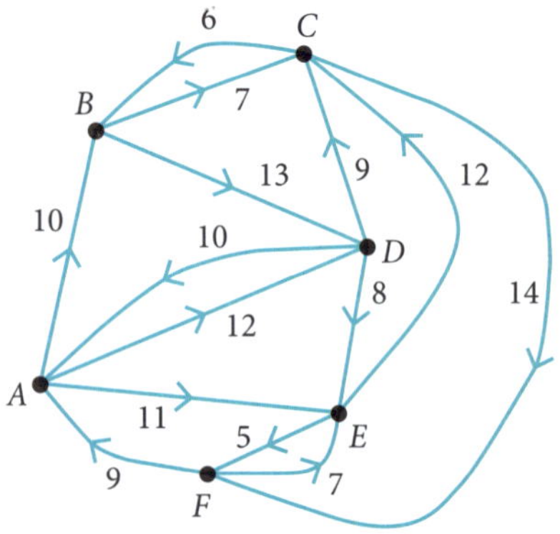

There are many slightly different possible representations.

→ PRACTICE

1 Draw a table to show the information in this network diagram.

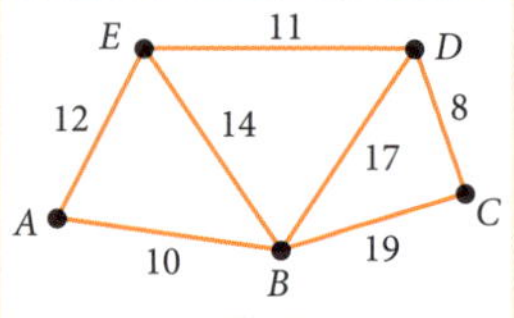

2 Show the information from this directed network diagram in a table.

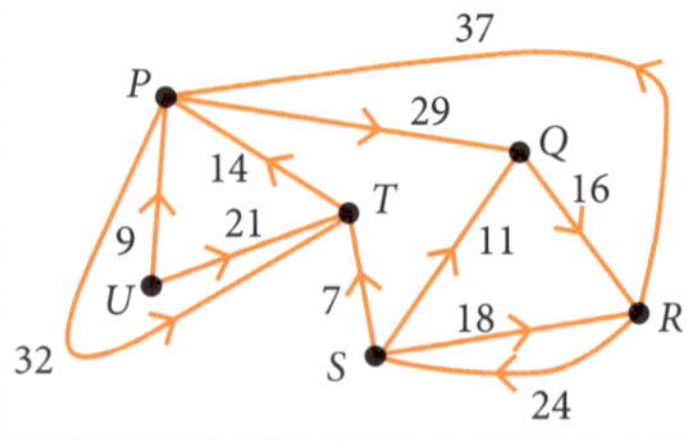

3 Construct a network diagram from this table.

		To					
		A	B	C	D	E	F
From	A	–	8	–	–	19	–
	B	–	–	9	13	21	–
	C	–	–	–	10	–	–
	D	17	15	–	–	11	–
	E	–	–	–	–	–	5
	F	12	–	–	–	–	–

4* a Construct a network diagram from this table.

	A	B	C	D	E	F	G
A	–	49	–	–	–	56	–
B	49	–	61	–	–	66	52
C	–	61	–	39	78	–	36
D	–	–	39	–	64	–	–
E	–	–	78	64	–	73	41
F	56	66	–	–	73	–	59
G	–	52	36	–	41	59	–

b If the table shows the time, in minutes, to travel between the various points, how long, in hours and minutes, will it take to travel from *A* to *C* via *F*, *E* and *D*?

Answers ➲ p. 174

Shortest paths

1 Trees

- ➔ **A tree is an undirected network, or part of a network**, that does not contain a cycle. In a tree, any two vertices are connected, at most, by exactly one path.
- ➔ **A spanning tree is a tree of an undirected network** that contains every vertex. A network might have many different spanning trees.
- ➔ **In a spanning tree**, the number of edges will always be one fewer than the number of vertices.

EXAMPLE 1

Consider this network.

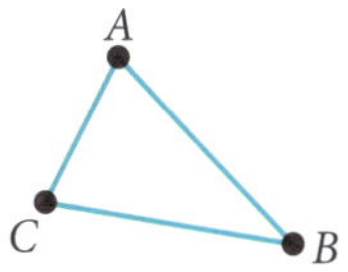

a **How many vertices are there?**

b **How many edges will any spanning tree have for this network?**

c **Show all the different possible spanning trees.**

Spanning trees use all the vertices but usually not all of the edges.

a There are three vertices.

b A spanning tree will have two edges.

c

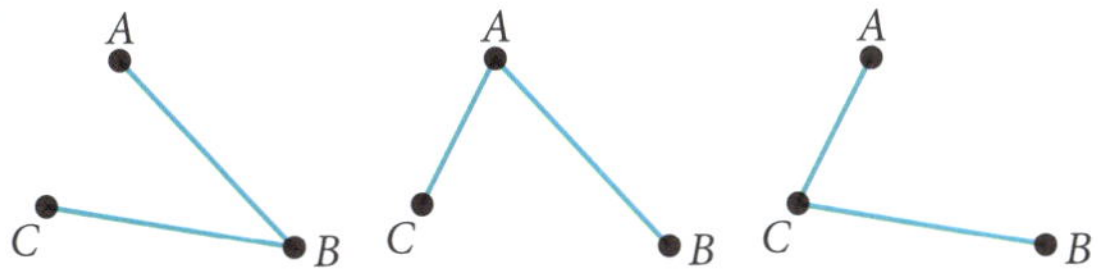

EXAMPLE 2

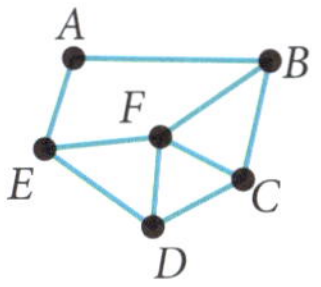

a **Briefly explain why this network is not a tree.**

b **How many edges will any spanning tree for this network have? Justify your answer.**

c **Draw three different possible spanning trees for the network.**

a The network is not a tree because it contains several cycles. There is not a single path linking all the vertices.

b Any spanning tree will have five edges. There are six vertices and any spanning tree will have one fewer edge than the number of vertices.

c

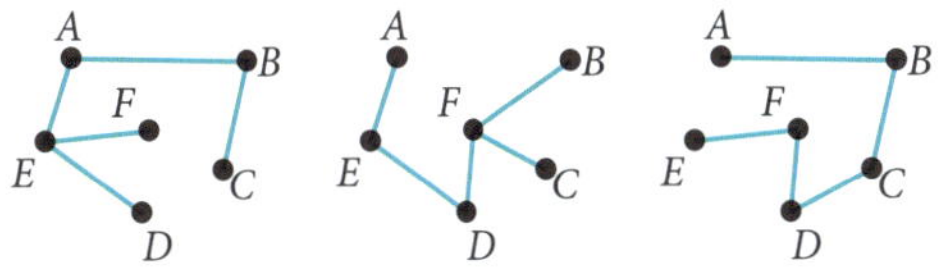

Four edges need to be removed but there must be one edge removed from each cycle and each vertex must remain connected. There are other possible answers.

➔ PRACTICE

1 **Consider this network.**

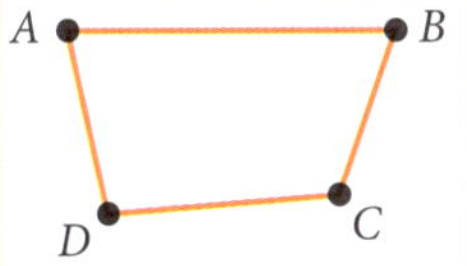

a **How many vertices are there?**

b **How many edges will any spanning tree have for this network?**

c **Show all the different possible spanning trees.**

2 **a** **Briefly explain why this network is not a tree.**

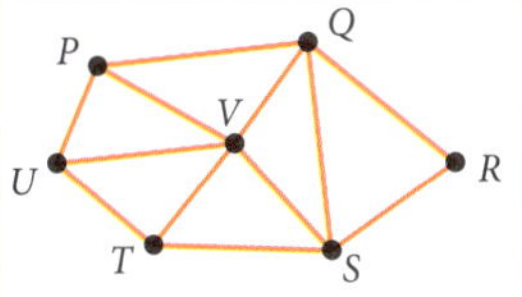

b **How many edges will any spanning tree for this network have? Justify your answer.**

c **Draw three different possible spanning trees for the network.**

3* **Draw a spanning tree for this network.**

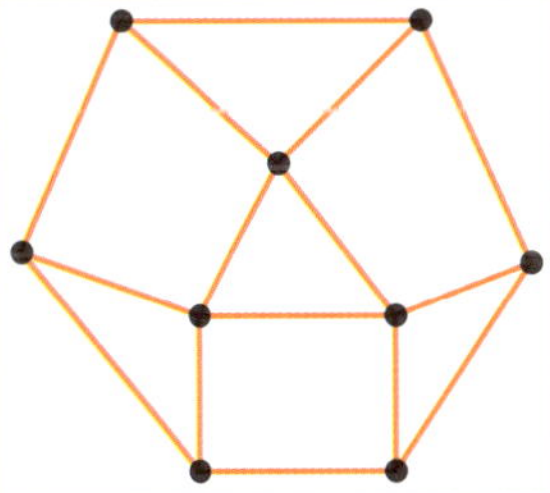

Answers ➲ pp. 174–175

2 Minimum spanning trees

- ➔ **The minimum spanning tree (or minimal spanning tree) of a network** that has weighted edges is the spanning tree for which the total of all the edges is least.
- ➔ **Prim's algorithm** is one method of **finding a minimum spanning tree.** Any vertex is chosen first and then connected vertices are added in turn, depending on the weight of their connecting edges.

Kruskal's algorithm is **another method of finding a minimum spanning tree**. The edges are chosen beginning with the one with the least weight and others are added in order of weight, provided they do not form a cycle.

EXAMPLE 1

This network shows the road distances, in kilometres, between three towns.

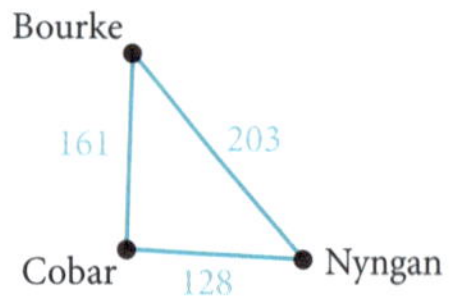

a **How many edges will a spanning tree have?**

b **Show the minimum spanning tree.**

a Two edges

b

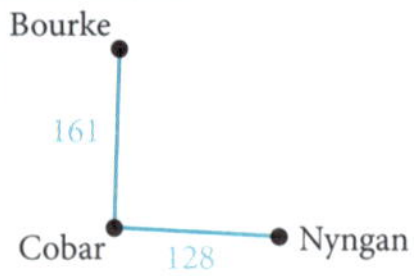

Simply leave out the longest edge.

EXAMPLE 2

Consider this network.

Use Prim's algorithm, beginning with vertex A, to find the minimum spanning tree.

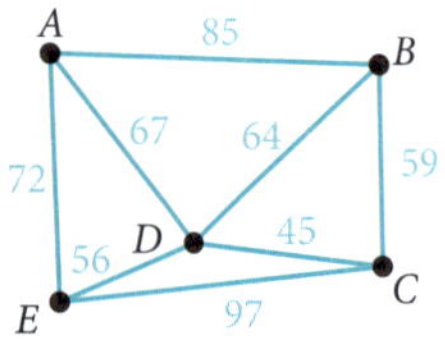

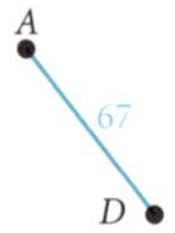

Begin with vertex *A* and then choose the vertex joined to *A* by the edge of least weight, *D*.

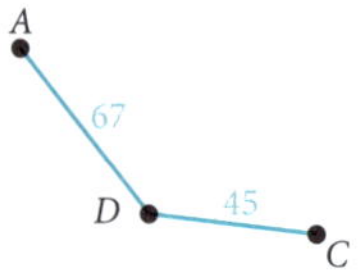

Then choose the vertex joined to either *A* or *D* by the edge of least weight, *C*.

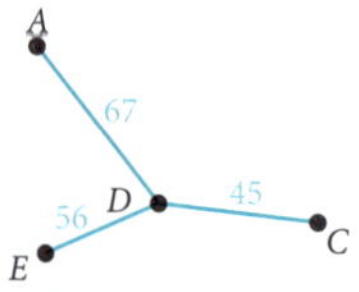

Then choose the vertex joined to either *A*, *D* or *C* by the edge of least weight, *E*.

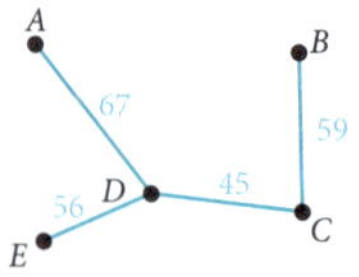

Then choose the final vertex, *B*, and the edge of least weight. The result will be the minimum spanning tree. The end result would be the same whichever vertex you begin with.

EXAMPLE 3

Use Kruskal's algorithm to find the minimum spanning tree for this network.

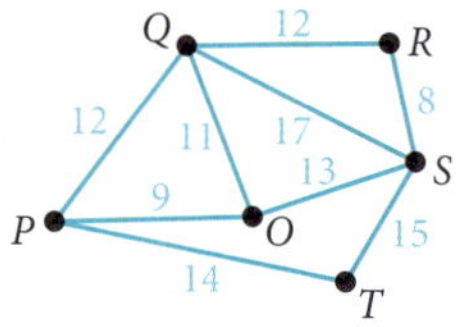

First choose the edge of least weight, 8.

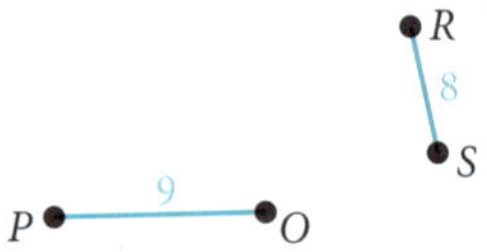

Then choose the edge of next least weight, 9. The edges do not have to adjoin.

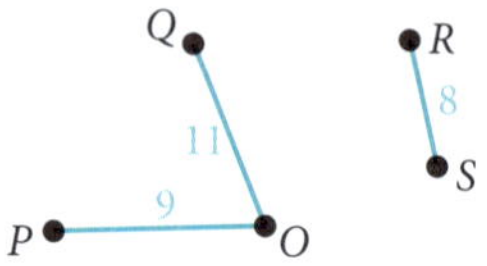

Choose the next edge, with next least weight, 11.

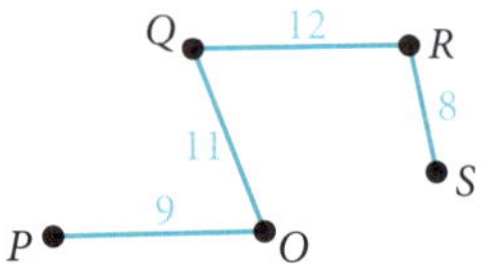

The next least weight is 12. There are two edges with weight 12 but *PQ* cannot be chosen because it would complete a cycle.

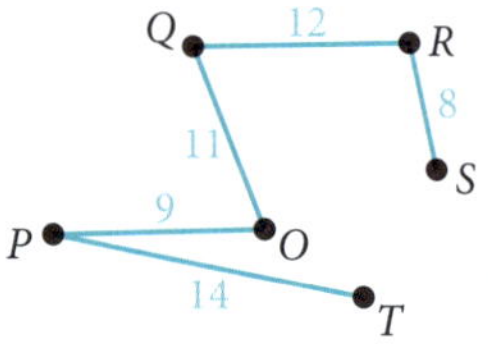

The next least weight is 13 but *OS* would complete a cycle. So the next least weight then is 14 and choosing *PT* completes the spanning tree as there are now five edges and all vertices are connected.

EXAMPLE 4

This table shows the cost, in dollars, to connect certain points with cables.

	A	B	C	D	E	F	G
A	0	320	–	–	–	350	–
B	320	0	460	–	–	560	430
C	–	460	0	530	590	–	370
D	–	–	530	0	390	–	–
E	–	–	590	390	0	610	280
F	350	560	–	–	610	0	470
G	–	430	370	–	280	470	0

a **Draw a network diagram from this table.**

b **Show that Prim's algorithm, beginning with vertex *A*, and Kruskal's algorithm give the same minimum spanning tree.**

c **What is the minimum cost to connect all the points?**

a

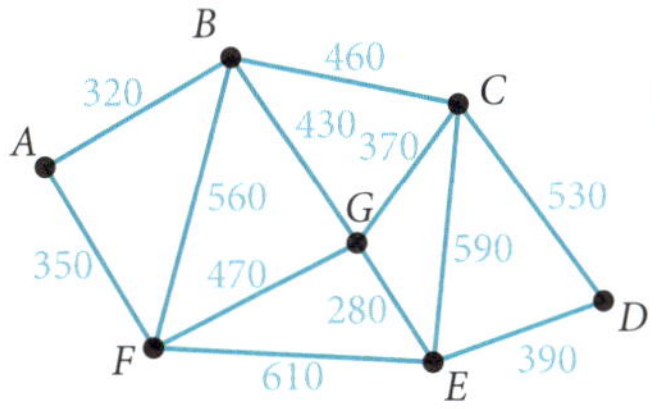

b Prim's algorithm

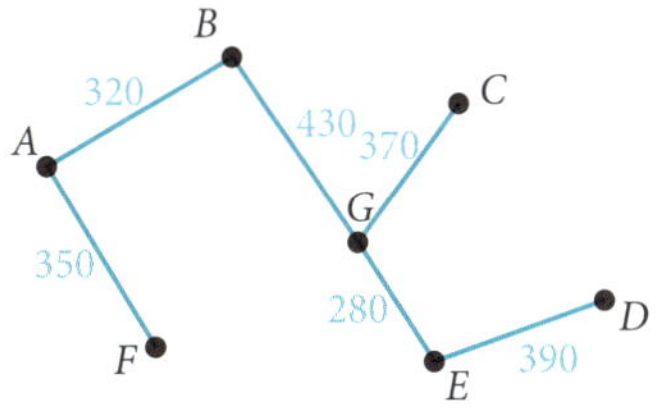

Beginning with vertex *A*, with Prim's algorithm you would add the vertices and connecting edges in this order: *B, F, G, E, C, D*.

Kruskal's algorithm

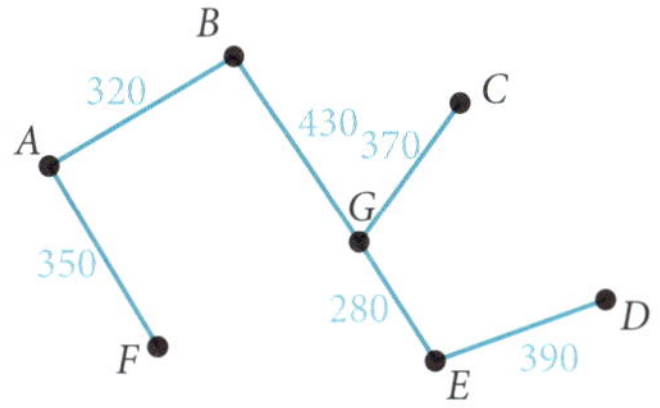

With Kruskal's algorithm the first edge is *GE* and then in order *AB, AF, GC, ED* and *BG*.

c Minimum cost

= $(350 + 320 + 430 + 370 + 280 + 390)

= $2140

EXAMPLE 5

The diagram shows a plan of a farm, with the position of a pump on the river and the distances to five water troughs that will be filled with water pumped from the river. Find the shortest length of pipes necessary to connect all the water troughs to the pump.

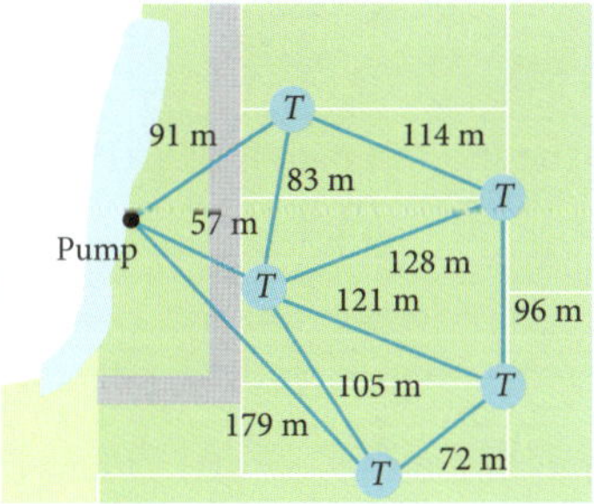

Consider the network:

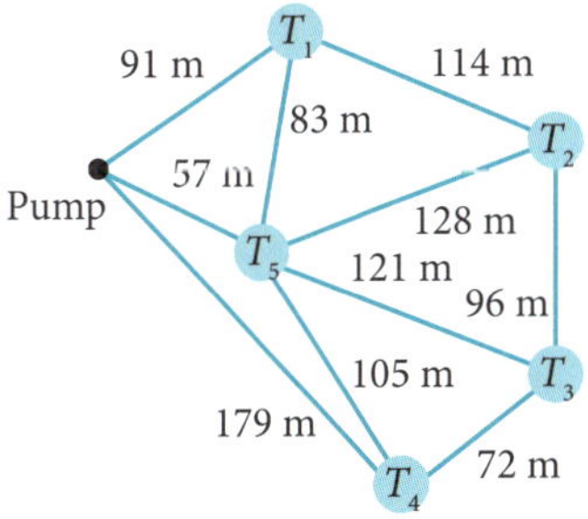

The minimum spanning tree is:

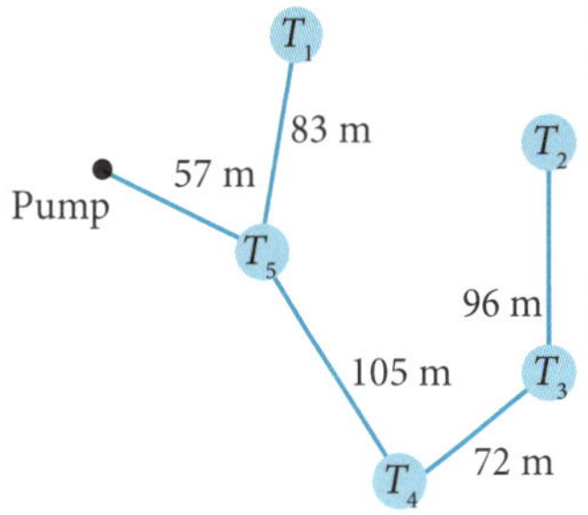

Minimum length of pipes

= (57 + 83 + 105 + 72 + 96) m

= 413 m

Finding the minimum spanning tree gives us the measurements that minimise the length. There are six vertices (one pump and five troughs) so there will be five edges.

PRACTICE

1 This network shows the road distances, in kilometres, between three towns.

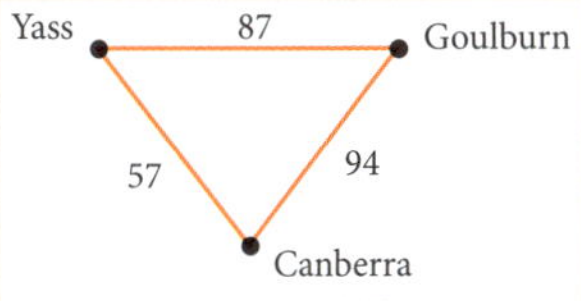

a How many fewer edges will a spanning tree have than shown here?

b Show the minimum spanning tree.

2 Consider this network.

Use Prim's algorithm, beginning with vertex *P*, to find the minimum spanning tree.

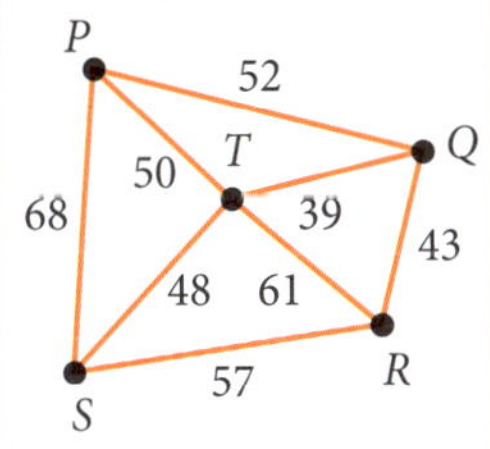

3 Use Kruskal's algorithm to find the minimum spanning tree for this network.

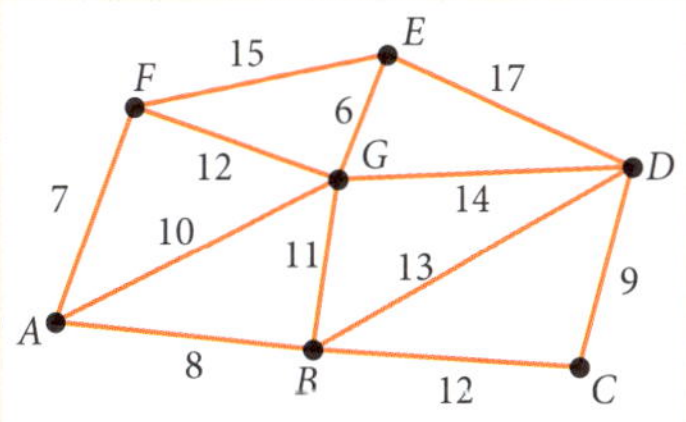

4 This table shows the distances, in metres, between certain points.

	A	B	C	D	E	F	G
A	–	540	–	495	–	–	–
B	540	–	480	482	–	–	513
C	–	480	–	–	–	575	506
D	495	482	–	–	329	–	378
E	–	–	–	329	–	368	417
F	–	–	575	–	368	–	439
G	–	513	506	378	417	439	–

a Draw a network diagram from this table.

b Show that Prim's algorithm, beginning with vertex *A*, and Kruskal's algorithm give the same minimum spanning tree.

c What is the minimum distance connecting all the points?

5 The diagram shows a plan of a subdivision, showing the position of electricity connections and the distances between them in metres. Find the shortest length of cables necessary to connect all the electricity points.

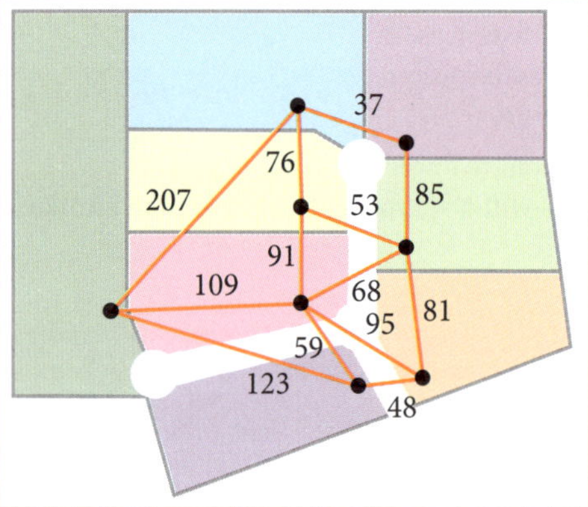

6* The diagram shows the cost to connect the rooms in a house to a vacuum system with ducting.

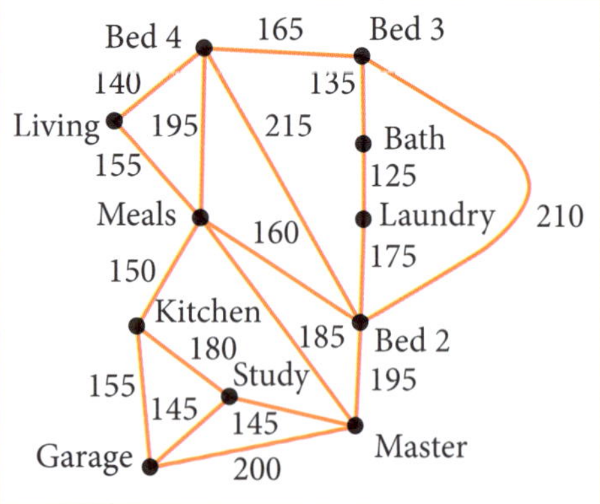

a Find the minimum spanning tree.

b Find the least cost to install the ducting.

Answers ➲ p. 175

3 Shortest paths

- **The shortest path between any two vertices** in a network can be found. It is the path of minimum weight.
- **The shortest path differs from the minimum spanning tree** because not all vertices must be included.
- **The shortest path can be found** by finding the least cumulative total for each step.
- **Dijkstra's algorithm is a method of finding the shortest path**. It works in a similar way to Prim's algorithm.

EXAMPLE 1

Find the shortest path from *A* to *J* in this network.

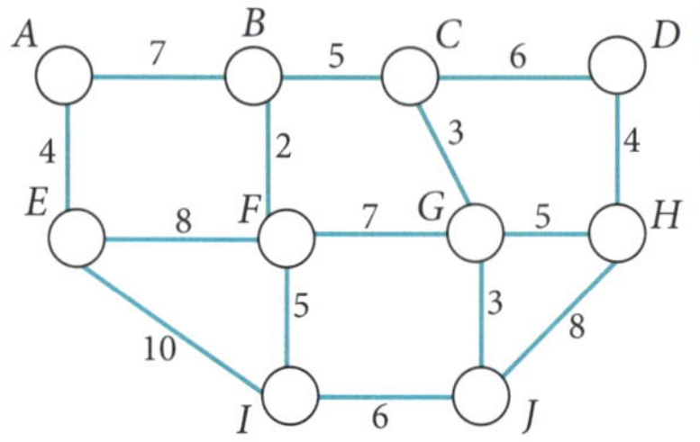

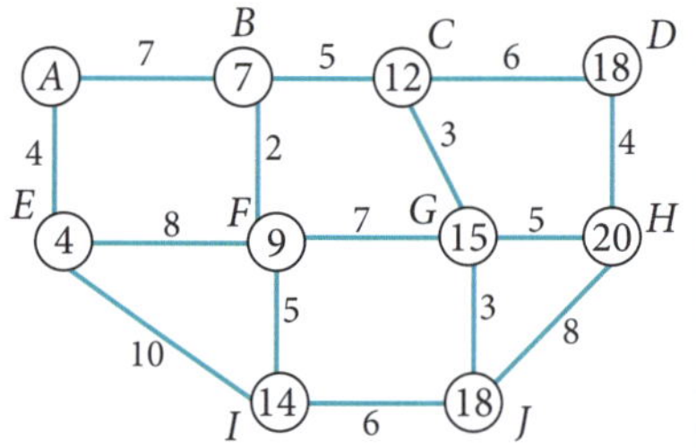

The shortest path is *ABCGJ*.

Begin at *A*. Fill in the distances one step from *A*, to *B* (7) and to *E* (4). Then fill in the total distances one step from either *B* or *E*, choosing the least value when there is more than one option. So *C* is 12. *F* is 9 because 7 + 2 = 9 is less than 4 + 8 = 12. *I* is 14 whether the path followed is *AEI* or *ABFI*. The minimum distance from *A* to *J* is 18.

EXAMPLE 2

Use Dijkstra's algorithm to find the shortest path from *A* to *J*.

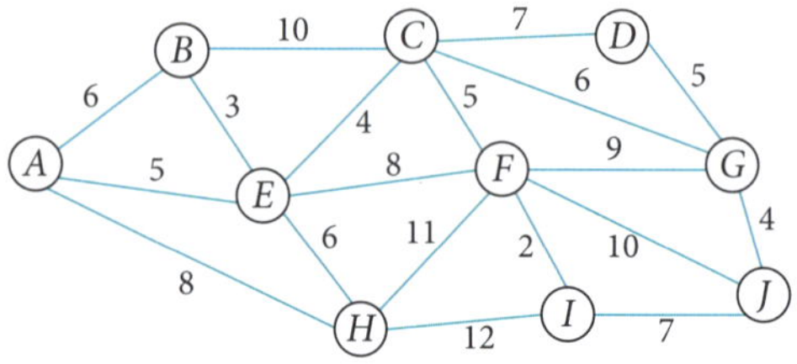

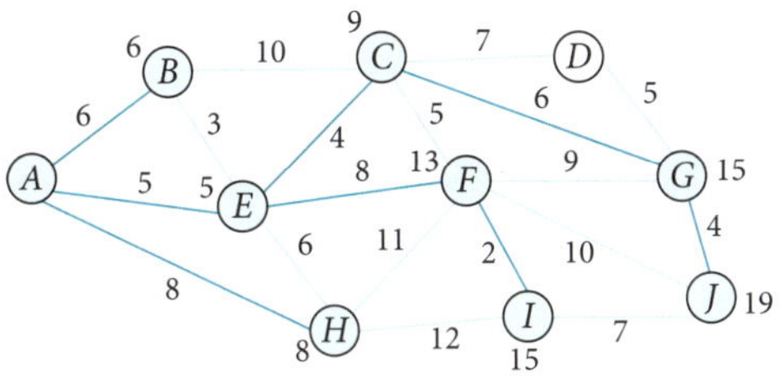

The shortest path is *AECGJ*.

Begin by shading the starting vertex, *A*. *E* is the vertex joined to *A* by the edge of least weight, so shade *E* and the path from *A* to *E* and place the weight beside the vertex. *B* is the vertex joined to either *A* or *E* by the least total weight from *A*, so shade *B* and write the total weight. *C* is the vertex joined to *A*, *B* or *E* by the edge of least total weight from *A*, followed by *F*, *I* or *G*, and finally *J*.

EXAMPLE 3

The diagram shows the times, in minutes, for a train to travel between certain points. Find the shortest possible time for a train to travel from *A* to *E*.

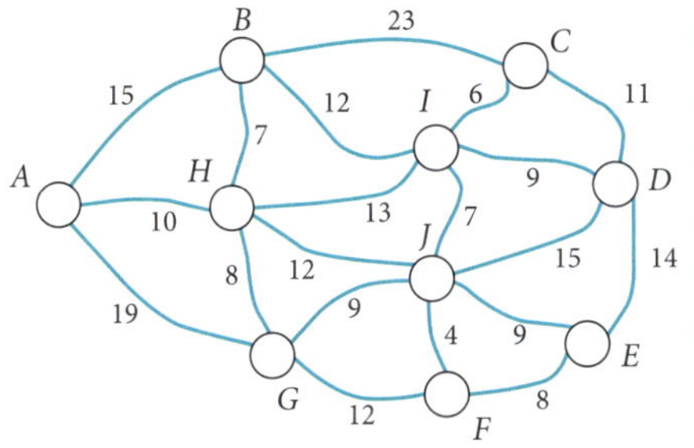

The shortest time is 31 minutes.

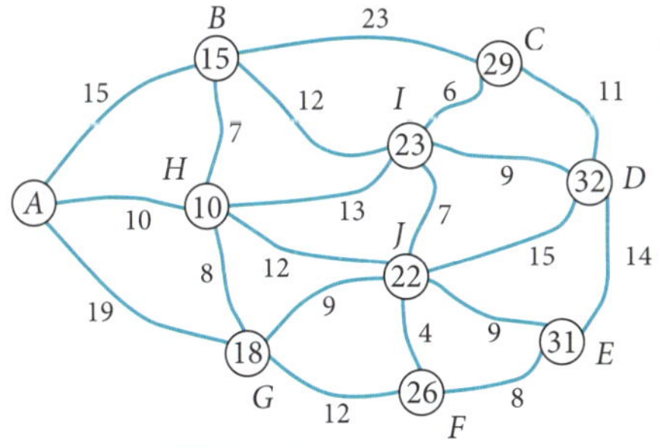

Travelling along the path *AHJE* gives the shortest time.

EXAMPLE 4

Find the shortest path from *A* to *G* in the network represented by this table.

	A	B	C	D	E	F	G
A	–	7	13	12	6	–	–
B	7	–	8	–	–	–	–
C	13	8	–	6	–	–	9
D	12	–	6	–	7	–	11
E	6	–	–	7	–	5	17
F	–	–	–	–	5	–	10
G	–	–	9	11	17	10	–

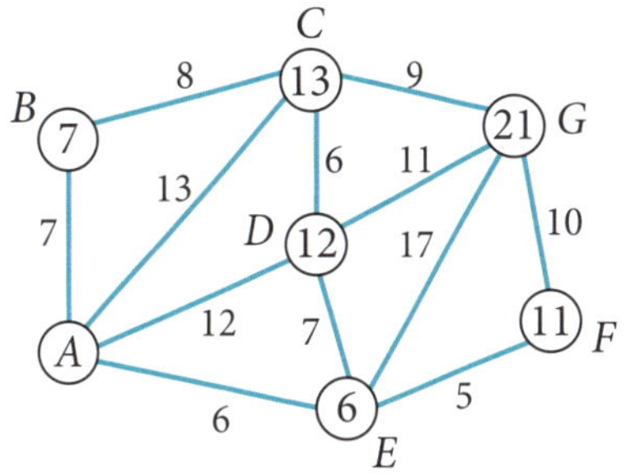

The shortest path is *AEFG*.

The shortest path doesn't have to be unique. There could be several paths all with the same minimum weight.

EXAMPLE 5

The diagram shows the cost, in dollars, to run cables from *A* to *B* along different routes.

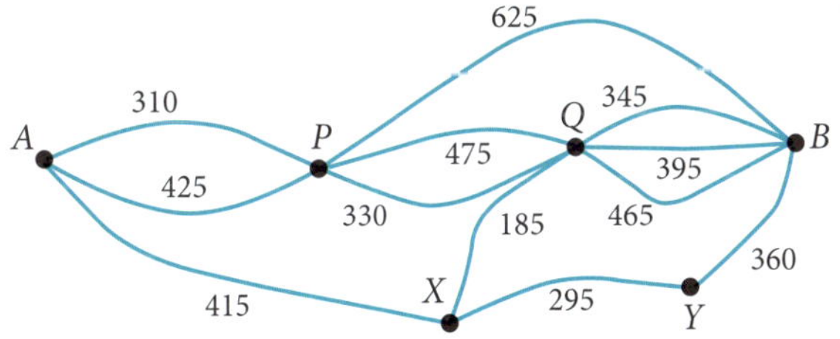

a **Find the cheapest cost.**

b **Suggest three possible reasons why the cheapest might not be the most appropriate route.**

a The cheapest cost is $935.

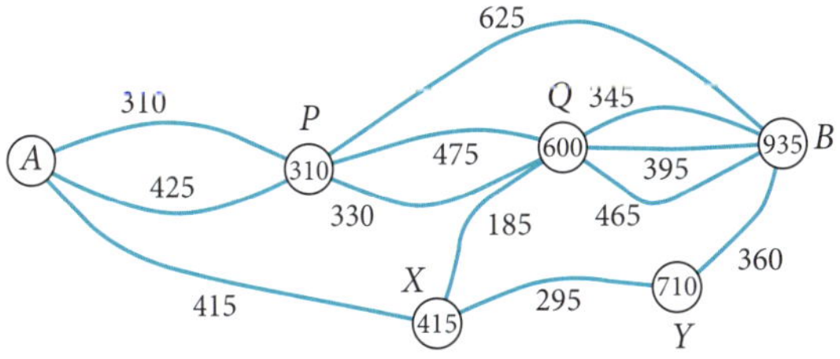

b Possible answers: It might not be the shortest path in distance. It might be cheapest to build but not to maintain. It might be better to run cables through more points in case they are required in the future.

> It is important to remember that different considerations might need to be taken into account when considering the 'best' path.

➔ PRACTICE

1 Find the shortest path from A to I in this network.

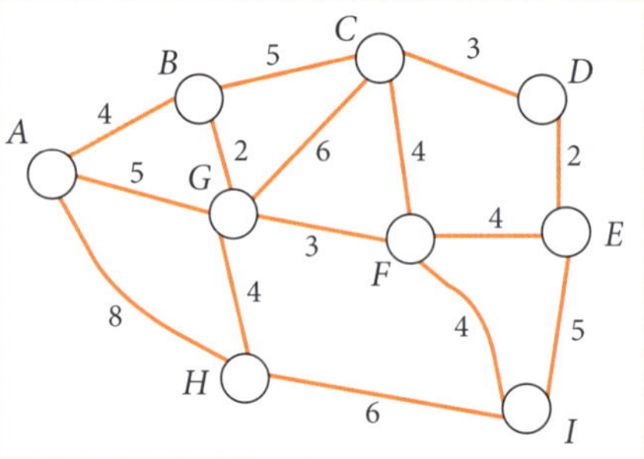

2 Use Dijkstra's algorithm to find the shortest path from A to I.

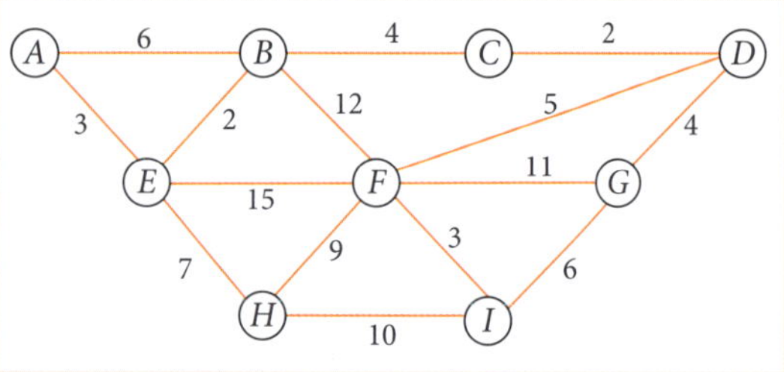

3 The diagram shows the timetabled times, in minutes, for a bus to travel between certain points. Find the shortest possible timetabled time for a bus to travel from A to E.

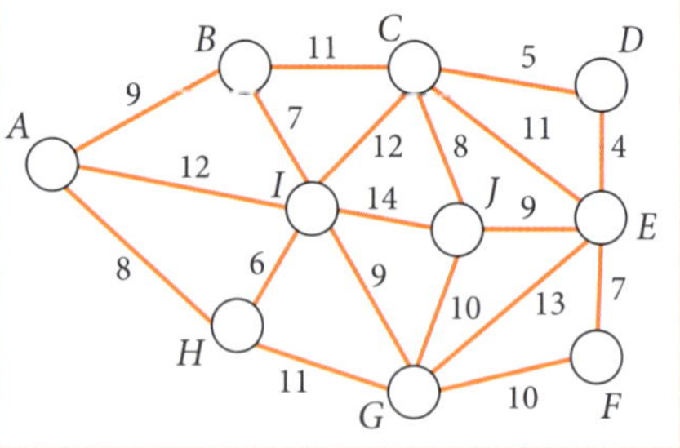

4 Find the shortest path from P to S in the network represented by this table.

	P	Q	R	S	T	U	V
P	–	13	–	–	–	11	9
Q	13	–	6	–	–	10	–
R	–	6	–	5	10	17	–
S	–	–	5	–	15	–	–
T	–	–	10	15	–	14	8
U	11	10	17	–	14	–	12
V	9	–	–	–	8	12	–

5 The diagram shows the cost, in thousands of dollars, to run cables from P to Q along different routes.

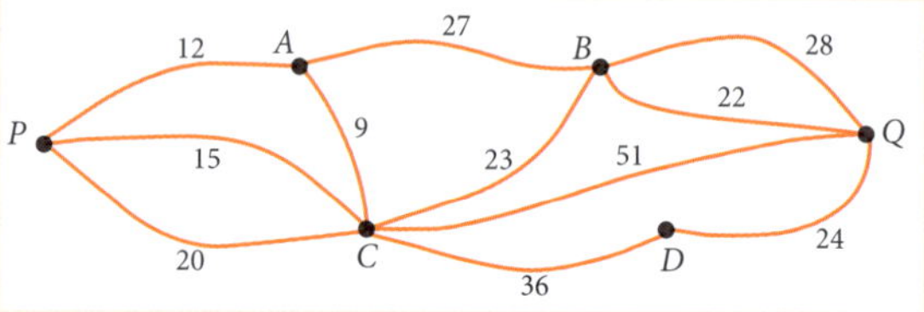

a Find the cheapest cost.

b Suggest three reasons (apart from lower cost) why the cheapest route might also be the most appropriate.

6* Andy drew this map which shows the distances between certain towns and also his average travelling time, in minutes, between the towns.

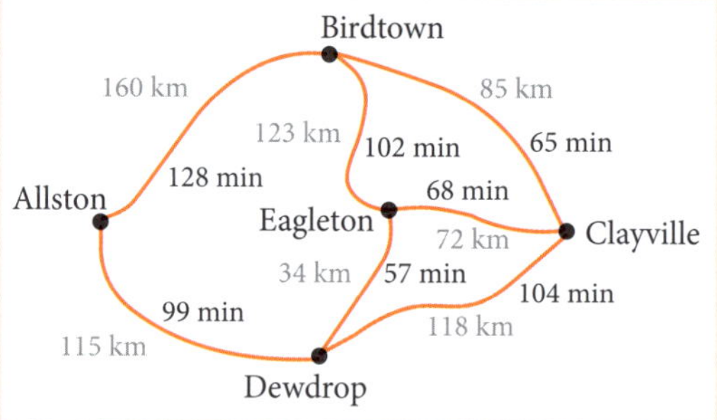

a Use a network diagram to find the shortest distance from Allston to Clayville.

b Use another network diagram to find the shortest average travelling time between Allston and Clayville.

c Which route would you recommend? Discuss.

Answers ➲ pp. 175–176

HSC EXAM-TYPE QUESTIONS

Total marks: 30

PART A

(Suggested time: 12 minutes)

(1 mark each)

1 What is the degree of vertex *B*?

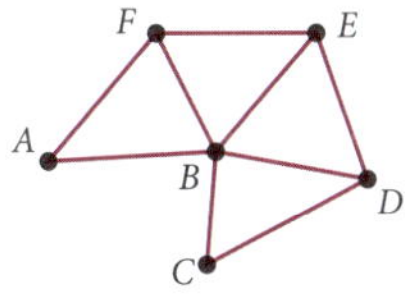

A 2
B 3
C 4
D 5

2 Referring to the network diagram in Question 1, how many edges will a spanning tree have?

A 5 B 6 C 7 D 8

3 Which of these networks is **not** traversable?

A
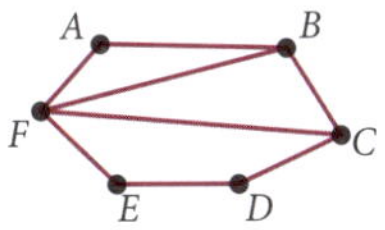

B
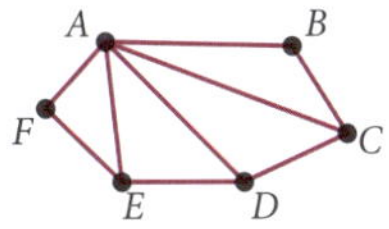

C
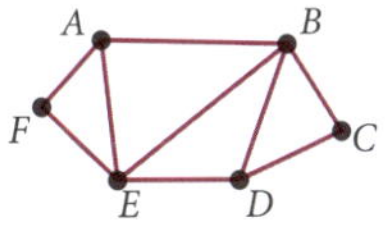

D
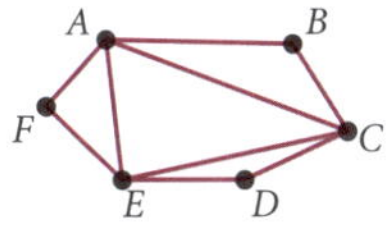

4 In how many different ways is it possible to travel from *A* to *E* without passing through any vertex more than once?

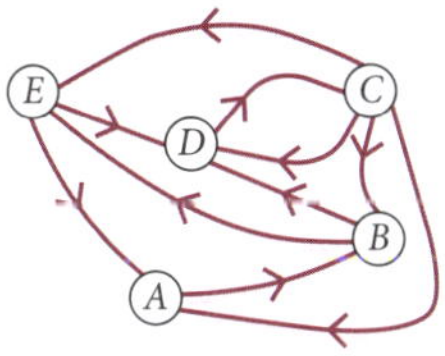

A 1
B 2
C 3
D 4

5 Which is not a walk from vertex *R* to vertex *T*?

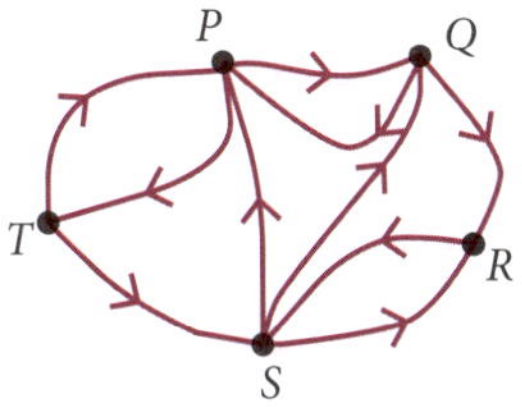

A *RSQPT* B *RSPT* C *RQSPT* D *RSPQPT*

6 Which diagram correctly represents the information in the table?

	P	*Q*	*R*	*S*	*T*
P	–	2	–	6	5
Q	2	–	4	–	–
R	–	4	–	3	1
S	6	–	3	–	8
T	5	–	1	8	–

A
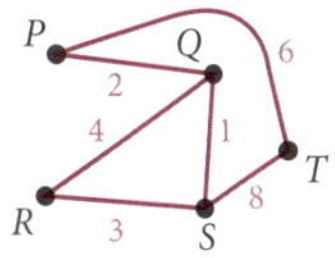

B
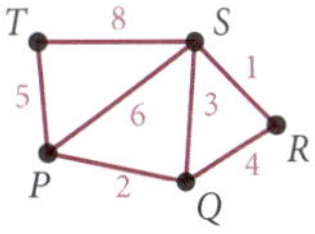

C
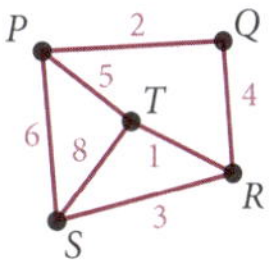

D
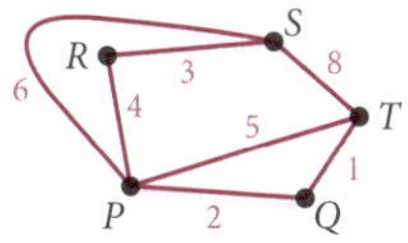

7 Which is a correct network diagram showing the connections between these houses?

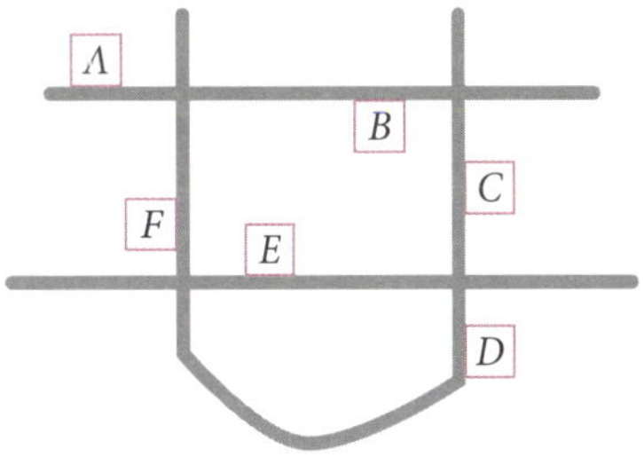

A
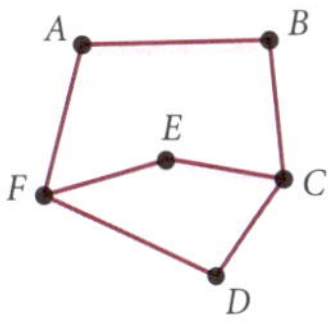

B
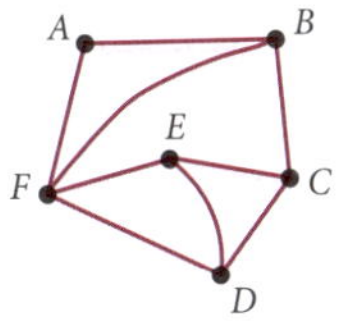

C
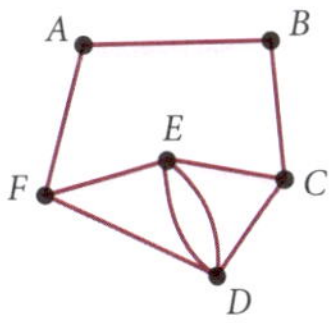

D
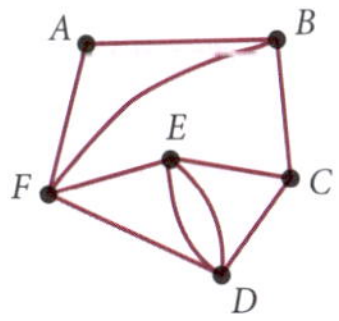

8 This diagram shows the times, in minutes, to run through certain points.

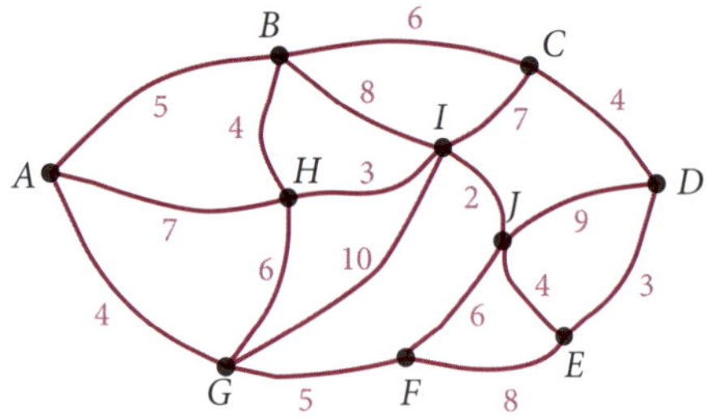

What is the shortest time to run from A to E?

A 16 min **B** 17 min
C 18 min **D** 19 min

PART B

(Suggested time: 33 minutes)

Show all working.

9 Consider this network.

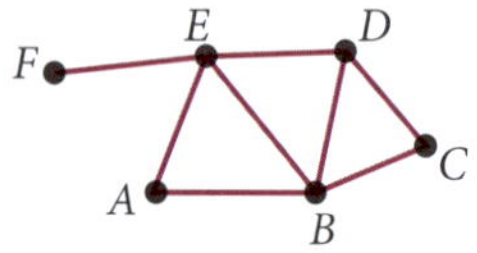

a How many vertices are there? (1 mark)
b How many edges are there? (1 mark)
c How many vertices have degree 3? (1 mark)
d Is this a connected network? Justify your answer. (1 mark)

10 Consider this network.

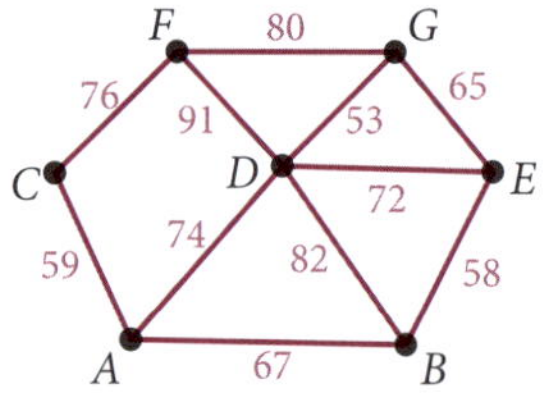

a Using Prim's algorithm, beginning at vertex A, explain why vertex C would be the first other vertex connected. (1 mark)
b Which vertex would be connected after vertex C using Prim's algorithm? (1 mark)
c Complete a spanning tree using Prim's algorithm. List the vertices in the order in which they are added. (2 marks)
d Using Kruskal's algorithm explain why edge DG would be chosen first. (1 mark)
e Which edge would be chosen after DG using Kruskal's algorithm? (1 mark)
f Complete a spanning tree using Kruskal's algorithm, listing the edges in the order in which they are added. (2 marks)

11 The diagram shows the cost to connect different hubs in a building.

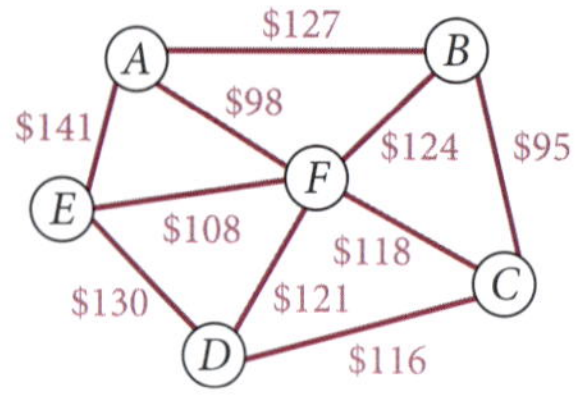

a Show the minimum spanning tree. (2 marks)
b What is the minimum cost of connecting the hubs? (1 mark)

12 The diagram shows distances in kilometres between certain towns.

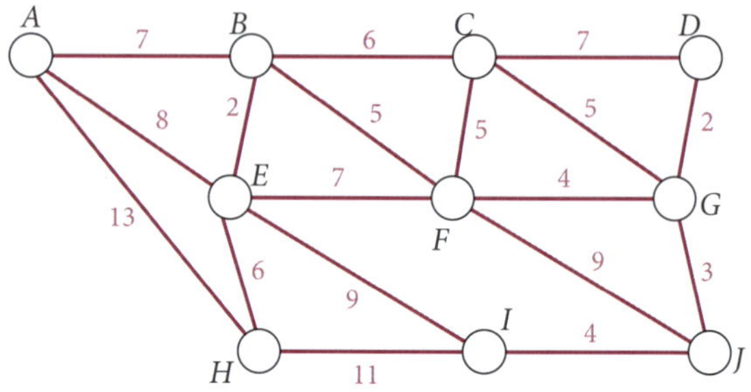

a What is the minimum distance from A to J? (1 mark)
b List the shortest path(s). (1 mark)
c When Ed travelled from A to J he needed to visit two towns on the way. The distance he travelled was 24 km. Which two towns did Ed need to visit? (1 mark)

13 This diagram shows the bus routes between certain suburbs and the cost of the trip.

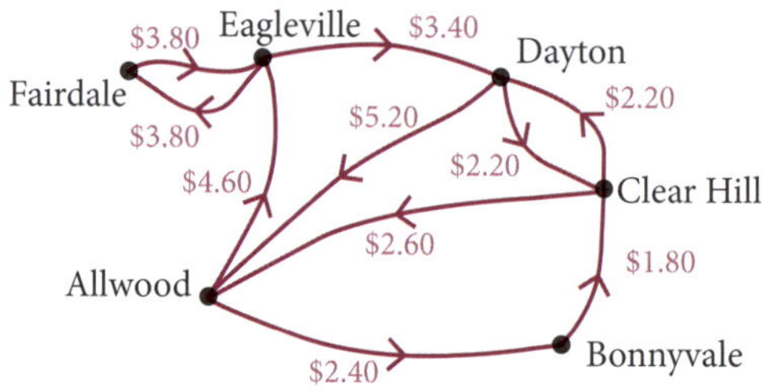

a What is the minimum cost to travel from Clear Hill to Eagleville? (1 mark)
b How much more does it cost to travel from Allwood to Dayton via Eagleville than via Bonnyvale and Clear Hill? (1 mark)
c An inspector needs to travel on all the routes to check them. Describe a trip he could take beginning and finishing at Allwood to travel each route exactly once. (2 marks)

PRACTICE

NETWORKS

1 Network definitions ➲ pp. 158–159

1 a There are 6 vertices.

b There are 8 edges.

c There are 4 different paths from *A* to *D*: *ABD*, *ABCD*, *AFD*, *AFED*.

2

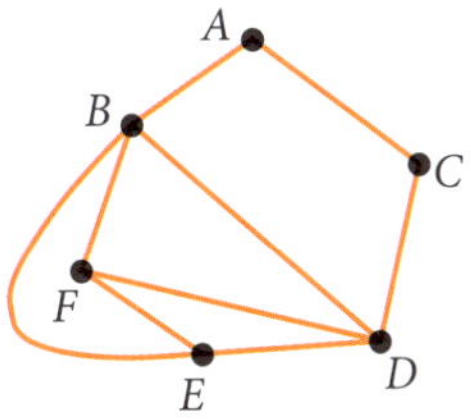

3

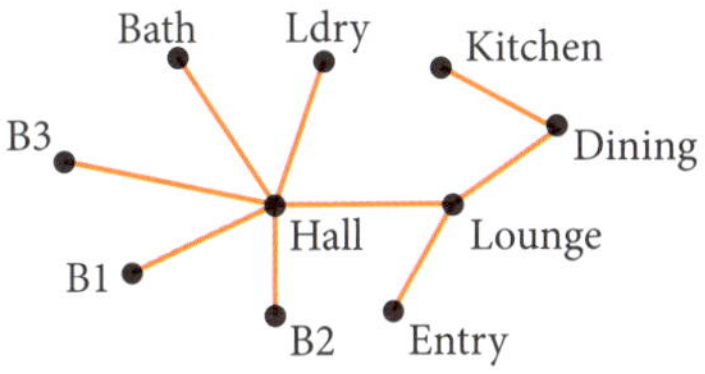

2 Terms and further definitions ➲ pp. 159–160

1 a **Yes**, it is a cycle. From any town (vertex) a walk leads to each of the other towns in succession and returns to the original town.

b **Brisbane and Tenterfield**

2 a **Every room (vertex) is connected to every other room.**

b **Porch, entry, kitchen, hall, bath**

c **Kitchen, dining, terrace, living, kitchen**

Kitchen, living, entry, kitchen

Kitchen, dining, terrace, living, entry, kitchen

3 a

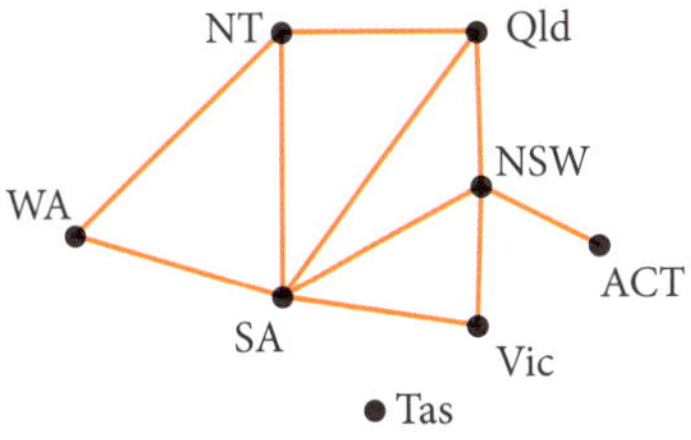

b **No**, it is not a connected network, because Tasmania is not connected to any of the other states.

c **ACT, NSW, Qld, NT or ACT, NSW, SA, NT or several other possible answers**

d **Qld, NT, SA, Qld**

Qld, NSW, SA, Qld

Qld, NT, WA, SA, Vic, NSW, Qld

or several other possible answers

3 Further terms ➲ pp. 161–162

1 a **4**

b **1** (vertex *F*)

c Sum = 4 + 4 + 3 + 4 + 4 + 2 + 5

= **26**

d

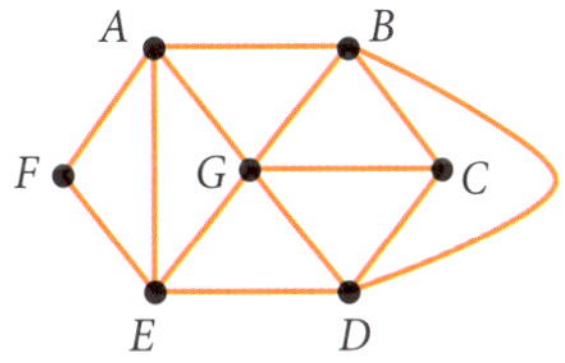

2 a ***A***

b ***B, C, D, A, E***

c **3**; From *A* to *D* then to *B* and *C* or from *A* to *E* then directly to *B* or through *D* to *B* and finally to C.

3 a **199 km** between Grafton and Armidale

b Distance = (97 + 67) km

= **164 km**

c From Tenterfield to Glen Innes to Grafton

= (91 + 159) km

= 250 km

From Tenterfield to Casino to Grafton

= (126 + 100) km

= 226 km

Difference = (250 – 226) km

= 24 km

So it is 24 km further from Tenterfield to Grafton via Glen Innes than via Casino.

4 a

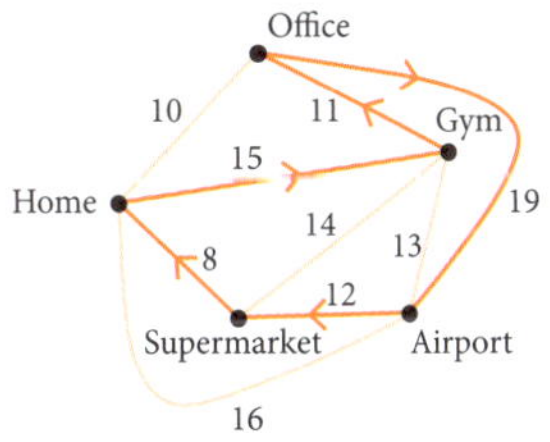

b Time = (15 + 11 + 19 + 12 + 8) min

= **65 min**

c Time = (10 + 11 + 14 + 12 + 16) min

= 63 min

Time difference = 2 min

d **The time difference is not great so Ken might prefer the slightly longer route because the roads are better or the traffic is less or he might wish to visit the various places in the order that he does.**

5 a **From Redhill to Bowville to Tulip**

b Directly from Vineyard to Bowville the time is 35 min.

From Vineyard to Tulip to Redhill to Bowville:

Time = (26 + 29 + 23) min

= 78 min

Time difference = (78 – 35) min

= 43 min

It is 43 minutes quicker to travel directly from Vineyard to Bowville than via Tulip and Redhill.

4 Applications ➲ p. 163

1 **One possibility: from the entrance to the falls, back to the entrance, then to the pool, back to the falls, then the lookout back to the pool, then to the camp. From the camp travel to the lookout and back by different routes and then return directly to the entrance.**

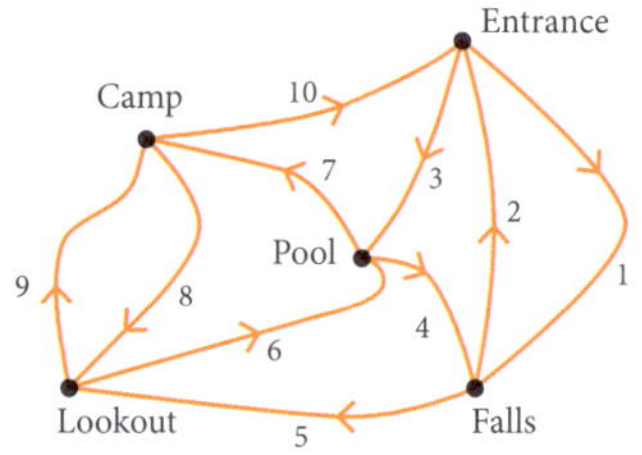

2 **It is possible because only two vertices have degree 3 and the other vertices are even.** The walk would need to begin (or end) in paddock D and end (or begin) in paddock F.

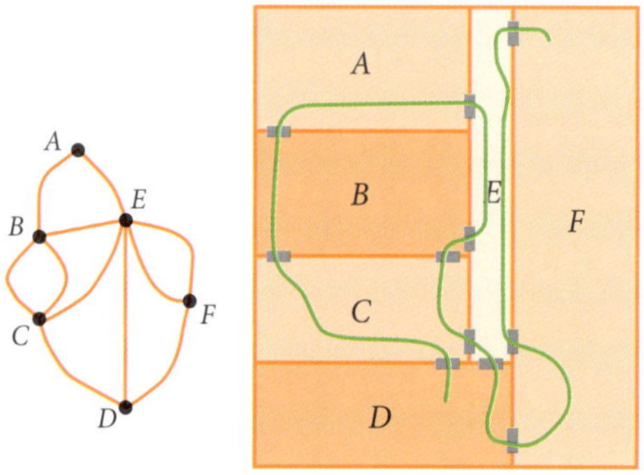

3 **One possible route is from Trent to Uptown, then Mayville, Blake, Everton, Richwood, Freda, Palmerton, Shottle, Oak Hill and return to Trent.**

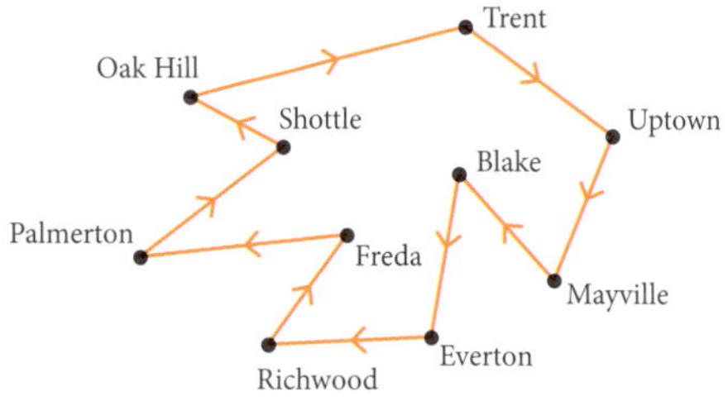

5 Tables and networks ➲ p. 164

1

	A	B	C	D	E
A	–	10	–	–	12
B	10	–	19	17	14
C	–	19	–	8	–
D	–	17	8	–	11
E	12	14	–	11	–

2

From\to	P	Q	R	S	T	U
P	0	29	–	–	32	–
Q	–	0	16	–	–	–
R	37	–	0	24	–	–
S	–	11	18	0	7	–
T	14	–	–	–	0	–
U	9	–	–	–	21	0

3 One possibility:

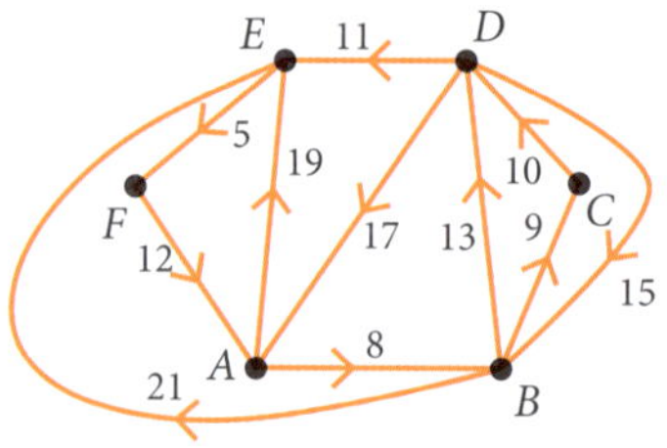

4 a

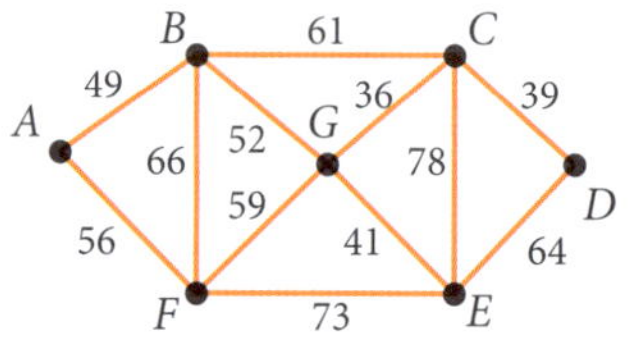

b Time = (56 + 73 + 64 + 39) min
= 232 min
= **3 h 52 min**

SHORTEST PATHS

1 Trees ➲ p. 165

1 a **4 vertices**

b **3 edges**

c

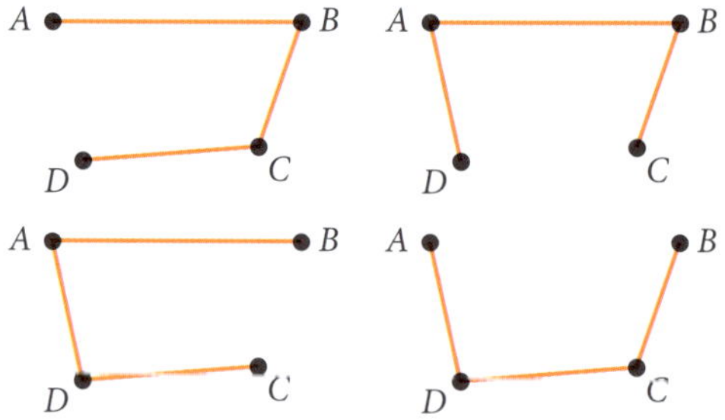

2 a **This network is not a tree because it contains cycles.** There is more than one path connecting many of the vertices.

b **Any spanning tree will have 6 edges.** There are 7 vertices and any spanning tree has one less edge than the number of vertices.

c Some possibilities are:

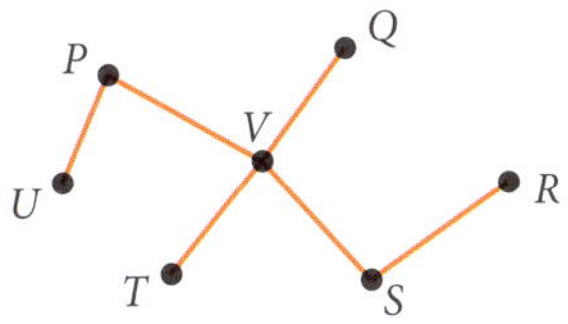

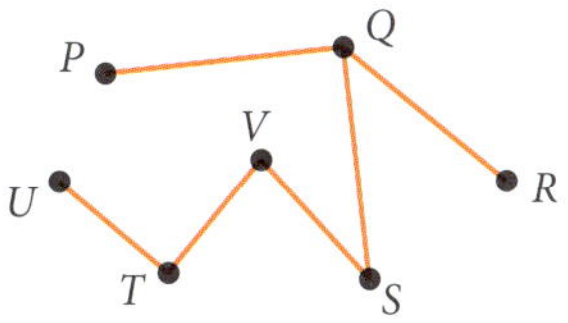

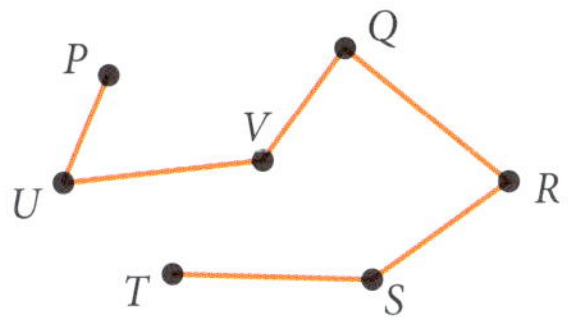

3 One possibility is:

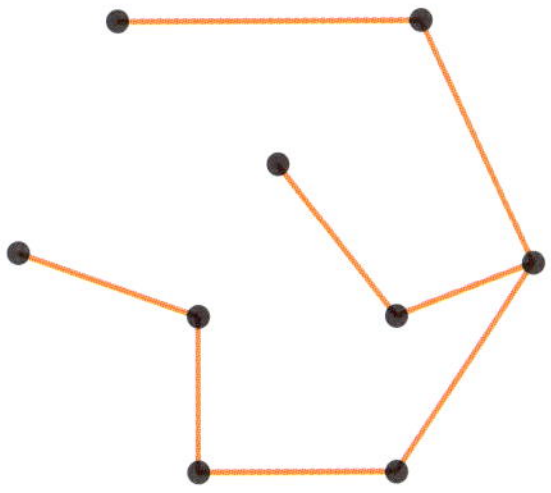

2 Minimum spanning trees ➲ pp. 167–168

1 a **One fewer edge**

b

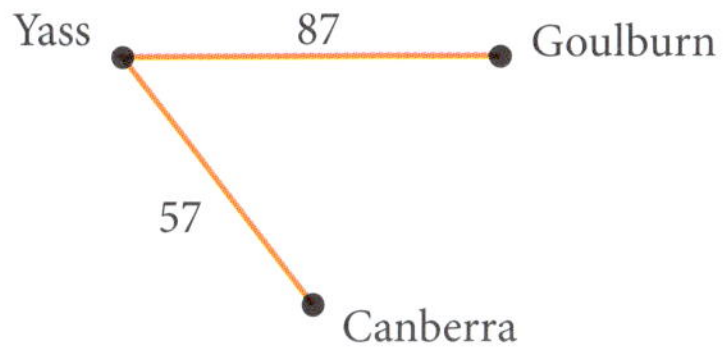

2 (Beginning at P, the vertices are added in this order: T, Q, R and S.)

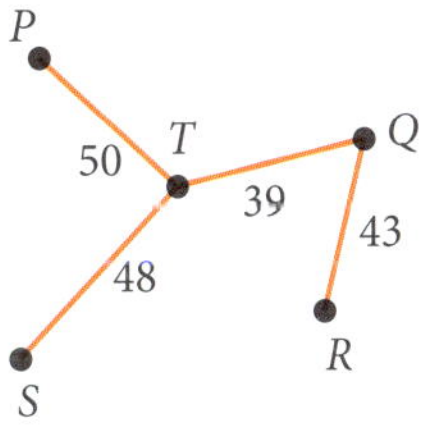

3 (Begin with edge EG, then add AF, AB, CD, AG and BC.)

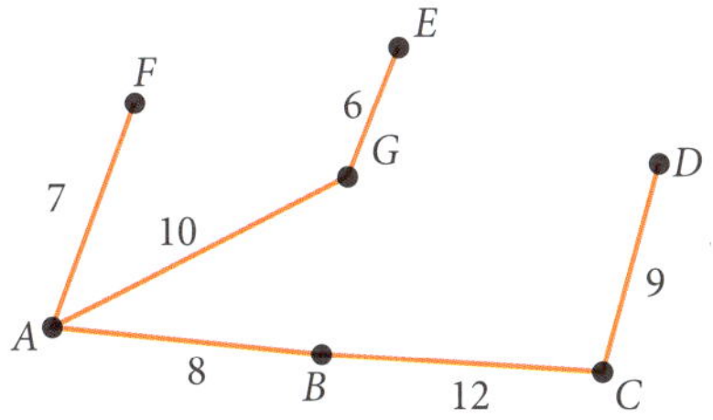

4 a

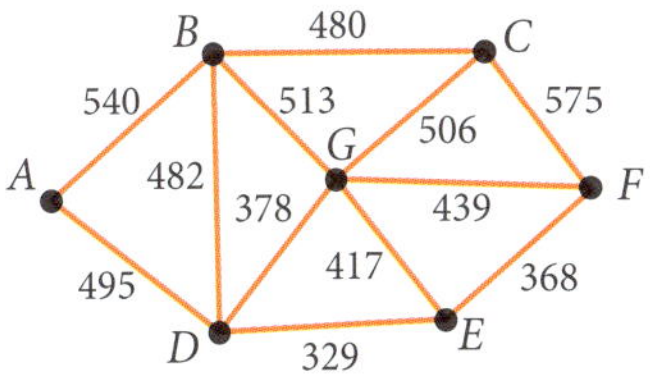

b For Prim's algorithm, vertices would be chosen in this order: *A, D, E, F, G, B, C*.

For Kruskal's algorithm, edges would be chosen in this order: *DE, EF, DG, BC, BD, AD*.

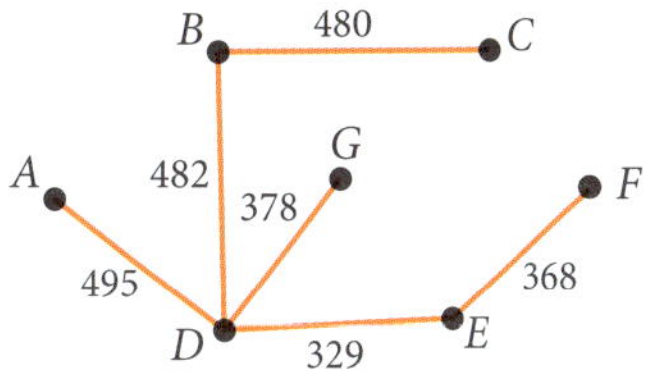

c Minimum distance

= (329 + 368 + 378 + 480 + 482 + 495) m

= 2532 m

= **2.532 km**

5 Shortest length = (37 + 76 + 53 + 68 + 59 + 48 + 109) m

= **450 m**

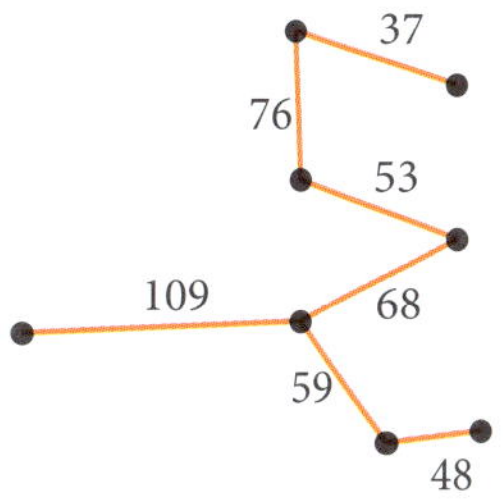

6 a

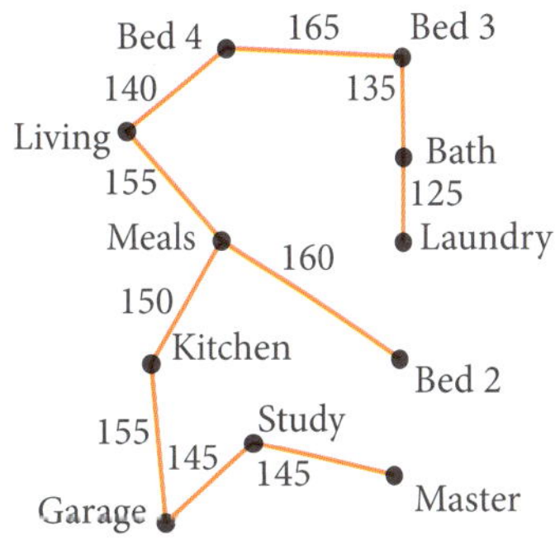

b Least cost

= $(125 + 135 + 165 + 140 + 155 + 160 + 150 + 155 + 145 + 145)

= **$1475**

3 Shortest paths ➲ p. 170

1 **The shortest path from *A* to *I* is *A, G, F, I*.**

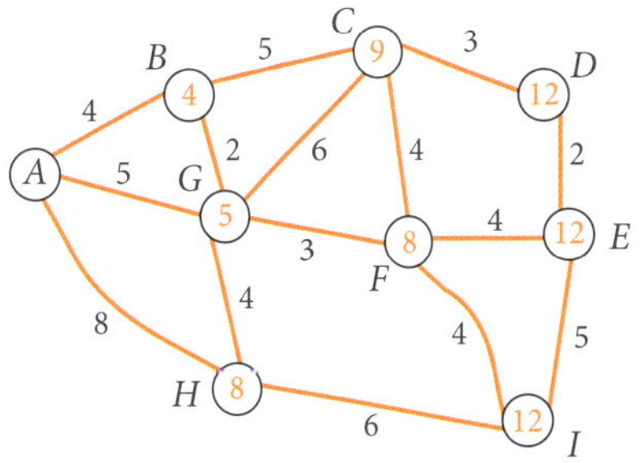

2 **The shortest path is *AEBCDFI*.**

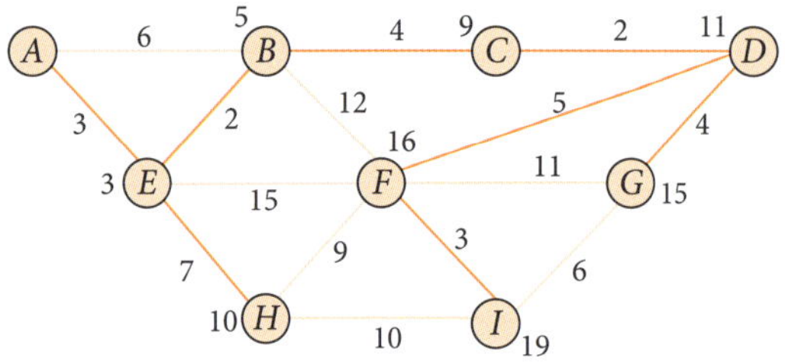

3 **The shortest possible timetabled time for a bus to travel from *A* to *E* is 29 minutes.**

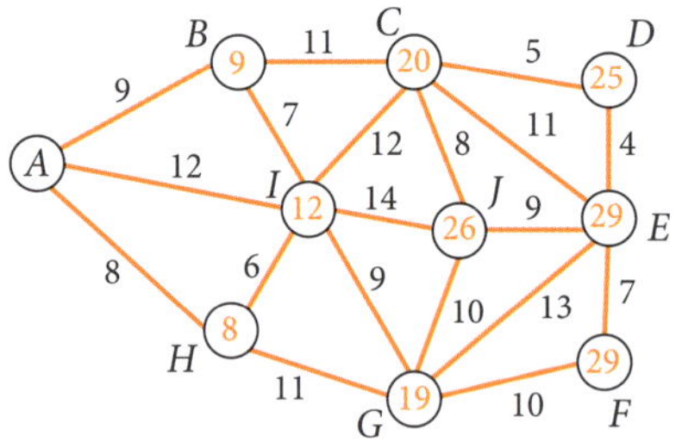

4 **The shortest path from *P* to *S* is *P, Q, R, S*.**

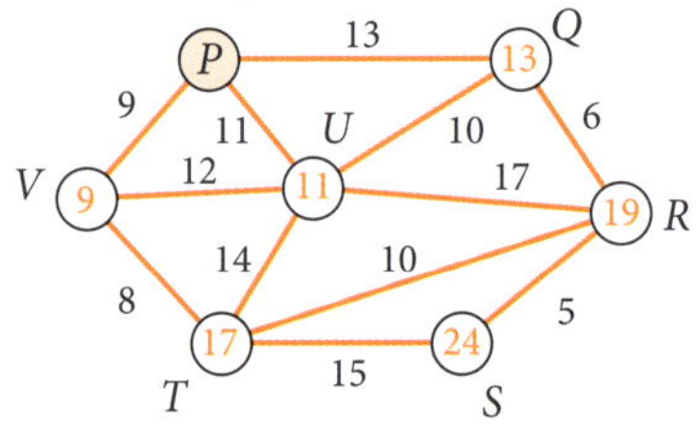

5 **a** **The cheapest cost is $60 000.**

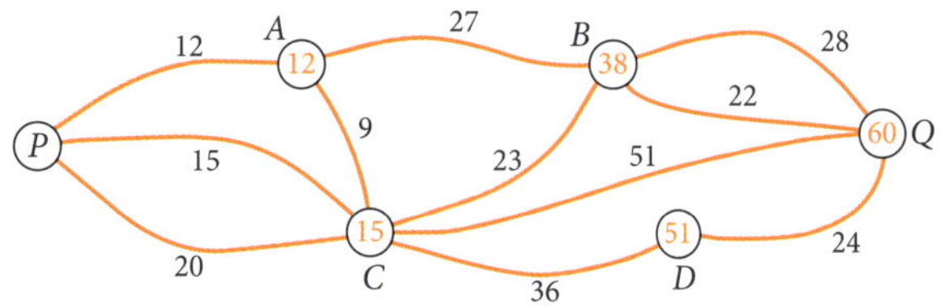

b **Possible answers: The cheapest route might also be the most appropriate because it might be the easiest to access. It might offer the best route taking into account future planning. It might be the most environmentally friendly. It might involve fewer disruptions or issues with property and roads.**

6 **a** **The shortest distance from Allston to Clayville is 221 km.**

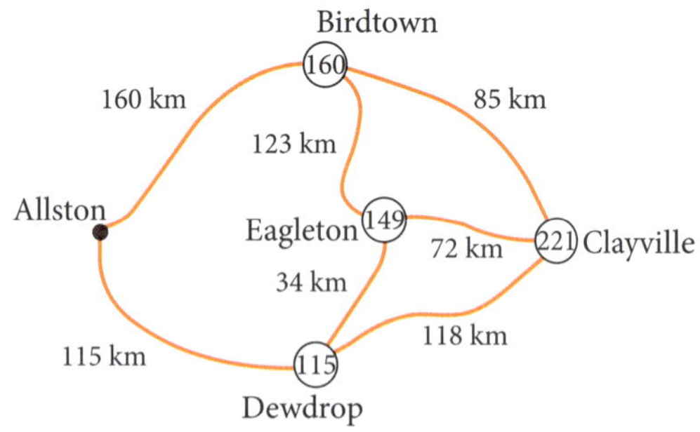

b Shortest travelling time = 193 min

= **3 h 13 min**

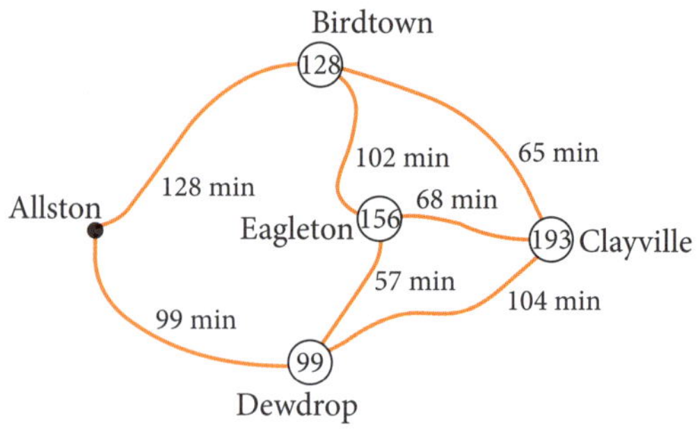

c **Without further information on the condition of the roads, the amount of traffic and Andy's plans it is not possible to make an informed choice for the recommended route.** If Andy's priority is minimising distance travelled he should travel through Dewdrop and Eagleton. If his priority is to minimise the time for the journey he should travel through Birdtown. One recommendation is to travel through Dewdrop and then straight to Clayville as that route is only 12 km further than the shortest distance and only 10 minutes longer than the shortest time.

HSC EXAM-TYPE QUESTIONS

Part A

1 **D.** The degree of vertex *B* is 5.

2 **A.** There are 6 vertices so a spanning tree will have 5 edges.

3 **B.** Network *B* has more than 2 odd vertices so it is not traversable.

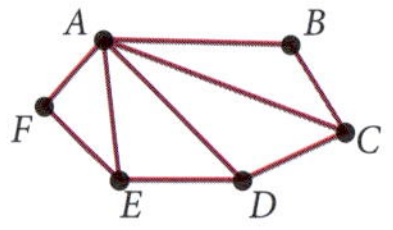

4 **B.** There are 2 ways to travel from *A* to *E* (from *A* to *B* to *E* or from *A* to *B* to *D* to *C* to *E*).

5 **C.** There is not a path from *R* to *Q*. So there cannot be a walk *RQSPT*.

6 **C.** The correct diagram is C.

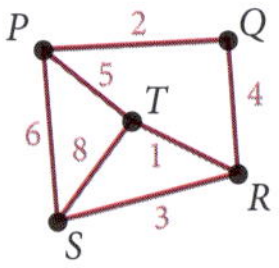

7 **D.** There must be a path from *B* to *F* and 2 paths linking *D* and *E*.

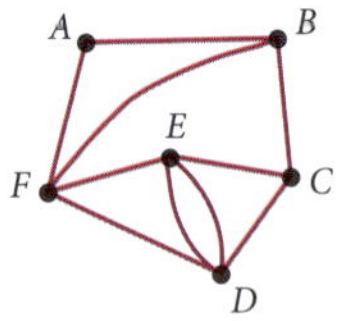

8 **A.** The shortest time from *A* to *E* is 16 min.

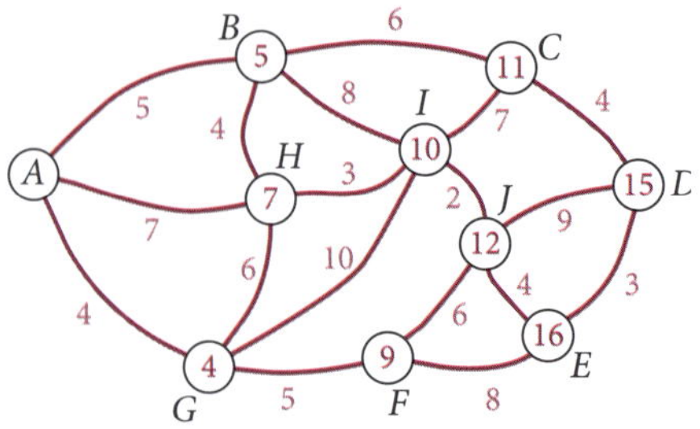

Part B

9 a There are 6 vertices. ✓

b There are 8 edges. ✓

c 1 vertex has degree 3 (vertex D) ✓

d **Yes, it is a connected network.** There is a path between any vertex and all the other vertices. ✓

10 a **Vertex C is the vertex connected to vertex A by the edge of least weight.** ✓

b B ✓

c A, C, B, E, G, D, F ✓

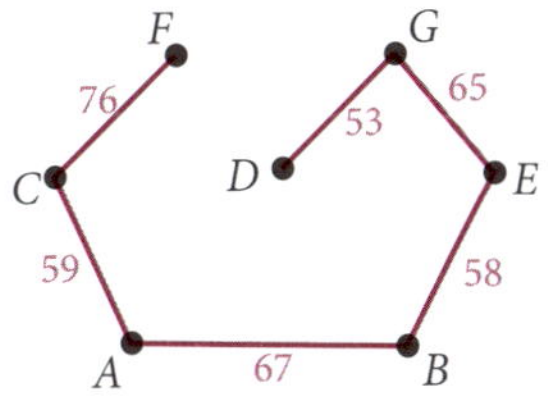

✓

d DG is the edge of least weight. ✓

e Edge BE ✓

f DG, BE, AC, EG, AB, CF ✓

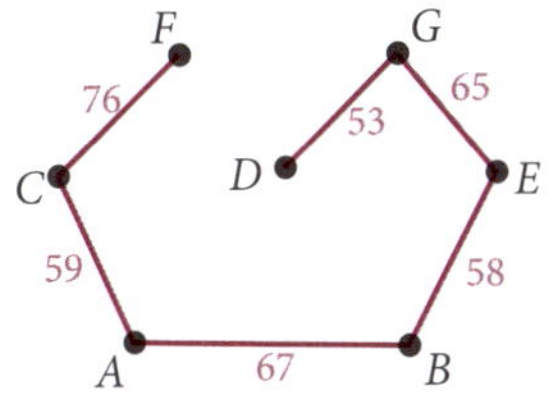

✓

11 a

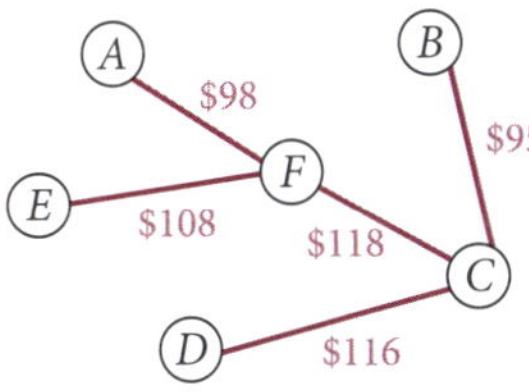

✓✓

b Minimum cost = $(98 + 108 + 95 + 118 + 116)
= **$535** ✓

12 a **The minimum distance from A to J is 19 km** ✓

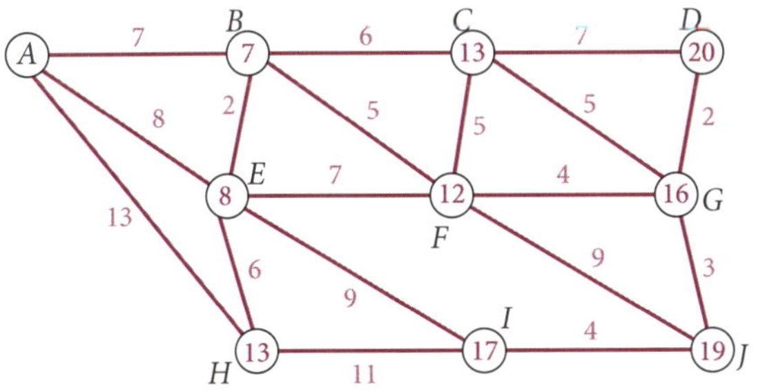

b **The shortest path is A, B, F, G, J.** ✓

c **Ed must have needed to visit E and F. From A to E to F to J is 24 km.** ✓

13 a Cost = $2.60 + $4.60
= **$7.20** ✓

b Allwood to Dayton via Eagleville:

Cost = $4.60 + $3.40
= $8.00

Allwood to Dayton via Bonnyvale and Clear Hill:

Cost = $2.40 + $1.80 + $2.20
= $6.40

Difference = $8.00 − $6.40
= $1.60

It costs $1.60 more to travel from Allwood to Dayton via Eagleville than via Bonnyvale and Clear Hill. ✓

c **One possible route: From Allwood to Eagleville, then to Fairdale and back to Eagleville. From Eagleville to Dayton, back to Allwood and then on to Bonnyvale and Clear Hill. Finally from Clear Hill to Dayton and back to Clear Hill and finish at Allwood.** ✓✓

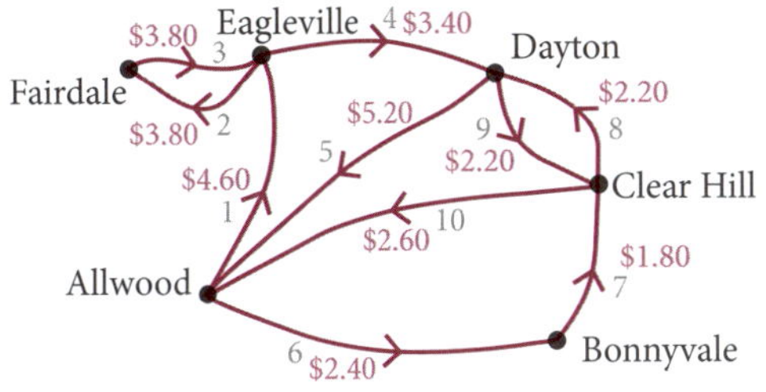

SAMPLE HSC EXAMINATION 1

Try to complete these papers as if they are the real thing. These are the instructions you need to follow in the HSC Exam:

General instructions
- Reading time: 10 minutes
- Working time: 2 hours
- Write using black pen.
- NESA approved calculators may be used.
- A reference sheet is provided at the back of this book.
- For questions in Section II, show relevant mathematical reasoning and/or calculations.

Total marks: 80
Section I: 10 marks
Section II: 70 marks

Section I: 10 Marks

- **Attempt Questions 1–10.**
- **Allow about 15 minutes for this section.**

1 Which of these could be a travel graph?

A 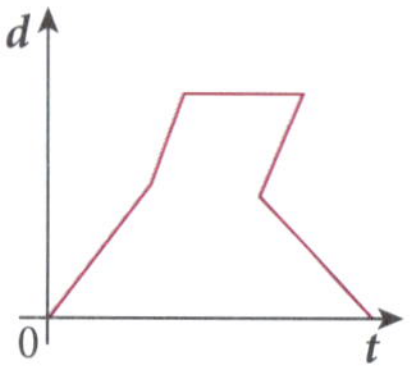

B 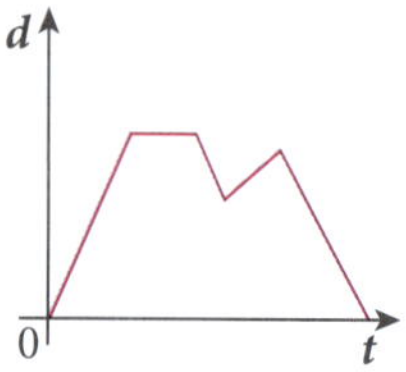

C 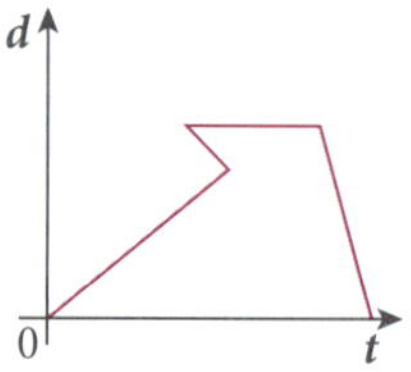

D 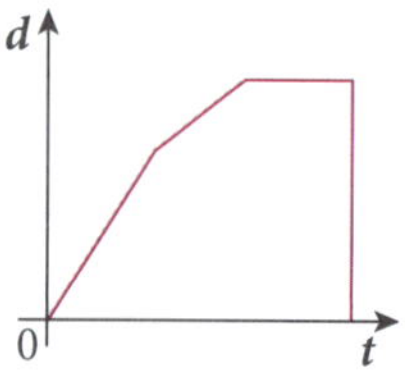

2 What is the ratio of 2 hours to 20 minutes in simplest form?

A 1 : 10 **B** 10 : 1 **C** 1 : 6 **D** 6 : 1

3 Which of these triangles would NOT be right-angled when drawn to scale? (All measurements are in metres.)

A 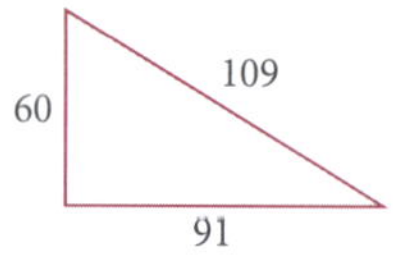

B 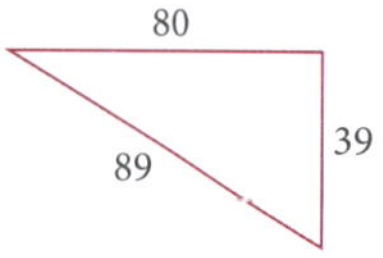

C 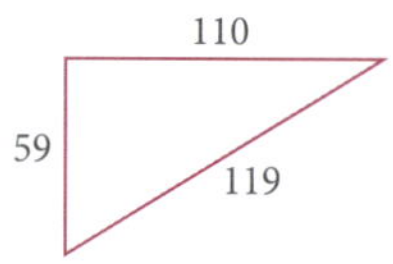

D 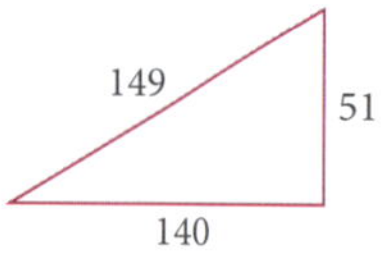

4 What true bearing is the same as a compass bearing of S65°E?

A 065° **B** 115° **C** 155° **D** 245°

5 In this network diagram, how many vertices have degree 3?

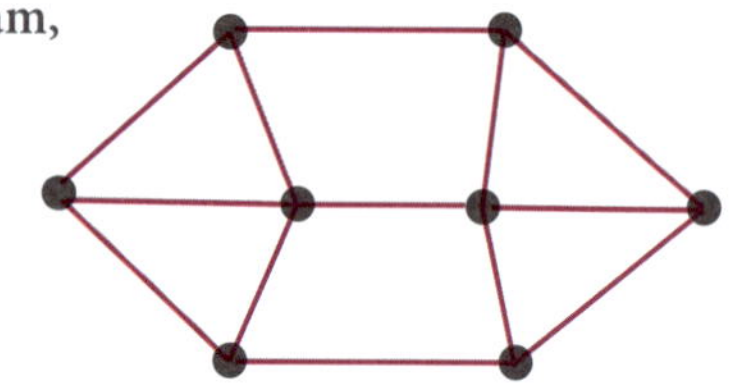

A 4 **B** 5
C 6 **D** 7

6 How much interest is earned, to the nearest dollar, when $5000 is invested for 2 years at 8% p.a. interest, compounded quarterly?

A $858 **B** $5832 **C** $202 **D** $832

7 Which could be the graph of $y = \frac{12}{x}$ $(x > 0)$?

A 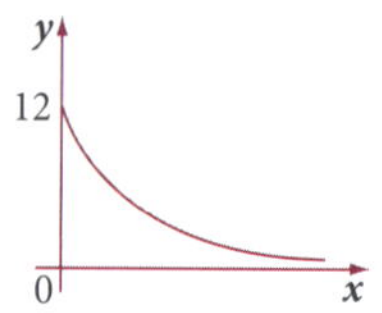

B 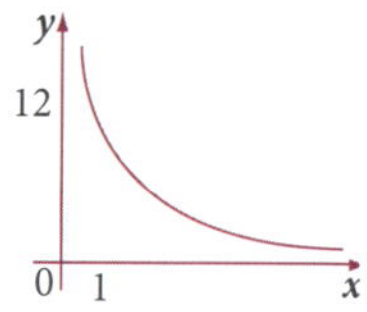

C 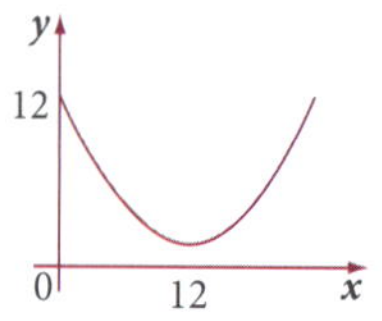

D 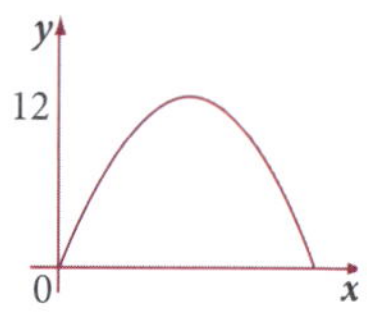

8 Ed's car uses 7.5 L/100 km of petrol. If petrol costs $1.25 per litre, how far could he travel on $60 worth of fuel?

A 360 km **B** 562.5 km **C** 640 km **D** 1000 km

9 Using Kruskal's algorithm to find the minimum spanning tree for this network, which edge would be chosen third?

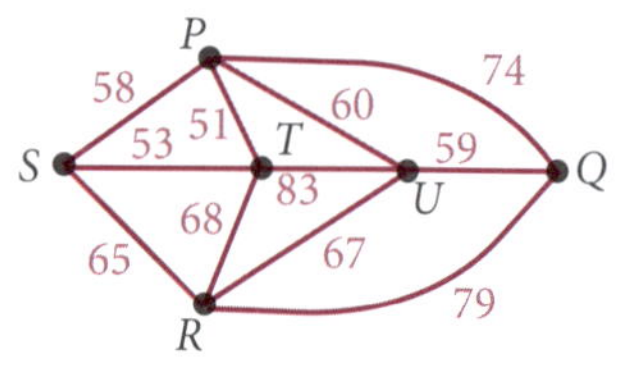

A *SP* **B** *SR*
C *PU* **D** *UQ*

10 Olive's orange juice is sold in four different sizes. Which is the best value?

A 450 mL for $1.10 **B** 600 mL for $1.60
C 1.5 L for $3.75 **D** 2 L for $5.20

Section II: 70 marks

- Attempt Questions 11–27.
- Allow about 1 hour and 45 minutes for this section.

Question 11 (5 marks)

A financial institution published the following table of home loan repayments per $1000 borrowed at various interest rates over various periods.

Monthly repayments per $1000 borrowed			
Period (years)	Interest rate		
	6%	7.5%	9%
7	$14.61	$15.34	$16.09
12	$9.76	$10.55	$11.38
15	$8.44	$9.27	$10.14
25	$6.44	$7.39	$8.39

a How much would the monthly repayments be on a $70 000 loan over 12 years at 6% p.a. interest? (1 mark)

b How much interest would be paid on a $110 000 loan over 25 years at 9% p.a. interest? (2 marks)

c Brock pays $613.60 per month on his loan. If the loan is over 7 years at 7.5% p.a., how much did Brock borrow? (1 mark)

d Rebel was originally offered a loan of $87 000 over 15 years at 7.5% p.a. Before she accepted the loan, interest rates rose. She finally took the loan over 15 years at 9% p.a. interest. How much extra were Rebel's repayments? (1 mark)

Question 12 (4 marks)

Yvonne makes pies to sell at a stall each weekend. The graph shows the cost and income for different numbers of pies.

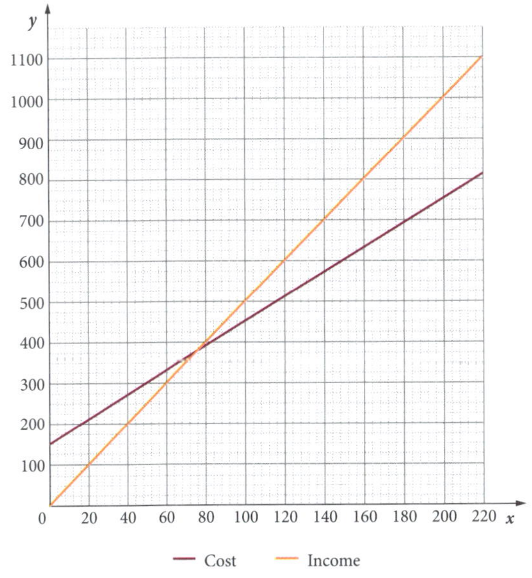

a How many pies must Yvonne sell to break even? (1 mark)

b How much profit will Yvonne make on a weekend when she sells 200 pies? (1 mark)

c What is the y-intercept of the cost line? What does that represent? (1 mark)

d What is the gradient of the income line? What does that represent? (1 mark)

Question 13 (3 marks)

The scatterplot shows the number of car accidents over a certain period of time in particular regions and the number of bottle shops in those regions.

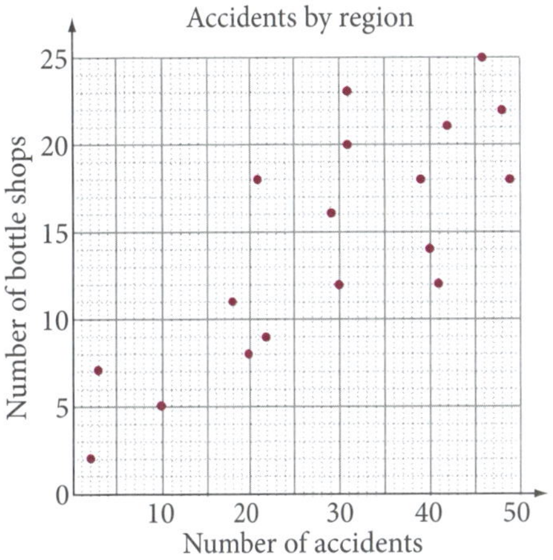

a How many accidents occurred in the region that has 20 bottle shops? (1 mark)

b Gary is a community leader in one of the regions. After studying the data he proposed that bottle shops should be banned or restricted to reduce the number of accidents. Explain why Gary is wrong. (2 marks)

Question 14 (2 marks)

A recipe for shortbread biscuits uses flour and icing sugar in the ratio 8 : 1. If 2 cups of flour makes 30 biscuits, find the amount of icing sugar needed to make 75 biscuits. (2 marks)

Question 15 (3 marks)

A caravan purchased for $54 000 in January 2016 is depreciated using the declining balance method at 15% per year.

a How much will the caravan lose in value in the first year? (1 mark)

b What will be its approximate value in January 2025? (1 mark)

c Use your calculator and a trial-and-error method to determine after how many years the value of the caravan will first fall below $5000. (1 mark)

Question 16 (3 marks)

These are the ingredients in a recipe for a meatloaf that will serve 6 people.

1 kg mince	2 eggs
2 small onions	1 teaspoon oregano
$\frac{1}{2}$ cup tomato sauce	$1\frac{1}{2}$ teaspoons salt
$\frac{1}{2}$ teaspoon pepper	

a What is the ratio, in simplest form, of oregano to salt? (1 mark)

b How many eggs would be used to make enough meatloaf for 15 people? (1 mark)

c How much sauce should be used if $4\frac{1}{2}$ kg of mince is used? (1 mark)

Question 17 (8 marks)

This diagram shows the costs, in dollars, to connect different points in a complex with cables.

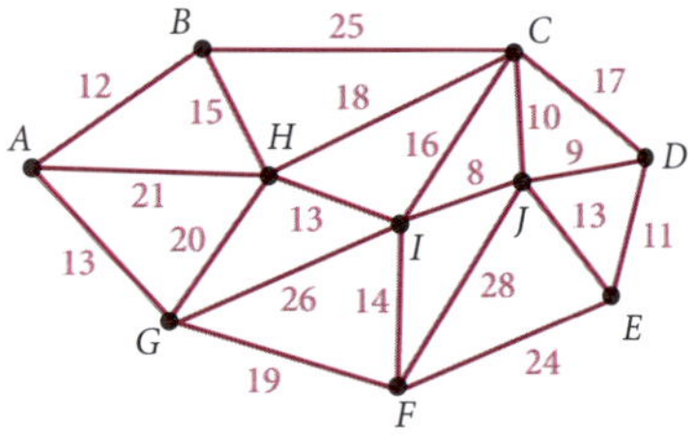

a Find the minimum spanning tree for the network. (2 marks)

b What is the minimum cost to connect all the points? (1 mark)

c What is the shortest path from A to E, and what is its length? (3 marks)

d Is the length of the shortest path the same as the minimum cost? Explain why or why not. (2 marks)

Question 18 (2 marks)

Tim made a study of the arm length (from elbow to wrist) and leg length (from knee to ankle) of some students.

a Tim found the equation of the least squares line of best fit to be $y = 1.1x + 16.4$. Using this equation, what would be the leg length of someone whose arm length is 20 cm? (1 mark)

b Tim's leg length was 56 cm. What would you predict his arm length to be? (1 mark)

Question 19 (5 marks)

The population, in millions, of a country is given by the equation $P = 9(1.04)^t$ where t is the time in years since 2010. The graph of $P = 9(1.04)^t$ is shown below.

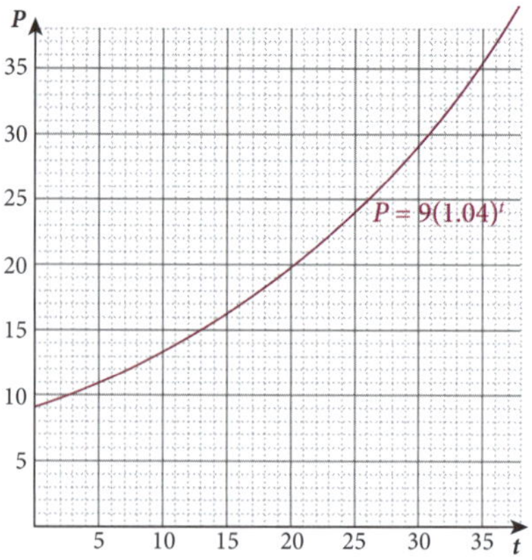

a What type of function is this an example of? (1 mark)

b What was the population of the country in 2010? (1 mark)

c In about what year will the population reach 30 million? (1 mark)

d The population will double in about how many years? Will this always be the case? Justify your answer. (2 marks)

Question 20 (4 marks)

The maximum heart rate is 220 – age. Jeremy is 37 years old and his maximum heart rate is three times his resting heart rate.

a How many times will Jeremy's heart beat in a day at his resting rate? (2 marks)

b Jeremy's target heart rate when exercising is between 65% and 75% of his maximum heart rate. What is Jeremy's target heart rate range? (2 marks)

Question 21 (4 marks)

a Complete the table of values for $T = \frac{100}{S}$.

S	1	2	4	5	10	20	25	50	100
T									

(1 mark)

b Sketch the graph of $T = \frac{100}{S}$ $(S > 0)$. (2 marks) c The rule $T = \dfrac{100}{S}$ can be used to give the time (T) in seconds to travel 100 m at a given speed (S) in metres per second.

Use the graph to find the approximate time to travel 100 m at 15 m/s. (1 mark)

Question 22 (5 marks)

Paul wanted to investigate whether there were differences between the performances of two classes X and Y. He made this stem-and-leaf plot from scores in a test.

Class scores

X		Y
9 3	4	6
7 7 2 0	5	1 4 8
9 8 5 4 1	6	0 3 5 6
8 6 5 3 2	7	1 2 4 7 8
9 6 5 3	8	2 4 4 7 9 9
4 1	9	0 3 5

a At what stage of the statistical investigation process would Paul prepare this? (1 mark)

b What is the range of scores in class X? (1 mark)

c What is the median score in class Y? (1 mark)

d Paul concluded that there was a difference between the performance of the two classes. Do you agree? Justify your answer. (2 marks)

Question 23 (4 marks)

a When the angle of elevation of the sun is 41°, a tree casts a shadow of length 20.7 m. How tall is the tree? Give the answer to the nearest metre. (2 marks)

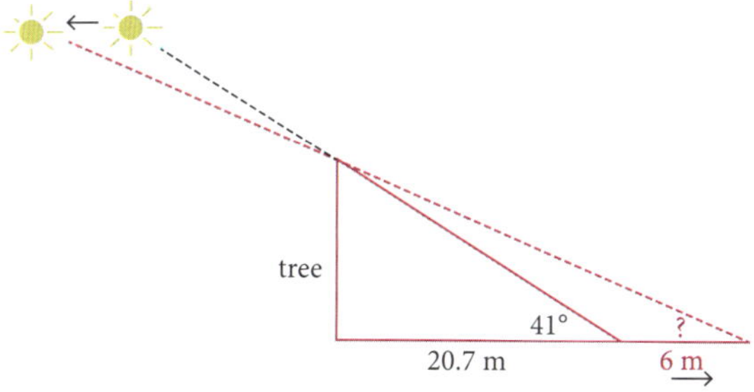

b When the shadow is 6 m longer, what is the angle of elevation then? (2 marks)

Question 24 (3 marks)

A house was valued at \$620 000. The value increased at a rate of 12% p.a. over 3 years.

a What was the value of the house at the end of the 3 years? (1 mark)

b At the same time another house increased in value from \$410 000 to \$560 000. At what rate did that house increase? (2 marks)

Question 25 (2 marks)

A 1.5 m high post casts a shadow 3.8 m long on level ground. At the same time a pole casts a shadow 17.7 m long.

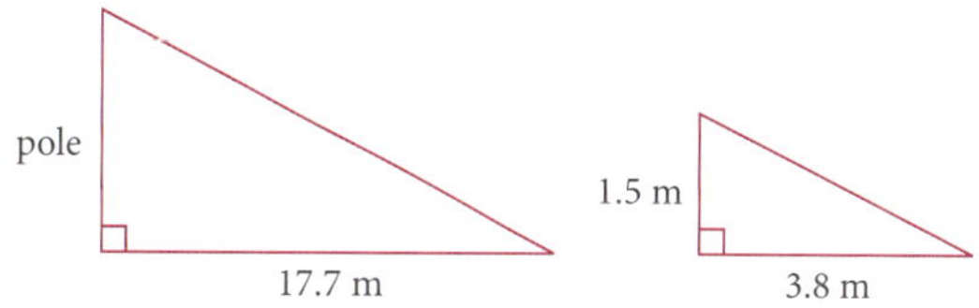

How high is the pole? (2 marks)

Question 26 (8 marks)

This is a scale drawing of a courtyard.

a By measurement and calculation, what is the exact length of the southern side? (1 mark)

b Sketch a plan (not to scale) of the courtyard, showing all the actual measurements. (2 marks)

c The courtyard is to be paved with square pavers of side edge 600 mm. Find the total cost to pave the courtyard if the pavers cost \$24.75 each. (2 marks)

d If, instead, rectangular pavers 600 mm by 400 mm were used, which way should they be laid? (1 mark)

e What is the difference in price to pave the courtyard if the rectangular pavers cost \$16.50 each? (2 marks)

Question 27 (5 marks)

Tilly's credit card has no interest-free period and the compound interest rate is 19.71% p.a. charged daily from, and including, the date of purchase until, and including, the date of repayment. Find the total amount that Tilly must repay on 10 June to clear the debt if there was no outstanding balance and she uses the card to buy groceries for \$324 on 17 May and pays \$586 for a car service on 28 May. (5 marks)

SAMPLE HSC EXAMINATION 2

Try to complete these papers as if they are the real thing. These are the instructions you need to follow in the HSC Exam:

General instructions

- Reading time: 10 minutes
- Working time: 2 hours
- Write using black pen.
- NESA approved calculators may be used.
- A reference sheet is provided at the back of this book.
- For questions in Section II, show relevant mathematical reasoning and/or calculations.

Total marks: 80

Section I: 10 marks

Section II: 70 marks

Section I: 10 Marks

- **Attempt Questions 1–10.**
- **Allow about 15 minutes for this section.**

1 Which expression will give the value of x?

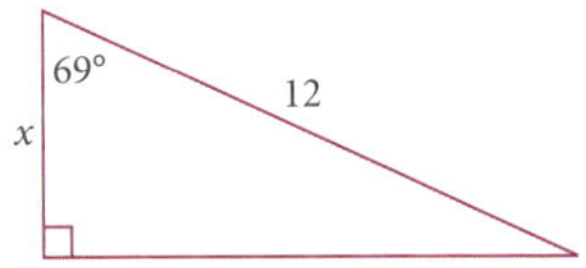

A 12 cos 69° **B** 12 sin 69°

C $\dfrac{12}{\sin 69°}$ **D** $\dfrac{12}{\cos 69°}$

2 How much compound interest, to the nearest dollar, is earned when $30 000 is invested at 9% p.a. interest, compounded monthly, for 7 years?

A $18 900 **B** $33 677

C $24 841 **D** $26 196

3 What is the equation of the line?

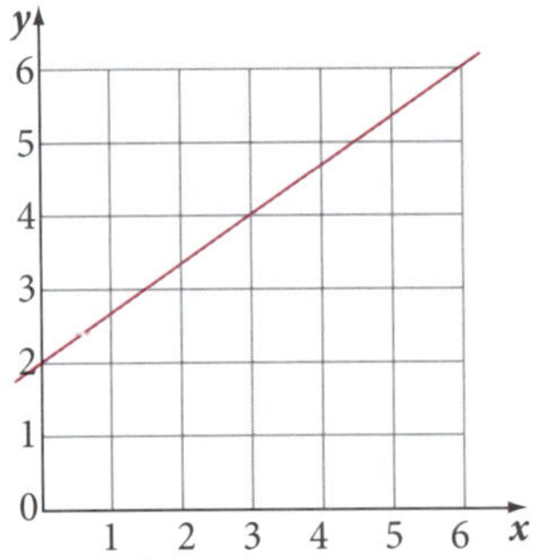

A $y=\dfrac{3}{2}x+2$ **B** $y=2x+\dfrac{3}{2}$

C $y=\dfrac{2}{3}x+2$ **D** $y=2x+\dfrac{2}{3}$

4 On a scale drawing 7.5 cm represents a real distance of 150 km. What is the scale?

A 1 : 20 000 000 **B** 1 : 2 000 000

C 1 : 200 000 **D** 1 : 20 000

5 Guy runs around a semicircular track. Which graph best represents the distance Guy has covered over time?

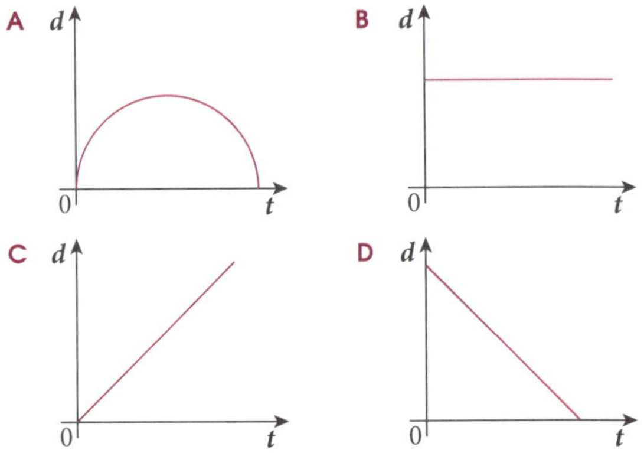

6 Prim's algorithm, beginning with vertex P, will be used to find the minimum spanning tree for this network. Which vertex will be added last?

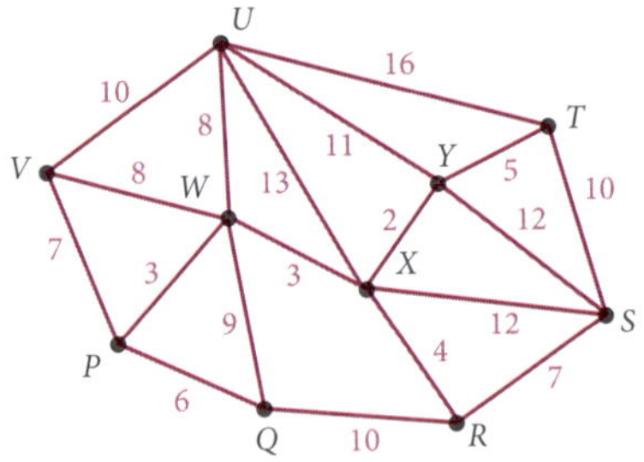

A Y **B** U **C** S **D** T

7 Which is the best fuel consumption?

A 586 km on 41 L **B** 654 km on 44 L

C 735 km on 53 L **D** 812 km on 58 L

8 What is the length, in metres correct to one decimal place, of side QR?

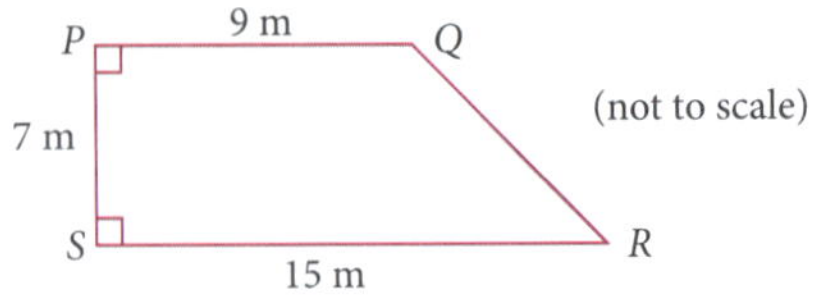

A 9.2 **B** 10.8 **C** 11.4 **D** 16.6

9 Joanne borrows \$16 000 and repays it with 60 monthly repayments of \$332. What annual flat rate of interest is charged?

A 4.9% B 5.4% C 24.5% D 24.9%

10 Mary ran 400 m in 54 seconds. What was her average speed in km/h (to the nearest km/h)?

A 6 B 7 C 26 D 27

Section II: 70 marks

- Attempt Questions 11–27.
- Allow about 1 hour and 45 minutes for this section.

Question 11 (2 marks)

Ava borrows \$12 000 over 3 years. The flat rate of interest charged on the loan is 7% p.a.

a Find the total amount of interest Ava will pay. (1 mark)

b If Ava repays the loan with three equal annual instalments, find the amount of each instalment. (1 mark)

Question 12 (8 marks)

Two different tradespeople charge different rates for their work. Each price is made up of a fixed charge and an hourly rate. The graph shows the cost for getting work done by each of the tradespeople for different hours.

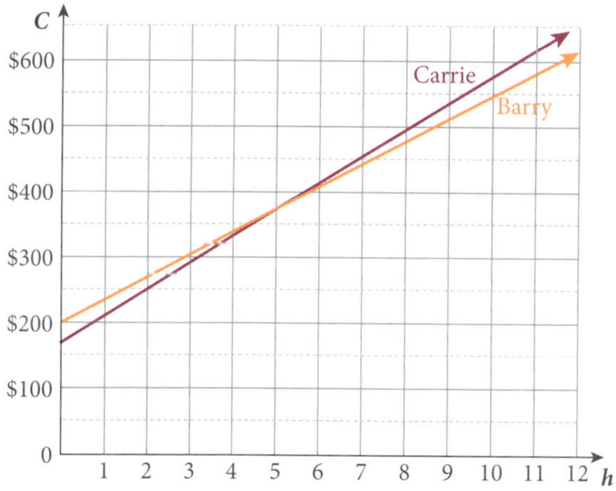

a How much does Barry charge as a fixed rate? (1 mark)

b How much will it cost for Carrie to work for 10 hours? (1 mark)

c At what time are the two costs the same? (1 mark)

d If you wanted to spend about \$500, who would be the most economical person to hire? (1 mark)

e What is the gradient of Barry's line and what does it represent? (2 marks)

f Write an equation for the amount, $\$C$, that Carrie charges per hour, h. (2 marks)

Question 13 (2 marks)

The diagram shows the cost to lay pipes to certain positions in a garden. Find the minimum cost to connect all the positions.

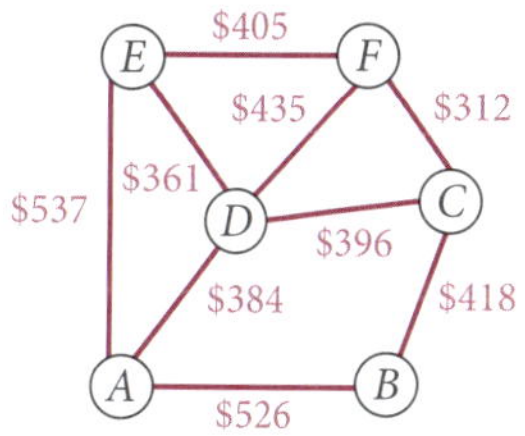

(2 marks)

Question 14 (1 mark)

An audience at the recording of a television show were surveyed. They listened to two songs and then were asked: 'Which of these two songs would you rank higher in a list of great songs?' Give a reason why this is not a good question. (1 mark)

Question 15 (2 marks)

A fruit drink is made by mixing ginger ale, lemonade and orange juice in the ratio 5 : 3 : 2.

a How much ginger ale should be used if 1 L of orange juice is used? (1 mark)

b How much lemonade is needed to make 7.5 L of the drink? (1 mark)

Question 16 (7 marks)

Corey collected some data on the value of vehicles at different ages.

a Corey produced a scatterplot from his data and drew a line of fit on it. How many vehicles did Corey have data for? (1 mark)

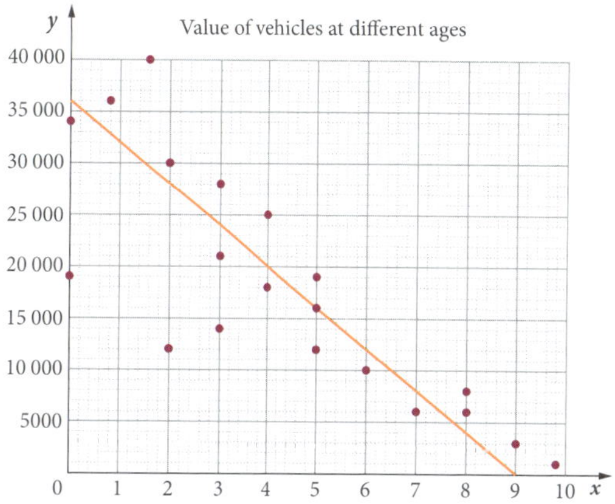

b What is the equation of Corey's line of fit? (2 marks)

c What value does the line of fit suggest for a vehicle that is $6\frac{1}{2}$ years old? (1 mark)

d What is the suggested age of a vehicle valued at $22 000? (1 mark)

e What restrictions must be placed on the equation of the line of fit? Give reasons for your answer. (2 marks)

Question 17 (4 marks)

The diagram shows the floor plan of a shed drawn to scale but with some measurements missing.

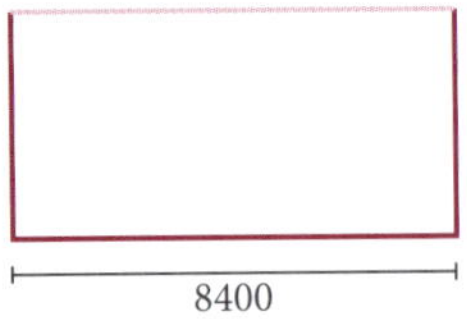

If 25 mm of rain falls on the roof, how many litres of water will be collected if it all flows into a tank? (4 marks)

Question 18 (3 marks)

a 75 mg of medication has been ordered for a patient. The medication is available as 5mg/100 mL. How many millilitres must the patient receive? (1 mark)

b The patient is to receive the medication through an IV set at 15 drops/mL. If the IV delivers 30 drops per minute, how long, in hours and minutes, will it take for the patient to receive the medication? (2 marks)

Question 19 (8 marks)

The diagram shows the graph of $y = 1000(a)^x$.

a What is the value of y when $x = 2$? (1 mark)

b For what value of x is y approximately 350? (1 mark)

The expression $y = 1000(a)^x$ can be used to model the depreciation of an asset, originally valued at $1000, under the declining balance method.

c How much will an asset originally valued at $1000 depreciate during the fourth year? (1 mark)

d What is the rate of depreciation? (1 mark)

e If the asset was being depreciated by the straight-line method, and the value was the same as the declining-balance method after 10 years, show the depreciation on the graph. (1 mark)

f How much depreciation is allocated per period under the straight-line method? (1 mark)

g For how many years could the straight-line method be used? Justify your answer. (2 marks)

Question 20 (3 marks)

A ramp 28 m long connects the car park level of a building to the first floor, which is 3.2 m above the car park level.

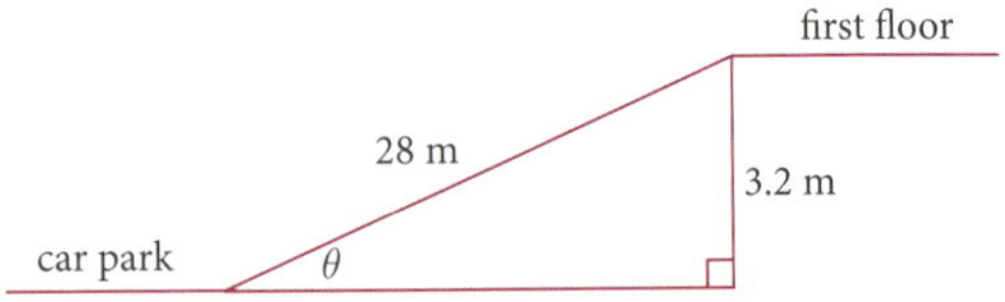

Find the size of the angle the ramp makes with the horizontal. Give the answer to the nearest minute. (3 marks)

Question 21 (4 marks)

a On 19 March Billy uses his credit card to get a cash advance of $2400. A fee is charged. The fee is the greater of $5 or 2.5% of the advance. How much is the fee that Billy must pay? (1 mark)

b Cash advances attract immediate interest at 18.6% p.a., compounded daily and charged from, and including, the date taken to, and including, the date repaid. If Billy repays the advance (plus fee) on 5 May, how much interest will he pay? (3 marks)

Question 22 (7 marks)

A paddock is bound on one side by a river as shown in the diagram.

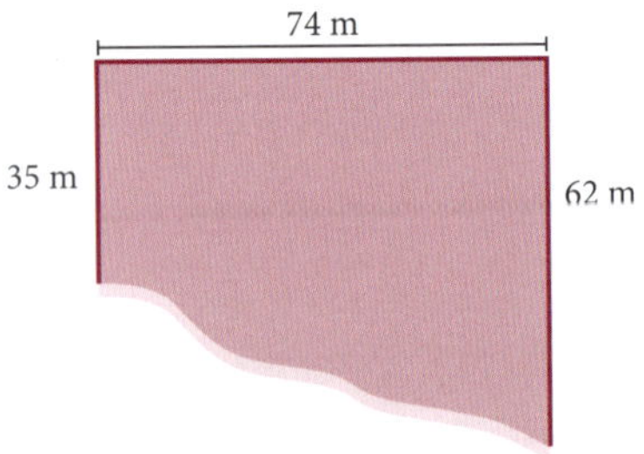

a Use the trapezoidal rule to find the approximate area of the paddock. (2 marks)

b Find the amount of seed required to plant the paddock at the rate of 120 kg/ha. Give the answer to the nearest kilogram. (2 marks) c After planting, fertiliser is to be applied providing nitrogen, phosphorus and sulphur in the ratio 10:7:5. If the paddock receives 8.6 kg of nitrogen, how many kilograms of phosphorus will it get? (1 mark)

d At what rate, in kg/ha, is sulphur being added to the paddock? (2 marks)

Question 23 (4 marks)

a Louise is intending to invest $20 000 for a term of 4 years. If she chooses to invest with Bud Bank she will earn 2.8% p.a. interest compounded quarterly. What will her investment be worth at the end of 4 years? Give the answer to the nearest dollar. (2 marks)

b If instead Louise chooses to invest with Bank Bods she will earn 2.7% p.a. interest compounded monthly. Which would be the best investment for Louise? Justify your answer with appropriate calculations. (2 marks)

Question 24 (5 marks)

a Draw a directed network diagram from this table which shows the distances in kilometres along bus routes linking different towns. (3 marks)

		To				
		A	**B**	**C**	**D**	**E**
From	**A**	0	10	–	14	11
	B	12	0	3	8	4
	C	–	2	0	9	–
	D	–	15	–	0	6
	E	–	7	–	5	0

b What is the length of the shortest path from D to A and what is that path? (2 marks)

Question 25 (2 marks)

A plan shows the dimension of a rectangular room to be 4150 mm long and 3225 mm wide. How many whole square metres of carpet would be needed to carpet the whole room? (2 marks)

Question 26 (3 marks)

Quilby is due east of Port Eck and 15 km due north of Rowland. The bearing of Port Eck from Rowland is 286°.

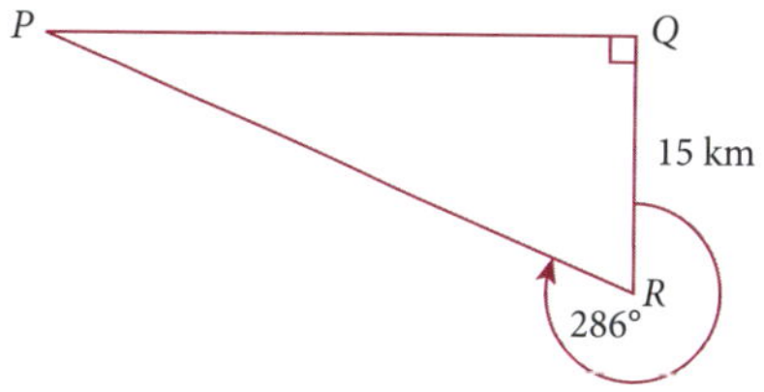

Find the distance from Rowland to Port Eck in kilometres to one decimal place. (3 marks)

Question 27 (5 marks)

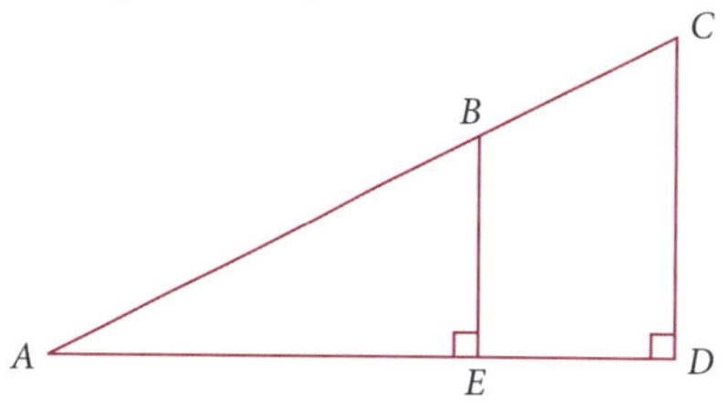

a Briefly explain why triangles ABE and ACD are similar. (1 mark)

b $BE = 7$ m, $CD = 10.5$ m and $ED = 5.5$ m. Find the length of AE. (2 marks)

c What is the length of AC? Give the answer to the nearest 10 centimetres. (2 marks)

SAMPLE HSC EXAMINATION 1

Section I

1 B. It is not possible to be in two or more places at the same time. So options A, C and D cannot be travel graphs.

2 D. 2 hours = 2 × 60 minutes
= 120 minutes

So the ratio is 120 : 20 or 6 : 1.

3 C. Try each option:

A: $60^2 + 91^2 = 11881 = 109^2$

B: $39^2 + 80^2 = 7921 = 89^2$

C: $59^2 + 110^2 = 15\,581 \neq 119^2$

D: $51^2 + 140^2 = 22\,201 = 149^2$

4 B. 180 − 65 = 115

So a compass bearing of S65°E is the same as a true bearing of 115°.

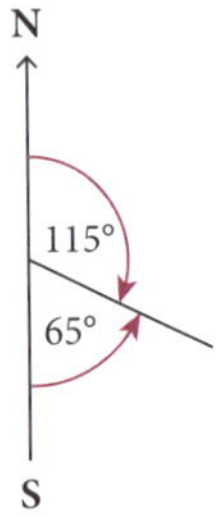

5 C. All except two vertices have degree 3.

So six vertices have degree 3.

6 A. $PV = \$5000$ $\quad r = 0.08 \div 4 = 0.02$ $\quad n = 2 \times 4 = 8$

$FV = PV(1 + r)^n$
$= \$5000 \times (1 + 0.02)^8$
$= \$5858.296\,90\ldots$
$= \$5858$ (nearest dollar)

Interest = \$5858 − \$5000
= \$858

7 B. $y = \dfrac{12}{x}$ is a hyperbola.

x cannot be equal to 0.

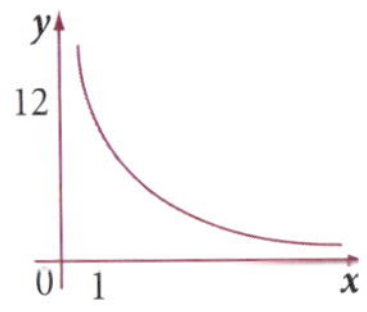

8 C. Amount of petrol = 60 ÷ 1.25
= 48 L

Distance = 48 ÷ 7.5 × 100 km
= 640 km

9 D. First edge is *PT* because it has the lowest weight, followed by *ST*. *SP* cannot be added because it would make a cycle. The edge of the next lowest weight is *UQ* so that would be chosen third.

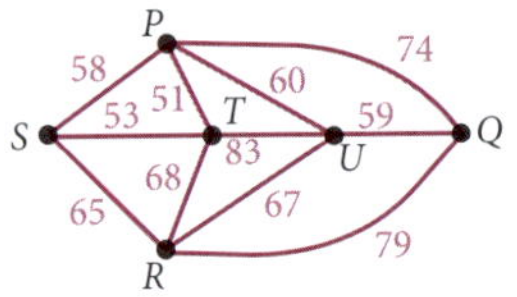

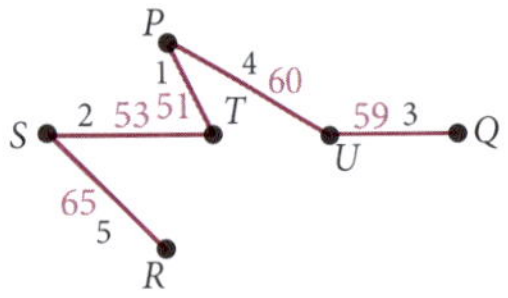

10 A. Try each option:

A: \$1.10 ÷ 0.45 L = \$2.444 44… per litre

B: \$1.60 ÷ 0.6 L = \$2.666 66… per litre

C: \$3.75 ÷ 1.5 L = \$2.50 per litre

D: \$5.20 ÷ 2 L = \$2.60 per litre

The best value is 450 mL for \$1.10.

Section II

11 a 12 years at 6%

Repayments = 70 × \$9.76
= **\$683.20** ✓

b 25 years at 9%

Repayments = 110 × \$8.39
= \$922.90 ✓

Total paid = \$922.90 × 25 × 12
= \$276 870

Interest = \$276 870 − \$110 000
= **\$166 870** ✓

c 7 years at 7.5%

Repayments = \$15.34 per \$1000

\$613.60 ÷ \$15.34 = 40

Brock borrowed **\$40 000.** ✓

d At 7.5%:

Repayments = 87 × \$9.27
= \$806.49

At 9%:

Repayments = 87 × \$10.14
= \$882.18

Extra repayments = \$882.18 − \$806.49
= **\$75.69** ✓

12 a From the graph, Yvonne must sell **75 pies** to break even. ✓

b If Yvonne sells 200 pies the cost is \$750 and the income is \$1000.

Profit = \$1000 − \$750
= **\$250** ✓

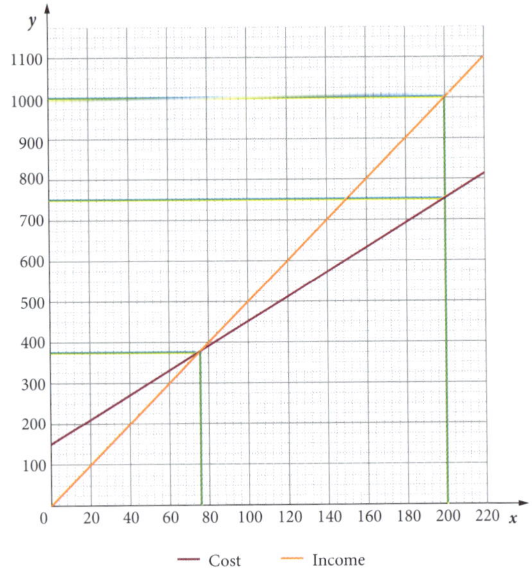

c **The y-intercept of the cost line is 150.** That represents the amount of fixed costs that Yvonne must pay. ✓

d The gradient of the income line is $\frac{100}{20} = 5$. It represents the amount that Yvonne receives for each pie. So she sells the pies for $5 each. ✓

13 a **31 accidents** occurred in the region that has 20 bottle shops. ✓

b **The bottle shops don't cause the accidents.** ✓

It might be, for example, that there are more bottle shops and more accidents in regions with higher populations. ✓

14 2 cups of flour make 30 biscuits.

Now $75 \div 30 = 2.5$

So $2.5 \times 2 = 5$ cups of flour make 75 biscuits. ✓

But the amount of flour is 8 times the amount of icing sugar.

So amount of icing sugar = $\frac{5}{8}$ **cup** ✓

15 a Lost value = 15% of $54 000

= 0.15 × $54 000

= **$8100** ✓

b From 2016 until 2025 is 9 years.

$S = V_0(1 - r)^n$

$= \$54\,000 \times (1 - 0.15)^9$

$= \$12\,507.315\ldots$

Its value in January 2025 will be about $12 500. ✓

c If $S = \$5000$,

$5000 = 54\,000(1 - 0.15)^n$

Using a calculator,

$54\,000(1 - 0.15)^{14} = 5549.562\,15\ldots$

$54\,000(1 - 0.15)^{15} = 4717.127\,83\ldots$

After 15 years, the value of the caravan will first be below $5000. ✓

16 a oregano : salt $= 1 : 1\frac{1}{2}$

$= 2 : 3$ ✓

b One quantity = 6 serves

Number of quantities = $15 \div 6$

= 2.5

Number of eggs = 2.5×2

= 5

So 5 eggs would be needed for 15 people. ✓

c Amount of sauce $= 4\frac{1}{2} \times \frac{1}{2}$ cup

$= 2\frac{1}{4}$ **cups** ✓

17 a

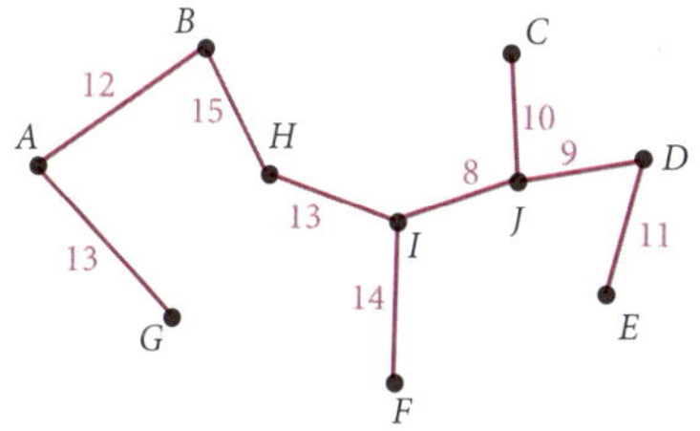

✓✓

b Minimum cost = $(13 + 12 + 15 + 13 + 14 + 8 + 10 + 9 + 11)

= **$105** ✓

c **The shortest path from A to E is $AHIJE$ and its length is $55.** ✓✓

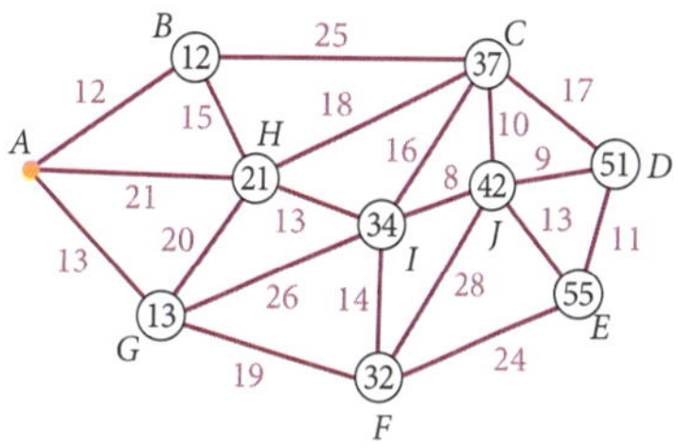

✓

d **No, the shortest path is only the least cost to connect A to E and does not include all the other vertices.** The minimum cost is the least cost to connect all the points in the complex. ✓✓

18 a $y = 1.1x + 16.4$

When $x = 20$,

$y = 1.1 \times 20 + 16.4$

$= 38.4$

So, using the equation, a person whose arm length is 20 cm would have a leg length of 38.4 cm. ✓

b $y = 1.1x + 16.4$

When $y = 56$,

$1.1x + 16.4 = 56$

$1.1x = 39.6$

$x = 36$

So a prediction for Tim's arm length is 36 cm. ✓

19 a **An exponential function. It is an example of exponential growth.** ✓

b **In 2010 the population was 9 million.** ✓

c The population will reach 30 million after about 31 years. **So in about the year 2041 the population will reach 30 million.** ✓

d **The population will double in about 18 years.** ✓

While the equation $P = 9(1.04)^t$ gives the population at time t, it will continue to double in about the same number of years as is the nature of exponential growth. If other factors influence the population it might change. ✓

20 a Maximum heart rate = 220 − 37

= 183

Resting heart rate = $183 \div 3$

= 61 bpm ✓

Times beating in a day = $61 \times 60 \times 24$

= 87 840

Jeremy's heart would beat 87 840 times in a day. ✓

b 65% of 183 = 118.95

75% of 183 = 137.25 ✓

So Jeremy's heart rate range would be between about 119 bpm and 137 bpm. ✓

21 $T = \frac{100}{S}$

a

S	1	2	4	5	10	20	25	50	100
T	100	50	25	20	10	5	4	2	1

✓

b

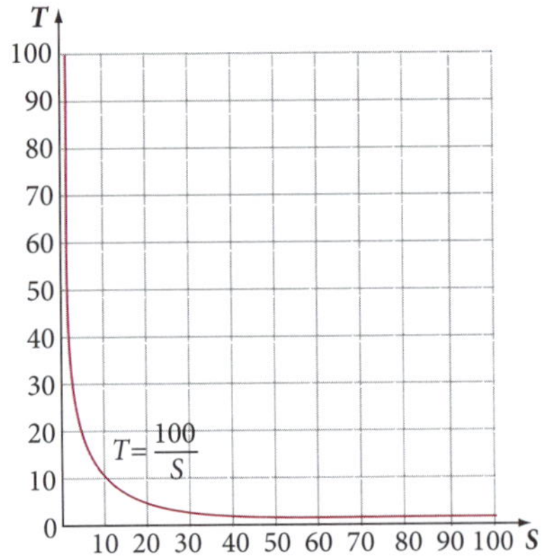

✓✓

c **Approximately 7 seconds** ✓

22 a **Displaying the data** ✓

b X: range = 94 − 43

= **51** ✓

c Y: median $= \frac{74+77}{2}$

= **75.5** ✓

d Yes, there does appear to be a difference in this test.

Class Y did better than class X with more higher scores and a higher median (75.5 compared to 70.5). ✓

The results in class X are **symmetrical but those in class Y are negatively skewed.** ✓

23 a Let h m be the height of the tree.

$\tan 41° = \frac{h}{20.7}$

$h = 20.7 \times \tan 41°$ ✓

$= 17.9942354\ldots$

$= 18$ (nearest unit)

The height of the tree is 18 m to the nearest metre. ✓

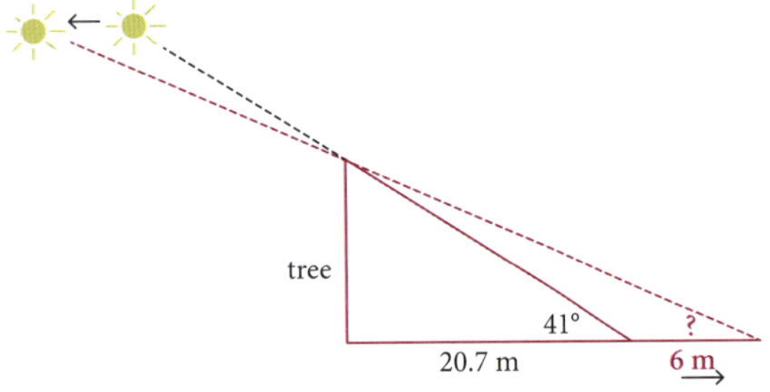

b The new shadow length is (20.7 + 6) or 26.7 m.

Let θ be the angle of elevation of the sun.

So $\tan \theta = \frac{17.994\ldots}{26.7}$ ✓

$= 33.9776614\ldots°$

$= 34°$ (nearest degree)

The angle of elevation of the sun is 34° to the nearest degree. ✓

24 a $PV = \$620\,000$, $r = 0.12$, $n = 3$

$FV = PV(1 + r)^n$

$= \$620\,000(1 + 0.12)^3$

$= \$871\,055.36$

The house would be valued at about $871 055 at the end of 3 years. ✓

b $FV = \$560\,000$, $PV = \$410\,000$, $n = 3$

$FV = PV(1 + r)^n$

$\$560\,000 = \$410\,000(1 + r)^3$

$(1 + r)^3 = \frac{\$560\,000}{\$410\,000}$ ✓

$1 + r = \sqrt[3]{\frac{\$560\,000}{\$410\,000}}$

$= 1.10951894\ldots$

$r = 0.10951894\ldots$

$= 0.11$ (2 d.p.)

The rate of increase was about 11%. ✓

25 Let the height of the pole be x m.

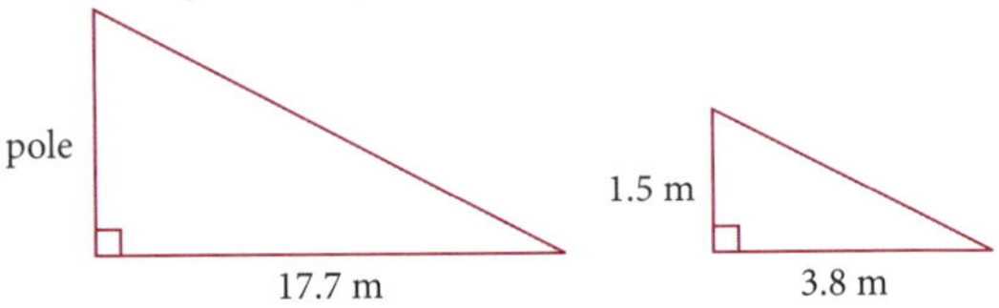

The triangles are similar (equiangular).

So $\frac{x}{1.5} = \frac{17.7}{3.8}$ ✓

$x = \frac{17.7}{3.8} \times 1.5$

$= 6.98684210\ldots$

$= 7.00$ (2 d.p.)

The pole is 7 m high. ✓

26 a The side measures 6 cm.

Actual size = 100 × 6 cm

= 600 cm

= **6 m** ✓

b

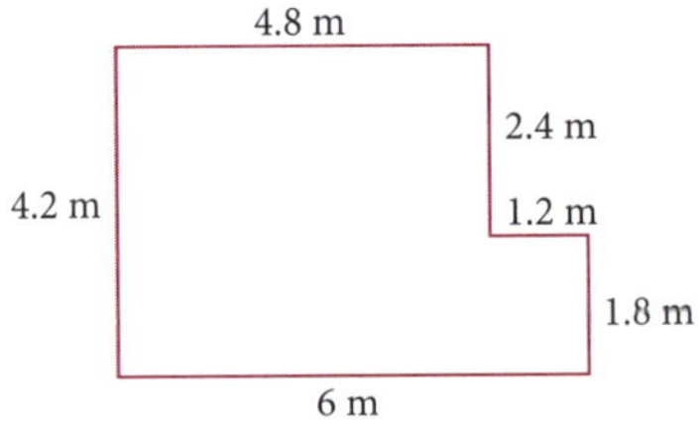

✓✓

c 600 mm = 0.6 m

Dividing the sides by 0.6 m we can see that 10 tiles will fit across the bottom side, and 8 across the top. The right side will have 4 tiles at the top and 3 at the bottom.

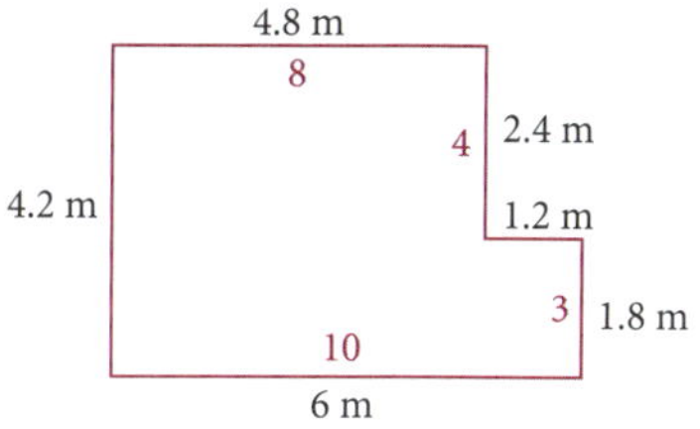

Number of tiles = 10 × 3 + 8 × 4
= 62 ✓

Cost = 62 × \$24.75
= **\$1534.50** ✓

d 6 and 4.8 are both divisible by 4 but 4.2 is not divisible by 4.
So the tiles need to be laid so that the shorter side is on the northern and southern sides and the longer side on the east and west. ✓

e There will be 4.8 ÷ 0.4 = 12 tiles across the top and 6 ÷ 0.4 = 15 across the bottom.

Number of tiles = 15 × 3 + 12 × 4
= 93 ✓

Cost = 93 × \$16.50
= \$1534.50

The cost would be exactly the same. ✓

27 19.71% p.a. = 0.054% per day

Groceries: \$324

From 17 May until 10 June inclusive is 25 days. ✓

$FV = PV(1 + r)^n$
= \$324 × $(1 + 0.000\,54)^{25}$
= \$328.402 461….
= \$328.40 (nearest cent) ✓

Car service: \$586

From 28 May until 10 June inclusive is 14 days. ✓

$FV = PV(1 + r)^n$
= \$586 × $(1 + 0.000\,54)^{14}$
= \$590.445 743…
= \$590.45 (nearest cent) ✓

Total to pay = \$328.40 + \$590.45
= **\$918.85** ✓

SAMPLE HSC EXAMINATION 2

Section I

1 A. $\cos 69° = \frac{x}{12}$

$x = 12 \cos 69°$

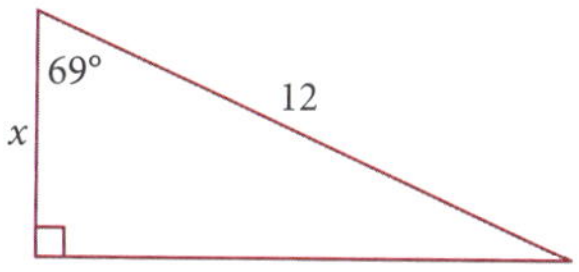

2 D. $FV = PV(1 + r)^n$

$= \$30\,000\left(1+\frac{0.09}{12}\right)^{7\times 12}$

= \$56 196 (nearest dollar)

I = \$56 196 − \$30 000
= \$26 196

3 C. Gradient = $\frac{2}{3}$

Vertical intercept = 2

Equation is $y = \frac{2}{3}x + 2$

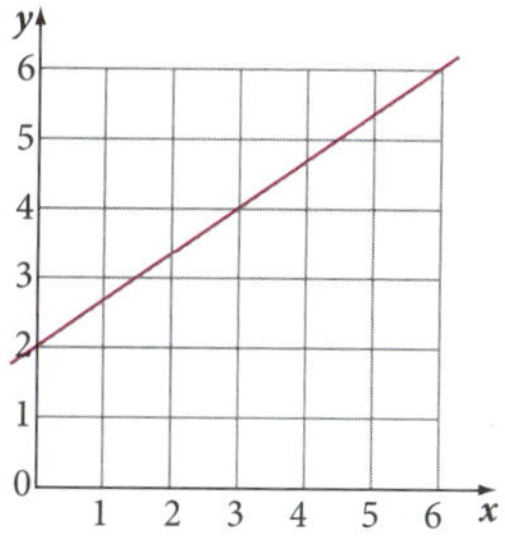

4 B. 7.5 cm represents 150 km.

Now 150 km = 150 000 m
= 15 000 000 cm.

So the scale is 7.5 : 15 000 000 = 1 : 2 000 000.

5 C. As Guy runs, the distance increases over time.

6 B. Vertices will be added in this order: *P, W, X, Y, R, T, Q, V* or *S, S* or *V* and finally *U*.

Vertex *U* will be added last.

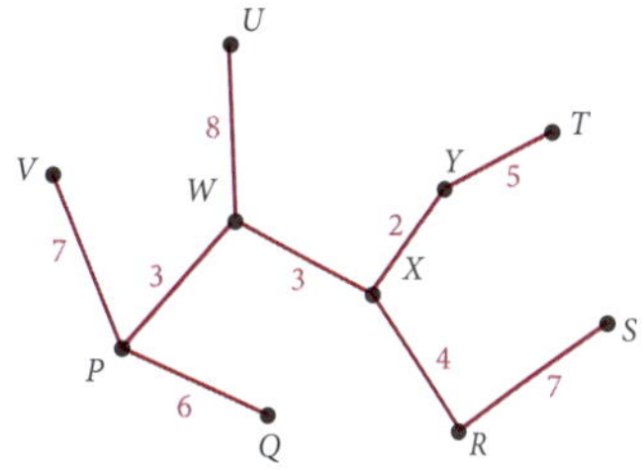

7 B. Consider each option:

586 km on 41 L = 6.996 587… L/100 km

654 km on 44 L = 6.727 828… L/100 km

735 km on 53 L = 7.210 884… L/100 km

812 km on 58 L = 7.142 857… L/100 km

The best fuel consumption is 6.727 828…. L/100 km, or 654 km on 44 L.

8 A. Let the length be x m.

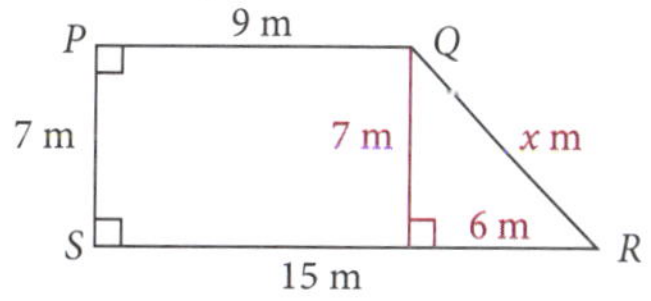

$x^2 = 7^2 + 6^2$
= 85

$x = \sqrt{85}$ $(x > 0)$
= 9.219 544 45…
= 9.2 (1 d.p.)

So the length of QR is 9.2 m to one decimal place.

9 A. Total repaid = 60 × \$332
= \$19 920

Interest = \$19 920 − \$16 000
= \$3920

Annual interest = \$3920 ÷ 5
= \$784

Annual interest rate = $\frac{\$784}{\$16\ 000} \times 100\%$
= 4.9%

10 **D.** Speed = 400 m in 54 s
= (0.4 ÷ 54) km in 1 s
= (0.4 ÷ 54 × 60 × 60) km in 1 hour
= 26.666 6666… km/h
= 27 km/h (nearest km/h)

Section II

11 **a** $I = Prn$
= $12 000 × 0.07 × 3
= **$2520** ✓

b Total to repay = $12 000 + $2520
= $14 520

Annual instalment = $14 520 ÷ 3
= **$4840** ✓

12 **a** **Barry charges $200 as a fixed rate.** ✓

b **It will cost $575 for Carrie to work for 10 hours.** ✓

c **The two costs are the same for 5 hours work.** ✓

d **Barry would work slightly longer than Carrie for $500 so Barry would be the most economical.** ✓

e Gradient = $\frac{375-200}{5}$
= **35** ✓

This represents Barry's hourly rate ($35/h). ✓

f y-intercept = 175

Gradient = $\frac{375-175}{5}$
= 40 ✓

So $C = 40h + 175$. ✓

13

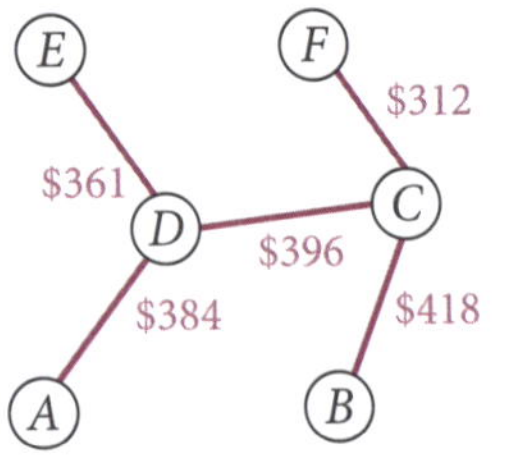

✓

Minimum cost = $(384 + 361 + 396 + 312 + 418)
= **$1871** ✓

14 **The question is ambiguous.** It could be taken to mean that the song thought best rated highest, i.e. closer to first. But it could also be taken to mean that the song thought worst rated highest as it had a larger number. So two people with opposing views could give the same answer. ✓

15 **a** 2 parts is 1 L

So 1 part is 0.5 L.

5 parts is 5 × 0.5 L = 2.5 L

So 2.5 L of ginger ale should be used. ✓

b Total parts = 5 + 3 + 2
= 10

So 10 parts is 7.5 L

1 part is 0.75 L

3 parts is 3 × 0.75 L = 2.25 L

So 2.25 L of lemonade is needed to make 7.5 L of the drink. ✓

16 **a** **20** ✓

b y-intercept = 36 000

gradient = $\frac{-36\,000}{9}$
= −4000 ✓

Equation is $\mathbf{y = -4000x + 36\,000}$ ✓

c **$10 000** ✓

d $\mathbf{3\frac{1}{2}}$ **years** ✓

e $\mathbf{x \geq 0}$ because a vehicle cannot have a negative age. ✓

$\mathbf{x \leq 9}$ because a vehicle cannot have a negative value. ✓

17 The length of the shed is 8400 mm or 8.4 m.

By measurement the width of the shed is half the length, so the width is 4.2 m. ✓

Area = 8.4 × 4.2
= 35.28 m^2 ✓

Rainfall = 25 mm
= 0.025 m

Volume = 35.28 × 0.025
= 0.882 m^3 ✓

Capacity = 0.882 × 1000 L
= **882 L** ✓

18 **a** Medication is 5 mg/100 mL.

The patient requires 75 mg.

Now 75 ÷ 5 = 15

So the patient requires 15 × 100 mL.

The patient must receive **1500 mL.** ✓

b Number of drops required = 15 × 1500
= 22 500 ✓

Time required = (22 500 ÷ 30) minutes
= 750 minutes

Now 750 ÷ 60 = 12.5.

12.5 h = 12 h 30 min

So the IV will take 12 hours and 30 minutes. ✓

19 **a** $y = 1000(a)^x$

When $x = 2$, $y \approx$ **810** ✓

b When $y = 350$, $x \approx$ **10** ✓

c When $x = 3$, $y \approx 730$

When $x = 4$, $y \approx 660$

Depreciation = $730 – $660
= $70

The asset would depreciate by approximately $70 in the fourth year. ✓

d In the first year the asset depreciates by $100.

Rate of depreciation = $\frac{\$100}{\$1000} \times 100\%$
= **10%** ✓

e

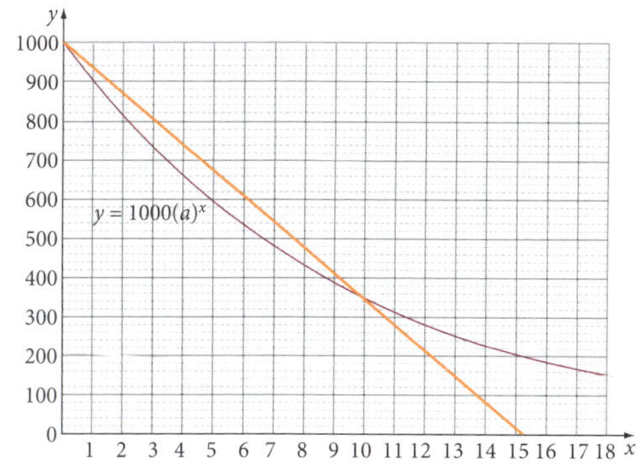

✓

f In 10 years:

Depreciation = \$1000 − \$350

= \$650

Depreciation per year is \$65. ✓

g The straight-line method could only be used for **15 years**. ✓

After that time, this method shows the value to be negative and that is not possible. ✓

20 $\sin\theta = \dfrac{3.2}{28}$ ✓

$\theta = \sin^{-1}\left(\dfrac{3.2}{28}\right)$ ✓

$= 6°\ 33'\ 44.74''$

$= 6°\ 34'$ (nearest minute) ✓

The ramp makes an angle of 6° 34' to the horizontal, to the nearest minute.

21 a 2.5% of \$2400 = 0.025 × \$2400

= \$60

Billy must pay a fee of \$60. ✓

b Number of days = (31 − 18) + 30 + 5

= 48 ✓

Total amount = \$2400 + \$60 = \$2460

$FV = \$2460 \times \left(1 + \dfrac{0.186}{365}\right)^{48}$

= \$2520.8985…

= \$2520.90 (nearest cent) ✓

Interest = \$2520.90 − \$2460

= **\$60.90** ✓

22 a $A = \dfrac{h}{2}(a + b)$

$= \dfrac{74}{2}(35 + 62)$ ✓

= 3589

The area of the paddock is about 3589 m^2. ✓

b 3589 m^2 = (3589 ÷ 10 000) ha

= 0.3589 ha ✓

Seed required = 0.3589 × 120 kg

= 43.068 kg

So the amount of seed required is 43 kg, to the nearest kilogram. ✓

c 10 parts is 8.6 kg

So 1 part is 0.86 kg

7 parts = 7 × 0.86 kg

= 6.02 kg

So 6 kg of phosphorus will be applied. ✓

d Amount of sulphur = 5 × 0.86 kg

= 4.3 kg ✓

Application rate = 4.3 kg for 0.3589 ha

$= \dfrac{4.3}{0.3589}$ kg/ha

= 11.981 0532… kg/ha

≈ 12 kg/ha

So sulphur would be added at the rate of 12 kg/ha. ✓

23 a $PV = \$20\,000$, $r = \dfrac{0.028}{4} = 0.007$, $n = 4 \times 4 = 16$

$FV = PV(1 + r)^n$

$= \$20\,000(1 + 0.007)^{16}$ ✓

= \$22 361.5304…

= **\$22 362** (nearest dollar) ✓

b $PV = \$20\,000$, $r = \dfrac{0.027}{12} = 0.002\,25$, $n = 4 \times 12 = 48$

$FV = PV(1 + r)^n$

$= \$20\,000(1 + 0.002\,25)^{48}$ ✓

= \$22 278.2519…

= \$22 278 (nearest dollar)

So the quarterly investment will be worth more, returning an extra \$84. ✓

24 a

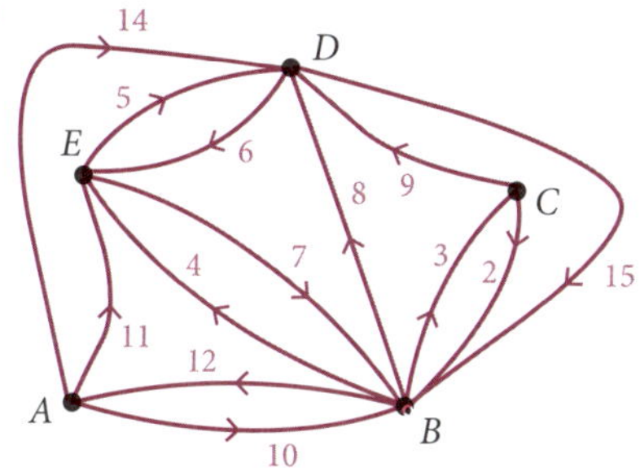

✓✓✓

b The length of the shortest path from *D* to *A* is **25 km**. ✓

The path is **from *D* to *E* to *B* to *A***. ✓

25 The room is 4.15 m long and 3.225 m wide. ✓

Area = 4.15 × 3.225

= 13.383 75 m^2

So 14 m^2 will be needed to carpet the whole room. ✓

26 Let the distance from Rowland to Port Eck be *x* km.

Now ∠*QRP* = 360° − 286°

= 74° ✓

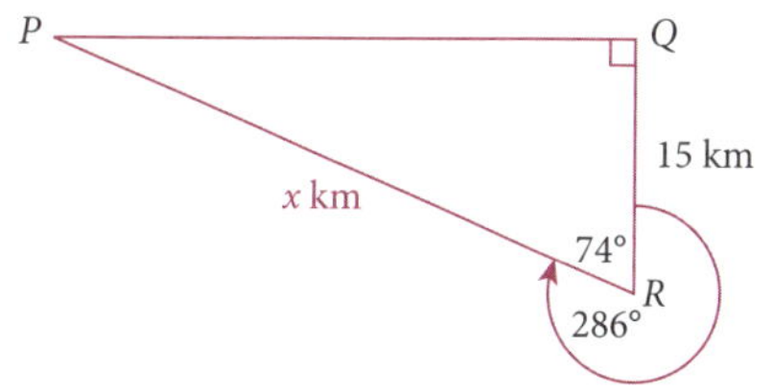

$\cos 74° = \dfrac{15}{x}$

$x = \dfrac{15}{\cos 74°}$ ✓

= 54.419 3291…

= 54.4 (1 d.p.)

So the distance from Rowland to Port Eck is 54.4 km, to one decimal place. ✓

27 a **The triangles are similar because they are equiangular.** They both have a right angle and they have a common angle. ✓

b Let the length of AE be x m.

So the length of AD will be $(x + 5.5)$ m.

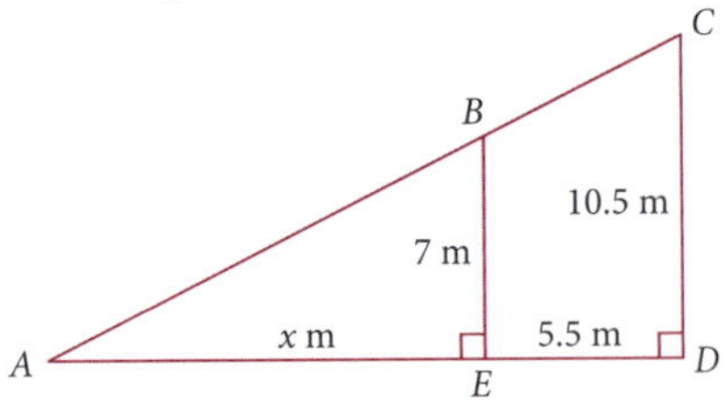

Now $\frac{AE}{AD} = \frac{BE}{CD}$

$$\frac{x}{x+5.5} = \frac{7}{10.5}$$ ✓

$$10.5x = 7(x + 5.5)$$

$$10.5x = 7x + 38.5$$

$$3.5x = 38.5$$

$$x = 11$$

So the length of AE is 11 m. ✓

c $AD = (11 + 5.5)$ m

$= 16.5$ m

$AC^2 = 10.5^2 + 16.5^2$

$= 382.5$

$AC = \sqrt{382.5}$ ✓

$= 19.557\,6072\ldots$

$= 19.6$ (1 d.p.)

The length of AC is 19.6 m to the nearest 10 centimetres. ✓

NSW GOVERNMENT

NSW Education Standards Authority

2020 HIGHER SCHOOL CERTIFICATE EXAMINATION

Mathematics Standard 1
Mathematics Standard 2

REFERENCE SHEET

Measurement

Limits of accuracy

Absolute error = $\frac{1}{2}$ × precision

Upper bound = measurement + absolute error

Lower bound = measurement − absolute error

Length

$$l = \frac{\theta}{360} \times 2\pi r$$

Area

$$A = \frac{\theta}{360} \times \pi r^2$$

$$A = \frac{h}{2}(a + b)$$

$$A \approx \frac{h}{2}(d_f + d_l)$$

Surface area

$$A = 2\pi r^2 + 2\pi rh$$

$$A = 4\pi r^2$$

Volume

$$V = \frac{1}{3}Ah$$

$$V = \frac{4}{3}\pi r^3$$

Trigonometry

$$\sin A = \frac{\text{opp}}{\text{hyp}}, \quad \cos A = \frac{\text{adj}}{\text{hyp}}, \quad \tan A = \frac{\text{opp}}{\text{adj}}$$

$$A = \frac{1}{2}ab\sin C$$

$$\frac{a}{\sin A} = \frac{b}{\sin B} = \frac{c}{\sin C}$$

$$c^2 = a^2 + b^2 - 2ab\cos C$$

$$\cos C = \frac{a^2 + b^2 - c^2}{2ab}$$

Financial Mathematics

$$FV = PV(1 + r)^n$$

Straight-line method of depreciation

$$S = V_0 - Dn$$

Declining-balance method of depreciation

$$S = V_0(1 - r)^n$$

Statistical Analysis

An outlier is a score
less than $Q_1 - 1.5 \times IQR$
or
more than $Q_3 + 1.5 \times IQR$

$$z = \frac{x - \mu}{\sigma}$$

Normal distribution

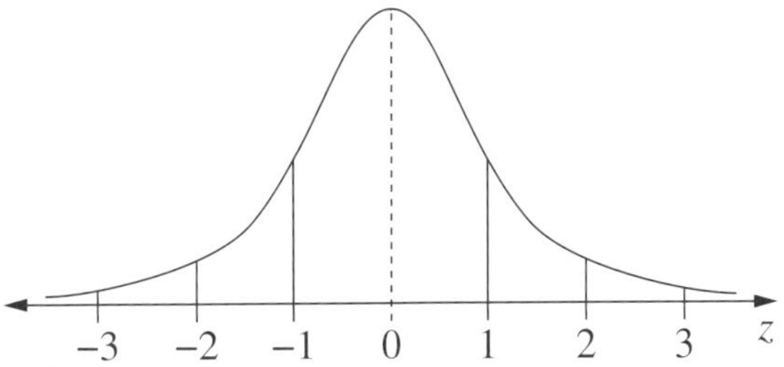

- approximately 68% of scores have z-scores between −1 and 1
- approximately 95% of scores have z-scores between −2 and 2
- approximately 99.7% of scores have z-scores between −3 and 3

INDEX

K

L

M

N

O

P

Q

R

S

T

U

V

W

X

Y

NOTES

NOTES

NOTES

ISBN 978 1 74125 689 5

Pascal Press
PO Box 250
Glebe NSW 2037
(02) 8585 4044
www.pascalpress.com.au

Publisher: Vivienne Joannou
Project editor: Rosemary Peers
Edited by Rosemary Peers
Answers checked by Peter Little
Cover and page design by Sonia Woo
Typesetting by Julianne Billington (lj Design)
Printed by Vivar Printing/Green Giant Press

Students
All care has been taken in the preparation of this study guide, but please check with your teacher or the NSW Education Standards Authority about the exact requirements of the course you are studying as these can change from year to year.

The validity and appropriateness of the internet addresses (URLs) in this book were checked at the time of publication. Due to the dynamic nature of the internet, the publisher cannot accept responsibility for the continued validity or content of these web addresses.

EXCEL YEAR 12 MATHS STANDARD 1 PASSCARDS

YOUR PATH TO SUCCESS!

This set of cards is designed for you to use as the final step in your revision program. The author has carefully selected the most important facts of the course for you to focus on just before your exam or test. You can use your ***Excel*** Year 12 Passcards:

✓ **ON THE GO.** If you are by yourself, read over each card again and again until you completely master its content. If you're with a friend, revise as a team by turning the bullet points into questions and quizzing each other on key points. Your answers will be there on the cards.

✓ **AT HOME.** Read each card thoroughly and make sure you understand all the points. You should also know more detailed information on each topic—if you are not completely sure of a topic, revise it in your ***Excel*** Year 11 and 12 study guides.

✓ **ON YOUR DIGITAL DEVICE.** Download a **FREE** digital copy of these cards at **www.pascalpress.com.au/free**

CONTENTS (2)

FORMULAE AND EQUATIONS — 1

Substitution and formulae

→ **When substituting into an algebraic expression**, the variable must be replaced with the given value every time it occurs. For example, when $x = 2$, $3x^2 - 4x = 3 \times 2^2 - 4 \times 2 = 4$.

→ **When solving equations** the same mathematical operation must be performed on both sides of the = sign.

→ **The subject of a formula is the variable before the = sign**. To change the subject, the formula is rearranged (in the same way as an equation is solved), so that the required variable appears before the = sign.

EXAMPLE: Make t the subject of the formula $v = u + at$.

Solution:
$$v = u + at$$
$$at = v - u$$
$$t = \frac{v-u}{a}$$

TOPIC: ALGEBRA

See ➲ *Excel* Year 11 Mathematics Standard pp. 4–5

LINEAR RELATIONSHIPS — 3

Gradient and y-intercept

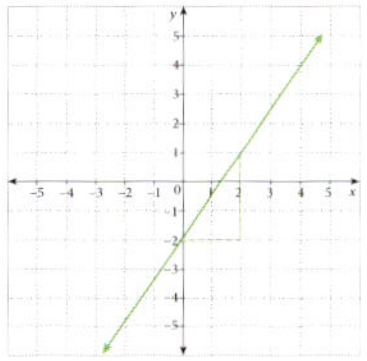

→ **The gradient of a line is its slope**. It is a measure of the steepness of the line.

→ **The gradient, m, of a line is given by**
$$m = \frac{\text{vertical change in position}}{\text{horizontal change in position}}.$$

→ **The gradient is usually expressed** in fractional form, if necessary as an improper fraction.

→ **The y-intercept, c, is where the line intersects** the vertical axis.

EXAMPLE: Find the gradient and y-intercept of the line in the graph.

Solution: $m = \dfrac{\text{vertical change in position}}{\text{horizontal change in position}} = \dfrac{3}{2}$; $c = -2$

TOPIC 1: ALGEBRA

See ➲ *Excel* Year 11 Mathematics Standard pp. 12–15

APPLICATIONS OF MEASUREMENT — 5

Perimeter, area and volume

→ **The perimeter and area might need to be found** for triangles, rectangles, parallelograms, trapeziums and circles plus composite shapes.

→ **Surface area is the total area of all the surfaces of a 3D object**. It is measured in square units.

→ **Volume is measured in cubic units** and is the amount of space contained by a 3D object.

→ **The volume and surface area** of prisms, cylinders, spheres and composite solids might be required.

→ **Conversions between volume and capacity might need to be made**. For example, 1 m^3 = 1000 L.

→ **Some formulae are given** on the reference sheet but others need to be known.

→ The **trapezoidal rule** approximates an **area with trapeziums**.

TOPIC: MEASUREMENT

See ➲ *Excel* Year 11 Mathematics Standard pp. 35–44

WORKING WITH TIME — 7

Working with time

→ **Timetables must be interpreted** and problems solved involving their use.

→ **Conversions between 12-hour and 24-hour time** might need to be made.

→ **Any position on the Earth's surface can be found** by its latitude and longitude.

→ **Time zones are designated areas all having the same time.** Differences between times in different zones all around the world might be calculated. Daylight saving needs to be taken into account when and where appropriate.

→ **Coordinated Universal Time** (UTC) is the time at the prime meridian (Greenwich in London).

→ **The International Date Line** (IDL) is an imaginary line that is the boundary between one day and the next.

→ **When travelling or determining the timing of events** in different countries, the differences in times must be taken into account.

TOPIC: MEASUREMENT

See ➲ *Excel* Year 11 Mathematics Standard pp. 56–63

MONEY MATTERS — 9

Budgeting and household expenses

→ **A budget, showing income balanced against expenses**, can be prepared to keep track of spending. Expenses include fixed costs and amounts available for discretionary spending.

→ **Household bills,** covering items like electricity, gas and water, need to be interpreted. They often contain a fixed charge and a rate for usage.

→ **On-road costs for the purchase of a vehicle** include all the extras that must be paid alongside the price of the vehicle. These might include stamp duty, registration and insurance, and loan costs if applicable.

→ **There are different types of insurance available** for vehicles, including compulsory third-party which is a condition of registration and covers injuries to people. Non-compulsory third-party insurance covers other vehicles or property belonging to others while comprehensive insurance covers your own vehicle and property as well as that of others.

→ **Running costs** for vehicles include things like fuel, servicing and tyres.

TOPIC: FINANCIAL MATHEMATICS

See ➲ *Excel* Year 11 Mathematics Standard pp. 75–82

DATA ANALYSIS — 11

Representing data

→ **Frequency distribution tables** are one way of displaying a lot of data in an organised form. They usually show the scores (x), frequency (f) and the product of the frequency with each score (fx). They might include a tally.

→ **Cumulative frequency** is where the frequency is accumulated or added up.

→ **Histograms** are displays of the data.

→ **A Pareto chart** is a type of graph that consists of both bars and a line graph. Variables shown by the bars are arranged in descending order. The line graph shows the cumulative total. With a Pareto chart the most important issues are highlighted.

→ **A stem-and-leaf plot** shows all the scores in a data set in a simple way even when the range is quite large. In a back-to-back stem-and-leaf plot two data sets are shown with a common stem. This allows us to easily compare the data.

TOPIC: STATISTICAL ANALYSIS

See ➲ *Excel* Year 11 Mathematics Standard pp. 92–100

DATA ANALYSIS — 13

Investigating data

→ **Outliers are members of a data set that are inconsistent** with the rest of the members. They can be identified by different methods; for example, the distance from the mean.

→ **A specific method of identifying outliers** are those scores that are more than 1.5 times the IQR above the upper quartile or more than 1.5 times the IQR below the lower quartile.

→ **The effect of outliers** on measures of central tendency can be explored.

→ **Graphical displays can be interpreted**. The number of modes or the shape of the graph give information about the data. For example, is the graph symmetrical or is it positively or negatively skewed?

→ **A box plot** gives an indication of the spread of scores using the five-number summary. If two box plots are drawn on the same scale it is easy to see the similarities and differences in location and spread.

TOPIC: STATISTICAL ANALYSIS

See ➲ *Excel* Year 11 Mathematics Standard pp. 107–108

CONTENTS (1)

ALGEBRA

MEASUREMENT

FINANCIAL MATHEMATICS

STATISTICAL ANALYSIS

ALGEBRA

CONTENTS (3)

STATISTICAL ANALYSIS

NETWORKS

FORMULAE AND EQUATIONS

Special formulae

2

- **Formulae exist to estimate the blood alcohol content** (BAC) after consuming alcohol. These are based on the number of standard drinks consumed, the time spent drinking and the weight of the person, and differ for males and females.
- **A standard drink is a drink that contains 10 g of alcohol**. Many alcoholic drinks contain more than one 'standard drink'.
- **A formula can also estimate the amount of time for the BAC to return to zero**.
- **It is important to remember that these formulae are only estimates** and extreme caution should be taken when applying them to actual situations.
- **Doses of medicine can also be determined using certain formulae**. For example, the dosage for a child might be worked out given the adult dose and age or weight of the child.

TOPIC: ALGEBRA

See ➲ *Excel* Year 11 Mathematics Standard pp. 6–8

APPLICATIONS OF MEASUREMENT

Practicalities of measurement

4

- **Metric units of measurement use standard prefixes**. Conversions can then be made between, for example, millimetres and metres, or litres and kilolitres.
- **Standard form (scientific notation) is often used with measurements**. In standard form, any number can be expressed as a number between 1 and 10 multiplied by a power of 10.
- **Error in a measurement** is the difference between the measured quantity and the actual quantity. Error might be due to human fallibility or it could be the result of the limitations of the measuring instrument.
- **The absolute error of a measurement** is half the smallest unit on the measuring device. The percentage error is the absolute error expressed as a percentage of the measurement.
- **The limits of accuracy are the upper and lower bounds**; that is, the measurement ± the absolute error.

TOPIC: MEASUREMENT

See ➲ *Excel* Year 11 Mathematics Standard pp. 24–33

APPLICATIONS OF MEASUREMENT

Units of energy

6

- **A calorie, cal, is a unit of energy** related to food.
- **The joule, J, is the metric unit of energy** (or work).
- **Power is the rate at which energy is changing**.
- **The watt, W, is a unit of power**. One watt is equal to one joule per second.
- **A watt hour (Wh or W-h) is a measurement of energy used**. Most often used to measure electricity, it is the energy used by an appliance drawing one watt for one hour.
- **A more common measurement is the kilowatt-hour** (kWh or kW-h) which is 1000 watt-hours. 1 kWh = 1000 Wh
- **Units of energy are used in solving problems** related to energy expended in activities or energy consumption related to food or to electricity.

TOPIC: MEASUREMENT

See ➲ *Excel* Year 11 Mathematics Standard pp. 45–46

MONEY MATTERS

Earning and managing money

8

- **In calculating profit and loss and GST**, percentage increases or decreases often need to be applied.
- **A salary is a fixed form of payment for a period of time**, often a year.
- **A wage is an amount paid at**, usually, an hourly rate.
- **Overtime, penalty rates and loadings** can be calculated from information about a wage or salary.
- **Commission is an amount paid for sales** and is usually a percentage of the sales made. Piecework is an amount paid for each item produced. Royalties are payments made for the use of material.
- **Income tax is an amount that must be paid to the government**. Taxable income is the income on which tax must be paid after allowable deductions have been made. In calculating tax a scale might be used and attention must also be made to the Medicare levy.

TOPIC: FINANCIAL MATHEMATICS

See ➲ *Excel* Year 11 Mathematics Standard pp. 70–75

DATA ANALYSIS

Classifying data

10

- **Data is information** that can be collected, arranged and analysed.
- **Samples taken from the population should be representative** of the population and can then be used to estimate characteristics of the population.
- **Samples might be random** (each member of the population has an equal chance of taking part), systematic or stratified (numbers of members from each subsection of the population are chosen in proportion).
- **Numerical (quantitative) data** is data that is a number or measurement. It might be discrete (that which can be counted) or continuous (that which is measured).
- **Categorical (qualitative) data** is data that can be categorised in some way, and might be nominal or ordinal. Ordinal data can be ranked or placed in order, while nominal data cannot.

TOPIC: STATISTICAL ANALYSIS

See ➲ *Excel* Year 11 Mathematics Standard pp. 90–91

DATA ANALYSIS

Exploring data

12

- **Measures such as the mean and median are measures of central tendency**. They give information about the location of scores.
- **The mean is the measure commonly known as the average**.
- **The median is the middle score when the scores are arranged in order**.
- **The mode is the score with the highest frequency**.
- **The range, interquartile range and standard deviation are measures of spread**.
- **The range is the difference between the highest and lowest score**.
- **Quartiles divide the data into four parts**. For example, the lower quartile, Q_1, is the number for which 25% of the scores are lower.
- **The interquartile range is the difference between the upper and lower quartiles**. $IQR = Q_3 - Q_1$
- **The standard deviation gives us information** about the position of most scores in relation to the mean.

TOPIC: STATISTICAL ANALYSIS

See ➲ *Excel* Year 11 Mathematics Standard pp. 101–106

RELATIVE FREQUENCY AND PROBABILITY

Language of probability

14

- **Outcomes of experiments** are the results that could possibly occur.
- **The sample space** is the set of all possible outcomes.
- **The probability that an event *A* will occur**, if outcomes are equally likely, is given by $P(A) = \dfrac{\text{number of favourable outcomes}}{\text{number of total outcomes}}$.
- $0 \le P(A) \le 1$
- **The complement of an event**, $P(\tilde{A})$, is the probability that the event does not occur. $P(\tilde{A}) = 1 - P(A)$.
- **In an experiment or trial the frequency of an event** is the number of times that event occurs. The relative frequency is found by dividing the frequency by the total frequency. It can be used to predict future outcomes.
- **Using a large number of trials** gives a more accurate relative frequency which in turn is more effective at future predictions.

TOPIC: STATISTICAL ANALYSIS

See ➲ *Excel* Year 11 Mathematics Standard pp. 119–121

RELATIVE FREQUENCY AND PROBABILITY 15

Tree diagrams

- **A tree diagram** can be used to trace the possible outcomes of a multi-stage experiment.
- **At each stage the sum of the probabilities** of all the branches emerging from the same position must be 1.
- **If the outcome of an event depends on the result of a previous event**, then the probabilities will be different at each stage and should be shown on the diagram.

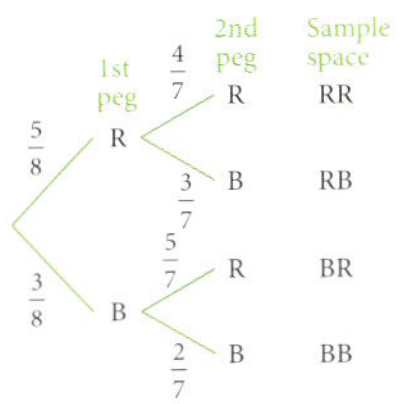

EXAMPLE: The diagram shows the possible results when 2 pegs are drawn from a bag holding 5 red and 3 blue pegs, without replacement.

TOPIC: STATISTICAL ANALYSIS See ➲ *Excel* Year 11 Mathematics Standard pp. 122–125

TYPES OF RELATIONSHIPS 17

Simultaneous equations

- **Two distinct straight lines will meet** at no more than one point. If parallel, the lines will not intersect but otherwise there will be one point of intersection.
- **Simultaneous equations** are two or more equations that have the same solution at the same time.
- **A pair of simultaneous equations** can be solved graphically by finding the point of intersection of the two lines.

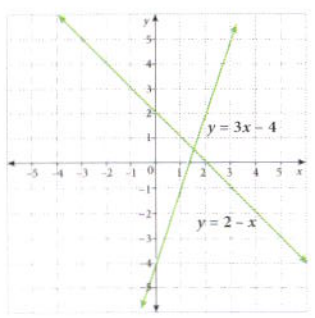

EXAMPLE:

Graph the lines $y = 3x - 4$ and $y = 2 - x$ and hence find the simultaneous solution.

Solution:

The simultaneous solution is $x = 1.5$ and $y = 0.5$.

Check the solution in both equations. If it is correct, the answer must be right.

TOPIC: ALGEBRA See ➲ *Excel* Year 12 Mathematics Standard 1 pp. 5–6

TYPES OF RELATIONSHIPS 19

Break-even points

- **When graphs are used to illustrate** the cost of producing goods and the return from selling those goods, then the point of intersection gives us the 'break-even' point.
- **When the cost is greater than the return** there will be a loss. When the return is greater than the cost there will be a profit. But when the cost and return are equal there is no profit or loss and the enterprise is said to break even.

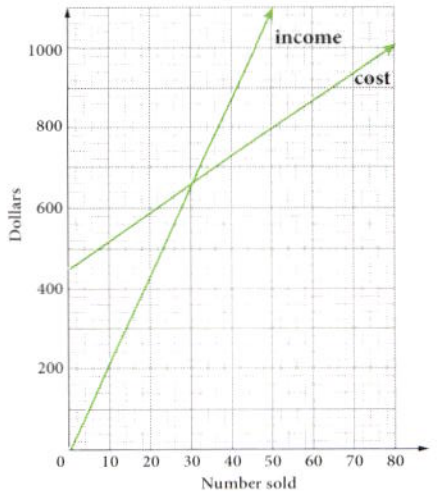

EXAMPLE: This graph shows the break-even point is when 30 items are sold.

TOPIC: ALGEBRA See ➲ *Excel* Year 12 Mathematics Standard 1 pp. 9–10

TYPES OF RELATIONSHIPS 21

Using descriptions

- **It may be possible to draw a graph** to illustrate a practical situation from a description.
- **The graph will be linear** (a straight line) when there is direct variation between variables.
- **The graph might be non-linear** (exponential for example) or some combination of different types.

EXAMPLE: Sketch a graph of the depth of water in this bucket as it is filled with water flowing into it at a constant rate.

Solution:

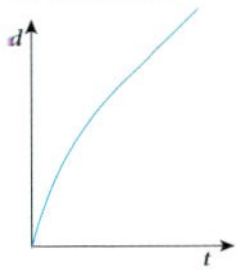

The depth will be increasing at a decreasing rate at first and later will increase at a constant rate.

TOPIC: ALGEBRA See ➲ *Excel* Year 12 Mathematics Standard 1 pp. 12–14

TYPES OF RELATIONSHIPS 23

Exponential curves

- **$y = a^x$ and $y = a^{-x}$ ($a > 0$, $a \neq 1$)** are exponential functions and their graphs are exponential curves.
- **The x-axis is an asymptote of both curves**. It is a line that the curves approach but do not intersect.
- **The y-intercept of each curve is 1** (because $a^0 = 1$).

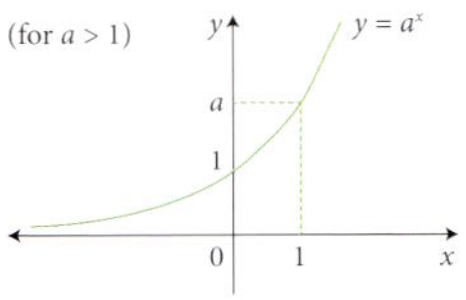

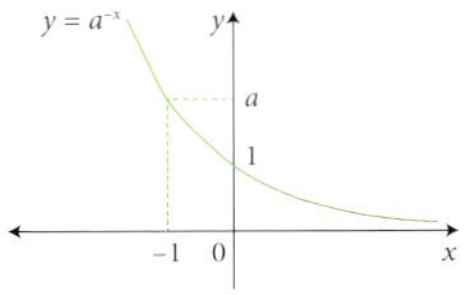

TOPIC: ALGEBRA See ➲ *Excel* Year 12 Mathematics Standard 1 pp. 16–19

TYPES OF RELATIONSHIPS 25

Exponential decay

- **For the curve $y = a^{-x}$ ($a > 1$)**, the y-values become very small very quickly. This is an example of exponential decay.
- **An equivalent form of the equation** of the curve is $y = a^x$ ($0 < a < 1$).
- **The mass of radioactive elements** over time and the value of items depreciating in proportion to their value are examples of exponential decay.

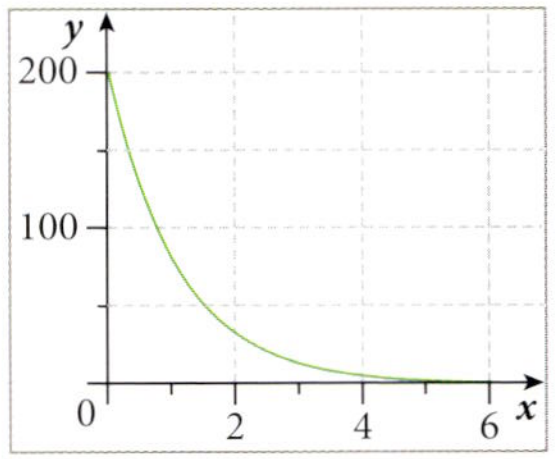

TOPIC: ALGEBRA See ➲ *Excel* Year 12 Mathematics Standard 1 pp. 16–19

TYPES OF RELATIONSHIPS 27

Hyperbolas

- $y = \frac{a}{x}$, where a is a constant, is a reciprocal function.
- **The graph of $y = \frac{a}{x}$** is a hyperbola.

  A hyperbola has two branches. The x and y axes are asymptotes of the curve.
- **Many models of physical phenomena** only use the positive branch of the hyperbola.

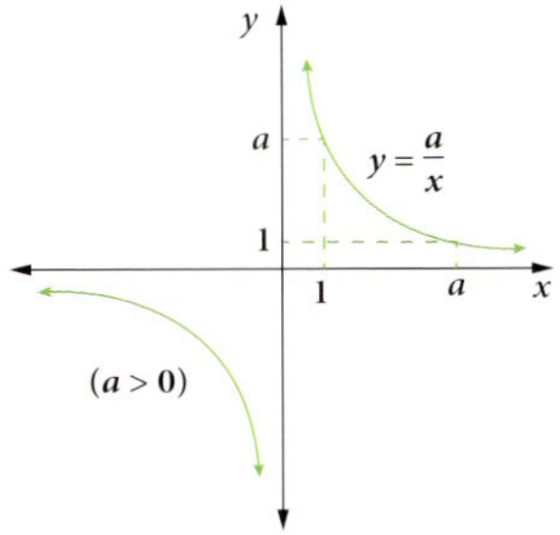

TOPIC: ALGEBRA See ➲ *Excel* Year 12 Mathematics Standard 1 pp. 19–21

TYPES OF RELATIONSHIPS 29

Further graphs

- **There are many types of graphs** that might be the graph of an expression which has practical significance.
- **Questions can be answered** by reading values from a graph or using an equation.
- **Values read from a graph** are often approximations, whereas an equation can give an exact answer.

EXAMPLE: From this graph we can see that $y \approx 17$ when $x = 1.5$ and $x \approx 2.4$ when $y = 70$.

When $x = 2.7$, $y \approx 100$ so $5(2.7)^3 \approx 100$, but using a calculator $5(2.7)^3 = 98.415$.

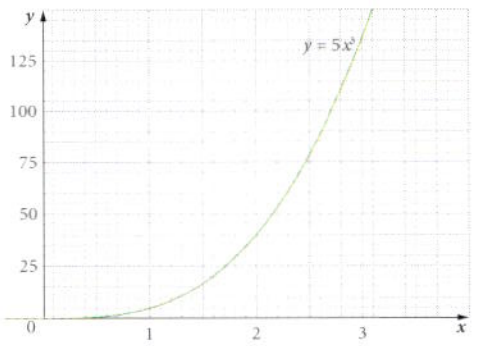

TOPIC: ALGEBRA See ➲ *Excel* Year 12 Mathematics Standard 1 pp. 21–23

RIGHT-ANGLED TRIANGLES 31

Right triangles and Pythagoras

- **A right-angled triangle** is any triangle that has one angle measuring 90°.
- **The hypotenuse is the longest side of a right-angled triangle.** It is the side opposite the right angle.
- **Upper-case letters (capitals) are used to name the angles and lower-case letters for the sides.** Opposite sides and angles use the same letter.
- **Pythagoras was an ancient Greek mathematician.** His theorem states: In any right-angled triangle, the square on the hypotenuse is equal to the sum of the squares on the other two sides. So $c^2 = a^2 + b^2$.

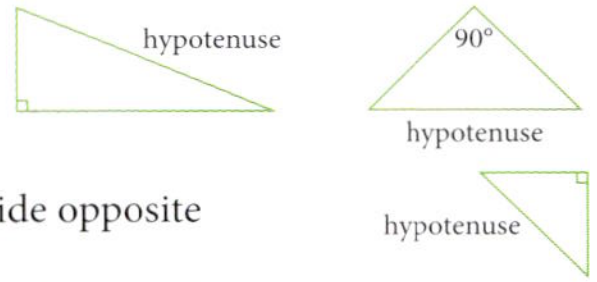

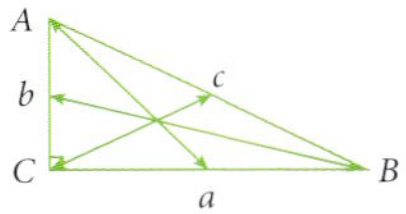

TOPIC: MEASUREMENT See ➲ *Excel* Year 12 Mathematics Standard 1 p. 38

TYPES OF RELATIONSHIPS
Constructing graphs
20

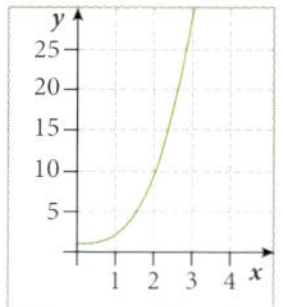

- **Graphs, either linear or non-linear** can be constructed from tables of values.
- **The drawing of a graph** gives a visual representation of some phenomena.

EXAMPLE: Complete a table of values for $y = x^3 + 1$ using values of x from 0 to 3 and sketch the curve for values of $x \geq 0$.

Solution:

x	0	0.5	1	1.5	2	2.5	3
y	1	1.125	2	4.375	9	16.625	28

TOPIC: ALGEBRA See ➲ *Excel* Year 12 Mathematics Standard 1 pp. 11–12

TYPES OF RELATIONSHIPS
Points of intersection
18

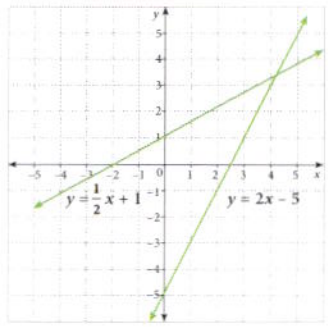

- **The simultaneous solution** is the only solution that works for both equations at the same time.
- **The point of intersection** gives the solution for which the two equations are equal.
- **When two equations that represent** models of physical phenomena are solved simultaneously, that solution has a practical significance.

EXAMPLE: The graph shows the lines $y = 2x - 5$ and $y = \frac{1}{2}x + 1$. What is the solution of the equation $2x - 5 = \frac{1}{2}x + 1$?

Solution: $x = 4$

Take care to answer the question that has been asked. Read the question carefully so that you know whether the whole simultaneous solution is required or only, say, the x-value.

TOPIC: ALGEBRA See ➲ *Excel* Year 12 Mathematics Standard 1 pp. 6–8

TYPES OF RELATIONSHIPS
Graphing straight lines
16

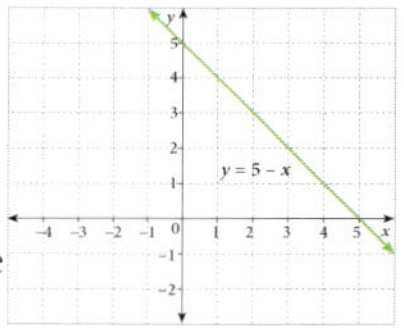

- **If x and y are related by the equation** $y = mx + c$ then they are in a linear relationship.
- **A linear function is a straight line** when graphed.
- **In many practical situations** one quantity will vary directly with another. This variation can be shown by the graph of a straight line.
- **A table of values can be drawn** up and points plotted to show the position of a line.
- **Graphing software can also be used** to draw lines.

EXAMPLE: Draw up a table of values and graph the line $y = 5 - x$.

Solution:

x	0	1	2
y	5	4	3

TOPIC: ALGEBRA See ➲ *Excel* Year 12 Mathematics Standard 1 pp. 4–5

TYPES OF RELATIONSHIPS
Inverse variation
26

- **If two quantities vary inversely** with one another, then one would become larger as the other becomes smaller.
- **If two quantities x and y vary inversely** with each other, then $y = \frac{a}{x}$ where a is a constant.

EXAMPLE: The cost per person, \$C, to go on a trip varies inversely with the number, n, of people travelling. If 30 people travel, the cost per person is \$25. What is the cost per person for 40 people?

Solution: $C = \frac{a}{n}$

When $n = 30$, $C = 25$

$25 = \frac{a}{30}$

$a = 750$

$\therefore C = \frac{750}{n}$

When $n = 40$, $C = \frac{750}{40} = 18.75$

The cost per person would be \$18.75.

TOPIC: ALGEBRA See ➲ *Excel* Year 12 Mathematics Standard 1 pp. 19–21

TYPES OF RELATIONSHIPS
Exponential growth
24

- **For the curve $y = a^x$ ($a > 1$)**, the y-values become very large very quickly. This is an example of exponential growth.
- **The growth of populations** and the growth of investments earning compound interest are examples of exponential growth.

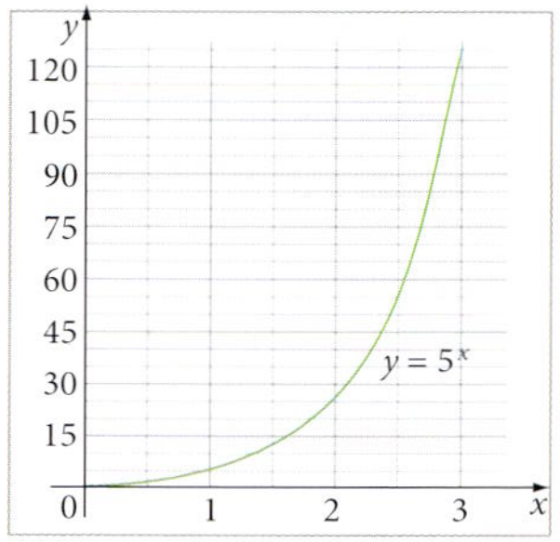

TOPIC: ALGEBRA See ➲ *Excel* Year 12 Mathematics Standard 1 pp. 16–19

TYPES OF RELATIONSHIPS
Further linear graphs of practical situations
22

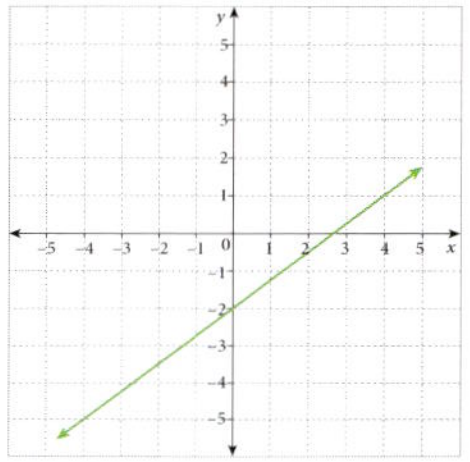

- **In the equation $y = mx + c$**, m is the gradient and c the y-intercept.
- **The meaning of the gradient** and of the y-intercept can be explained in the context of the relationship between the quantities.
- **If we are given information** about the relationship between variables, then a graph can be easily drawn. Using the gradient and y-intercept, the equation of the graph can be found.
- **In many practical situations**, one variable will vary in proportion to another. For example, the distance that is travelled at a fixed speed is directly proportional to the time taken to travel that distance.

TOPIC: ALGEBRA See ➲ *Excel* Year 12 Mathematics Standard 1 pp. 14–16

RIGHT-ANGLED TRIANGLES
Applying Pythagoras' theorem
32

- **Pythagoras' theorem can be used** to find the length of the hypotenuse or the length of one of the other sides of a right-angled triangle.
- **Pythagoras' theorem can also be used** to determine whether a triangle is right-angled.

EXAMPLE: Find the length of side AC to one decimal place.

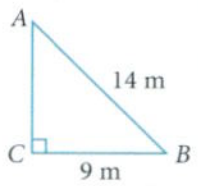

Solution:

$AC^2 = 14^2 - 9^2$

$= 115$

$AC = \sqrt{155}$

$= 10.7$ m (1 d.p.)

EXAMPLE: Is this triangle right-angled?

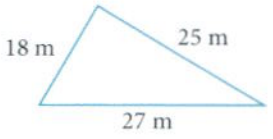

Solution: $27^2 = 729$

$18^2 + 25^2 = 324 + 625 = 949$

So $18^2 + 25^2 \neq 27^2$

The triangle is not right-angled.

TOPIC: MEASUREMENT See ➲ *Excel* Year 12 Mathematics Standard 1 pp. 38–39

TYPES OF RELATIONSHIPS
Limits and restrictions
30

- **Often limits need to be placed** on models of physical phenomena so they are valid.
- **Restrictions might be placed** on the independent variable.
- **Answers that cannot physically occur** are prevented from occurring.

EXAMPLE: The cost per person for a cruise varies inversely with the number of people taking the cruise.

$C = \frac{1350}{n}$ where n is the number of people travelling and C the cost in dollars. The minimum cost per person is \$22.50. What is the maximum number of people that can take the cruise?

Solution: $C = \frac{1350}{n}$

When $C = 22.5$,

$22.5 = \frac{1350}{n}$

$22.5n = 1350$

$n = 60$

The maximum number of people is 60.

TOPIC: ALGEBRA See ➲ *Excel* Year 12 Mathematics Standard 1 pp. 23–26

TYPES OF RELATIONSHIPS
Quadratic functions
28

- **A quadratic function** is any relationship of the form $y = ax^2 + bx + c$ ($a \neq 0$).
- **The graph of a quadratic function** is a parabola.
- **A parabola is a symmetrical curve** with a turning point that lies on the axis of symmetry.
- **The value c is the y-intercept.**

EXAMPLE: Graphing software has been used to graph $y = x^2 + 2x - 3$.

The axis of symmetry is $x = -1$.

The turning point is $(-1, -4)$.

The y-intercept is -3.

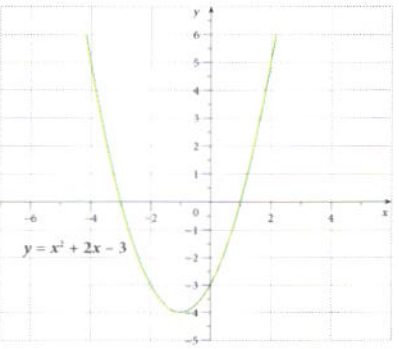

TOPIC: ALGEBRA See ➲ *Excel* Year 12 Mathematics Standard 1 pp. 21–23

RIGHT-ANGLED TRIANGLES
The trig ratios
33

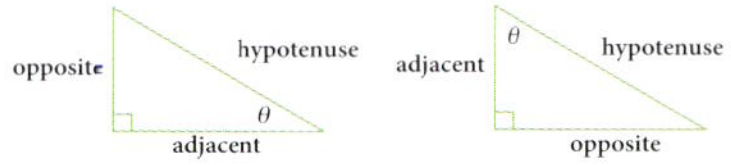

- **The hypotenuse is the side opposite the right angle in a right-angled triangle.**
- **The opposite and adjacent sides depend** upon the angle being considered.
- $\sin\theta = \dfrac{\textbf{opposite}}{\textbf{hypotenuse}}$ $\cos\theta = \dfrac{\textbf{adjacent}}{\textbf{hypotenuse}}$ $\tan\theta = \dfrac{\textbf{opposite}}{\textbf{adjacent}}$

EXAMPLE:
From this triangle, $\sin 28° = \dfrac{8}{17}$,

$\cos 62° = \dfrac{8}{17}$ and $\tan 28° = \dfrac{8}{15}$.

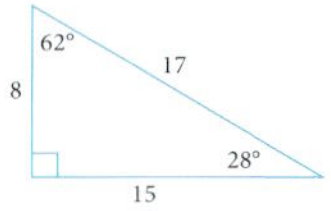

TOPIC: MEASUREMENT See ➲ *Excel* Year 12 Mathematics Standard 1 p. 39

RIGHT-ANGLED TRIANGLES
Using trigonometry to find angles
35

- **If two sides of a right-angled triangle are known**, then trigonometry can be used to find the size of either of the acute angles.
- **The answer might be rounded off** to the nearest degree or to the nearest minute.
- **There are 60 minutes in a degree and 60 seconds in a minute** so if there are 30 or more seconds, for example, we must round up to the next minute.

EXAMPLE: Find the size of θ to the nearest minute.

Solution: $\sin\theta = \dfrac{7}{13}$

$\theta = \sin^{-1}\left(\dfrac{7}{13}\right) = 32°34'44.29'' = 32°35'$ (nearest minute)

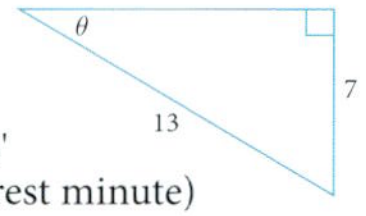

TOPIC: MEASUREMENT See ➲ *Excel* Year 12 Mathematics Standard 1 p. 40

RIGHT-ANGLED TRIANGLES
Angles of depression
37

- **An angle of depression** is formed when we look down.
- **Angles of depression** are always measured from the horizontal.
- **We might make use of the fact** that when alternate angles are formed by parallel lines, they are equal.

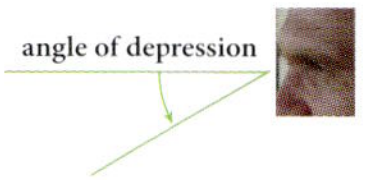

EXAMPLE: From the top of a cliff 92 m high, the angle of depression of a boat is 53°. How far is the boat from the base of the cliff?

Solution: Let d m be the distance from the cliff to the boat.

So $\tan 53° = \dfrac{92}{d}$

$d = \dfrac{92}{\tan 53°}$

$= 69.3269726\ldots$

$= 69$ (nearest unit)

The boat is about 69 m from the base of the cliff.

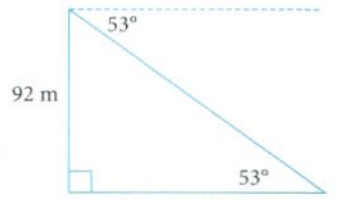

TOPIC: MEASUREMENT See ➲ *Excel* Year 12 Mathematics Standard 1 pp. 41–42

RIGHT-ANGLED TRIANGLES
Compass bearings
39

- **The points of the compass** should all be known.
- **Compass bearings are measured** from either north or south in either an easterly or westerly direction, e.g. N40°E N70°W S50°E S20°W

EXAMPLE: From a point X, a helicopter flew south-east to land at point Y. Leaving Y, it flew N60°E to point Z. Show the information on a diagram and find the size of $\angle XYZ$.

Solution: $\angle XYZ = 45° + 60°$
$= 105°$

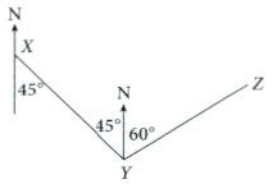

TOPIC: MEASUREMENT See ➲ *Excel* Year 12 Mathematics Standard 1 pp. 42–44

RIGHT-ANGLED TRIANGLES
Navigational methods
41

- **Navigational methods are ways of determining position** in the world and directions for movement.
- **The Global Positioning System (GPS)** uses satellites to give exact positions.
- **Stars have been used to navigate** for a long time. First Nations Australians used storylines that linked landscape with the night sky.
- **Compass bearings or true bearings might be used**. It is important to be able to convert between them.

EXAMPLE: Write S37°W as a true bearing.

Solution: Due south is a true bearing of 180°.

So, as a true bearing, S37°W is 180° + 37° or 217°.

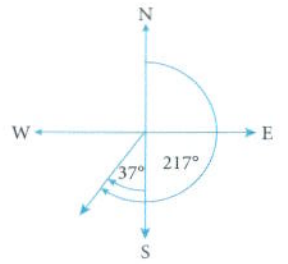

TOPIC: MEASUREMENT See ➲ *Excel* Year 12 Mathematics Standard 1 p. 42

RATES
Using rates
43

- **Rates compare quantities of different kinds.**
- **A rate is usually expressed** as the amount of one item for each unit of the other.
- **There are many examples** of different types of rates, most with practical applications.

EXAMPLE: Lawn seed is recommended to be sown at the rate of 40 g/m^2. How many kilograms of seed would be needed for a 350 m^2 lawn?

Solution: Seed required
$= 350 \times 40$ g
$= 14\,000$ g
$= 14$ kg

TOPIC: MEASUREMENT See ➲ *Excel* Year 12 Mathematics Standard 1 p. 53

RATES
Costs involving rates
45

- **The costs associated with different types of work** are often quoted as rates.
- **A cost for a tradesperson** might involve a fixed amount; for example, a call-out fee, plus a rate per a period of time.
- **There might be a travel component** if the person has to travel some distance to reach the job.

EXAMPLE: A technician charges a call-out fee of \$80 plus a rate of \$110 per hour or part thereof. How much would she charge for a job that takes 2 hours 25 minutes?

Solution: 3 hours will be charged for.

Total cost = \$80 + 3 × \$110
= \$410

Although the time is closer to 2 hours than 3 hours, part of the third hour has been taken up so 3 hours are charged for.

TOPIC: MEASUREMENT See ➲ *Excel* Year 12 Mathematics Standard 1 pp. 54–55

RATES
Travel graphs
47

- **A travel graph usually shows distance against time.**
- **Horizontal lines** usually show that the journey has been paused.
- **Average speed** can be calculated from the information shown in the graph.
- **The fastest average speed** is when the line is steepest.

EXAMPLE: In this graph, Ed left home at 10 am and hiked 6 km in the first 2 hours before stopping for an hour. He then hiked a further kilometre in an hour and stopped for half an hour. He hiked home at an average speed of 2 km/h arriving home at 6 pm.

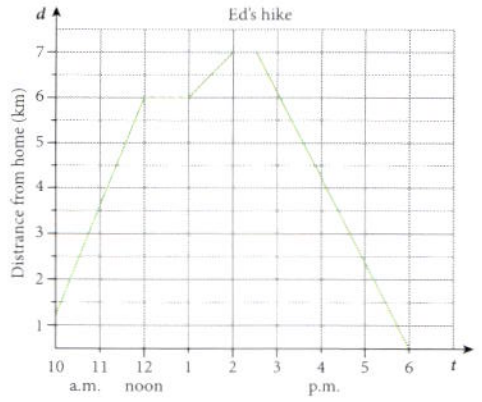

TOPIC: MEASUREMENT See ➲ *Excel* Year 12 Mathematics Standard 1 pp. 55–56

RATES
Fuel consumption
49

- **Fuel consumption is usually given** in litres used per 100 km travelled.
- **The lower the amount of fuel used,** the better the fuel consumption.
- **Fuel consumption is often quoted** differently for highway travel and city travel.

EXAMPLE: A car uses 42 L of petrol to travel 560 km. What is its fuel consumption?

Solution:
560 km = 5.6 × 100 km

Fuel consumption
= (42 ÷ 5.6) L/100 km
= 7.5 L/100 km

EXAMPLE: A car has a fuel consumption of 7.2 L/100 km. How far, to the nearest 10 km, will it travel on a full tank of 68 L?

Solution: Distance travelled
= (68 ÷ 7.2) × 100 km
= 944.44444… km
= 940 km (nearest 10 km)

TOPIC: MEASUREMENT See ➲ *Excel* Year 12 Mathematics Standard 1 pp. 56–57

RIGHT-ANGLED TRIANGLES 38
Solving problems involving trigonometry

- **When solving problems, draw a large clear diagram**, even if one is supplied, and mark all the given information.
- **Two or more right-angled triangles** might be involved.
- **Either trigonometry** or **Pythagoras' theorem** (or a combination of both) might be required.

EXAMPLE:
Find the length of AB.

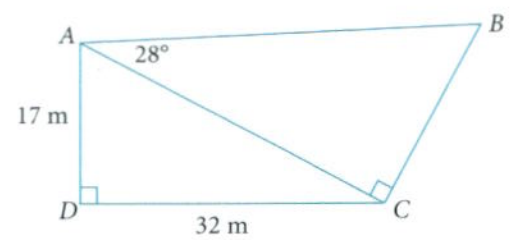

Solution: In ΔACD,
$AC^2 = 17^2 + 32^2 = 1313$
$AC = \sqrt{1313} = 36.2353418\ldots$ m
In ΔABC, $\cos 28° = \frac{AC}{AB}$

So $AB = \frac{AC}{\cos 28°} = 41.0390629\ldots$
The length of AB is about 41 m.

TOPIC: MEASUREMENT See ➲ *Excel* Year 12 Mathematics Standard 1 pp. 41–45

RIGHT-ANGLED TRIANGLES 36
Angles of elevation

- **An angle of elevation** is formed when we look up.
- **Angles of elevation** are always measured from the horizontal.

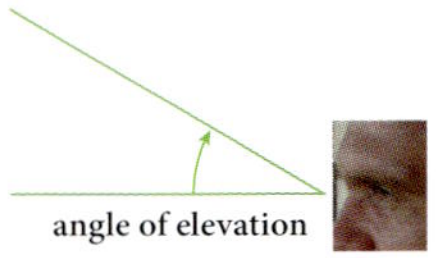

EXAMPLE: A person who is 1.65 m tall casts a shadow, on level ground, 2.3 m in length. Find the angle of elevation of the sun.

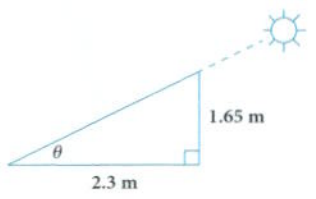

Solution:
$\tan\theta = \frac{1.65}{2.3}$
$\theta = \tan^{-1}\left(\frac{1.65}{2.3}\right)$
$= 36°$ (nearest degree)

TOPIC: MEASUREMENT See ➲ *Excel* Year 12 Mathematics Standard 1 pp. 41–42

RIGHT-ANGLED TRIANGLES 34
Using trigonometry to find sides

- **If one side and one of the acute angles of a right-angled triangle are known**, then trigonometry can be used to find either of the other two sides.
- **First identify the trig ratio** that links the known side with the required side.
- **Then form an equation and solve.**

EXAMPLE: Find the value of x.

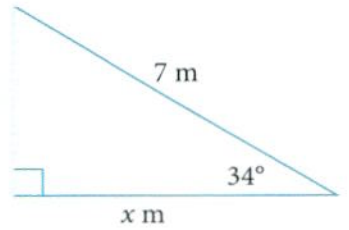

Solution: $\cos 34° = \frac{x}{7}$
$x = 7\cos 34°$
$= 5.8$ (1 d.p.)

We know the hypotenuse and want to find the side that is adjacent to the angle, so the ratio is cos.

TOPIC: MEASUREMENT See ➲ *Excel* Year 12 Mathematics Standard 1 pp. 39–40

RATES 44
Best buys

- **The price of an item** is often expressed as a rate.
- **Finding the best buy** is an application of rates.
- **Best buys are determined** by comparing unit prices. The best buy has the cheapest unit price.

EXAMPLE: Which is the best buy: 600 mL for \$3.30, 1.5 L for \$7.95 or 2.5 L for \$13.50?
Solution: 600 mL for \$3.30 = \$5.50 per kg (\$3.30 ÷ 0.6)
1.5 L for \$7.95 = \$5.30 per kilogram (\$7.95 ÷ 1.5)
2.5 L for \$13.50 = \$5.40 per kilogram (\$13.50 ÷ 2.5)
The cheapest rate is \$5.30 per kilogram.
The best buy is 1.5 L for \$7.95.

TOPIC: MEASUREMENT See ➲ *Excel* Year 12 Mathematics Standard 1 pp. 53–54

RIGHT-ANGLED TRIANGLES 42
Problems involving bearings

- **Problems involving bearings usually involve trigonometry** but might also require some simple geometry.
- **Draw a large clear diagram** and mark all the given information.
- **When drawing a diagram, use the correct directions** and measurements roughly to scale.

EXAMPLE: Q is 23 km from P on a bearing of 070°. R is 39 km from Q on a bearing of 160°. Find the bearing of R from P.

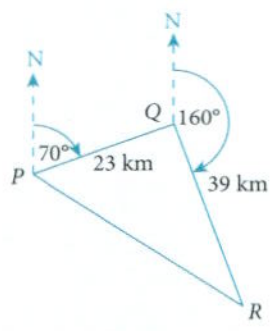

Solution:
$\angle PQR = 70° + 20° = 90°$
$\tan P = \frac{39}{23}$; $P = \tan^{-1}\left(\frac{39}{23}\right) = 59°$ (nearest degree)
So the bearing of R from P is 70° + 59° or 129°.

TOPIC: MEASUREMENT See ➲ *Excel* Year 12 Mathematics Standard 1 pp. 44–45

RIGHT-ANGLED TRIANGLES 40
True bearings

- **True bearings are measured** in a clockwise direction from north.
- **They are always given using three digits**, e.g. the bearing of P from Q is

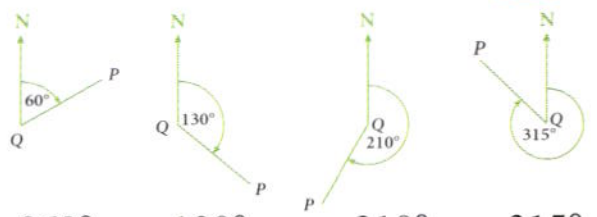

060° 130° 210° 315°

EXAMPLE: In the diagram B is due east of A, $\angle CAB = 30°$ and $\angle CBA = 40°$. So $\angle YBA = 90°$ and $\angle YBC = 90° - 40° = 50°$. $\angle XCB$ and $\angle YBC$ are co-interior angles and add to 180° so $\angle XCB = 130°$.
The bearing of B from C is 130°.

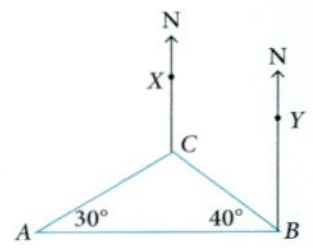

The bearing of P from Q and the bearing of Q from P are not the same. Read the question carefully to make sure you have the correct angle.

TOPIC: MEASUREMENT See ➲ *Excel* Year 12 Mathematics Standard 1 pp. 42–44

RATES 50
Heart rates

- **A person's heart rate** is also known as their pulse.
- **Heart rates are measured in beats per minute** (bpm).
- **The resting heart rate** is the person's usual heart rate when they are at rest. The normal range for adults is between 60 and 100 beats per minute, but can be as low as 40 bpm.
- **Heart rates can be affected** by such things as fitness levels, physical activity, medication, age, emotions and temperature.

EXAMPLE: Jai, who is 23 years old, has a resting heart rate of 82 bpm. At that rate, how many times will Jai's heart beat in a year?
Solution: Number of beats = 82 × 60 × 24 × 365
= 43 099 200

TOPIC: MEASUREMENT See ➲ *Excel* Year 12 Mathematics Standard 1 pp. 57–58

RATES 48
Conversions

- **It might be necessary to convert between rates**, e.g. changing m/s to km/h.
- **Each unit can be changed separately** or all units at once.
- **It is important to understand the process**, not just learn to multiply by a particular constant.

EXAMPLE: James ran 100 m in 15 s. What is that speed in kilometres per hour?
Solution: 100 m in 15 s = 1000 m in 150 s
= 1 km in 150 s
= 1 km in 2.5 min
= 2 km in 5 min
= 24 km in 60 min
So 100 m in 15 s = 24 km/h.

TOPIC: MEASUREMENT See ➲ *Excel* Year 12 Mathematics Standard 1 p. 56

RATES 46
Speed

- **Average speed, S,** is the rate at which distance, D, is covered over time, T.
- $S = \frac{D}{T}$
- **Average speed differs from instantaneous speed**, which is the actual speed at any given time.

EXAMPLE: A car travels 384 km in 5 hours and 20 minutes. What is its average speed?
Solution:
Speed = $(384 \div 5\frac{1}{3})$ km/h
= 72 km/h

EXAMPLE: A train travels at 124 km/h for $2\frac{1}{4}$ h. How far will it travel in that time?
Solution:
Distance = 2.25 × 124 km
= 279 km

TOPIC: MEASUREMENT See ➲ *Excel* Year 12 Mathematics Standard 1 pp. 55–56

RATES
Target heart rate

51

- **A person's maximum heart rate** is said to be 220 – age of the person.
- **A target heart rate range** is the range of heart rate values aimed to be achieved during exercise.
- **Achieving a target heart rate range** maximises the benefits of exercise and not exceeding the range minimises the risk of overdoing it.

EXAMPLE: Jacqui, who is 20 years old, has a resting heart rate of 75 bpm. Her target heart rate range is between 60% and 70% of her maximum heart rate. Find her target heart rate range.

Solution: Maximum heart rate = 220 – 20 = 200

Now 60% of 200 is 120 and 70% of 200 is 140.

So Jacqui's target heart rate range is between 120 bpm and 140 bpm.

TOPIC: MEASUREMENT

See ➲ *Excel* Year 12 Mathematics Standard 1 pp. 57–58

RATES
Applications of rates

53

- **There are many applications of rates** in various fields; for example, agriculture, health and energy efficiency.
- **In practical applications, limits and restrictions** might need to be taken into account.
- **The reasonableness of answers** should always be considered.

EXAMPLE: Find the cost, to the nearest dollar, of using a 120-watt electric blanket for 6 hours a night for 90 nights if electricity costs 32.6 cents per kilowatt hour.

Solution: 120 W = 0.12 kW

Electricity used = 0.12 kW × 6 h × 90
= 64.8 kWh

Cost = 64.8 × $0.326 = $21.1248 = $21 (nearest dollar)

TOPIC: MEASUREMENT

See ➲ *Excel* Year 12 Mathematics Standard 1 pp. 60–61

SCALE DRAWING
Dividing in a given ratio

55

- **To divide in a given ratio** means to separate an amount into different parts so that the parts form the required ratio.
- **When dividing in a given ratio**, the amount is first divided by the total number of parts in the ratio and then the required share of the amount is found for each component of the ratio.
- **A quick check can be made** that the total shares add to the original amount.

EXAMPLE: Divide $3600 in the ratio 3 : 5.

Solution: Total parts = 3 + 5 = 8

Each part is $3600 ÷ 8 or $450.

The first share is 3 × $450 or $1350.

The second share is 5 × $450 or $2250.

TOPIC: MEASUREMENT

See ➲ *Excel* Year 12 Mathematics Standard 1 p. 69

SCALE DRAWING
Similar figures

57

- **Similar figures are the same shape** but not necessarily the same size.
- **The corresponding angles of similar figures are equal** and the corresponding sides are in the same ratio.
- **Two similar figures will differ by a scale factor** (enlargement factor or reduction factor).

EXAMPLE: A rectangle that was 14 cm long and 9 cm wide is now 91 cm long.

a What is the enlargement factor?
b How wide is the new rectangle?

Solution:

a Enlargement factor = 91 ÷ 14 = 6.5
b New width = 6.5 × 9 cm = 58.5 cm

TOPIC: MEASUREMENT

See ➲ *Excel* Year 12 Mathematics Standard 1 pp. 71–73

SCALE DRAWING
Applying similar triangles

59

- **If two triangles are similar**, then corresponding angles are equal and corresponding sides are in the same ratio.
- **There are many applications of similar triangles**; for example, the height of any object can be found (if the length of the shadow can be accessed) using a shadow stick.

EXAMPLE: A metre-high stick casts a shadow 78 cm long at the same time as a building casts a shadow 14.2 m long. How tall is the building?

Solution: Let x m be the height of the building.

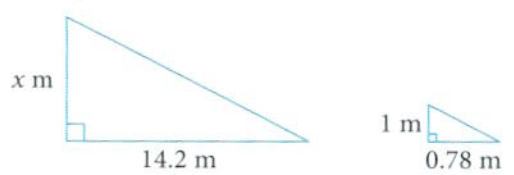

So $\frac{x}{1} = \frac{14.2}{0.78}$

$x = 18.205\,1282\ldots$

The height of the building is about 18.2 m.

TOPIC: MEASUREMENT

See ➲ *Excel* Year 12 Mathematics Standard 1 pp. 73–75

SCALE DRAWING
Maps and plans

61

- **Aerial photographs** are an example of scale diagrams. It is possible to find distances by calculating the scale from features of a photograph.
- **Tools in computer programs or online resources** can also provide information about distances and directions.

EXAMPLE: This is an aerial view of a paddock that is bordered on one side by a creek. The measurements were determined using a computer program. Find the approximate area of the paddock.

Solution: $A \approx \frac{h}{2}(d_f + d_l)$

$= \frac{118}{2}(101 + 147)$

$= 14\,632 \text{ m}^2$

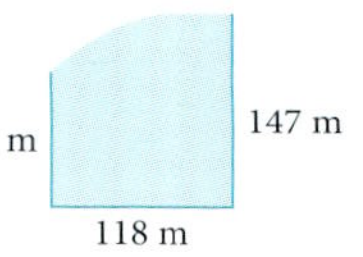

TOPIC: MEASUREMENT

See ➲ *Excel* Year 12 Mathematics Standard 1 pp. 75–80

SCALE DRAWING
Elevations and site plans

63

- **An elevation view shows how the building will look.** Elevations might be numbered or named for the direction in which they face.
- **A site plan shows the position of any buildings** and other features of the land.

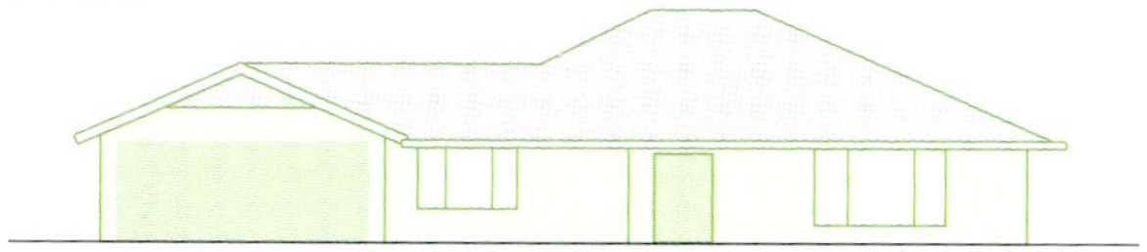

North Elevation

EXAMPLE: The front door and garage of this house open to the north. The position of the windows and the roof line can clearly be seen.

TOPIC: MEASUREMENT

See ➲ *Excel* Year 12 Mathematics Standard 1 pp. 78–80

INVESTMENT
Simple interest

65

- **Interest is the amount paid for the use of money.** It is the amount earned on an investment or paid for a loan.
- **Simple interest is where the amount paid at regular intervals** is the same over the whole period of the investment or loan.
- **Simple interest is calculated using the formula $I = Prn$** where P is the principal or initial amount of money, r is the interest rate per period of time expressed as a decimal and n is the number of time periods.

EXAMPLE: Find the amount of simple interest paid on an investment of $12 000 for 6 years at 7% p.a. simple interest.

Solution:
$P = \$12\,000$, $r = 0.07$, $n = 6$

$I = Prn$
$= \$12\,000 \times 0.07 \times 6$
$= \$5040$

TOPIC: FINANCIAL MATHEMATICS

See ➲ *Excel* Year 12 Mathematics Standard 1 pp. 91–92

INVESTMENT
Compounding periods

67

- **p.a. means per annum or per year.**
- **Often, interest is compounded** over a period other than a year.
- **The compounding period** and the interest rate over the compounding period must be used in any calculations.

EXAMPLE: If an amount is invested for 5 years with interest at 6% p.a. compounded quarterly, then interest is added to the amount four times every year. The compounding period is 5 × 4 quarters or 20 quarters and the interest rate is (6 ÷ 4)% per quarter or 1.5% per quarter.

Although r is expressed as a decimal it can be left in a form such as $r = \frac{0.06}{4}$.

TOPIC: FINANCIAL MATHEMATICS

See ➲ *Excel* Year 12 Mathematics Standard 1 pp. 92–93

SCALE DRAWING 56
Solving problems involving ratios

- **Ratios might be expressed as fractions** and problems solved using properties of fractions.
- **The unitary method can be used with ratios**. First find the value for one item and then the value for the required shares.
- **Given the ratio of all parts** that make up a whole, the percentage composition can be found.

EXAMPLE: If the ratio of gold to silver to bronze medals is 5:7:8 find the percentage composition.

Solution: $5 + 7 + 8 = 20$; Gold: $\frac{5}{20} \times 100\% = 25\%$;

Silver: $\frac{7}{20} \times 100\% = 35\%$; Bronze: $\frac{8}{20} \times 100\% = 40\%$

The percentage composition is 25% gold, 35% silver and 40% bronze.

SCALE DRAWING 54
Ratios

- **A ratio compares quantities of the same kind**.
- **Order is important in a ratio**. A ratio of 3 : 5 is different to a ratio of 5 : 3.
- A **ratio is expressed in simplest form** if expressed using whole numbers that have no common factors.
- **The units used in a ratio must be the same**. For example, the ratio of 20 g to 1 kg is not 20 : 1. It is 20 : 1000 or, when simplified, 1 : 50.

EXAMPLE: A box holds 15 red, 12 blue and 9 white balls. What is the ratio of blue to red to white?

Solution: Required ratio = 12 : 15 : 9
= 4 : 5 : 3

RATES 52
Blood pressure

- **Blood pressure measures the pressure of blood in the arteries**.
- **The unit of pressure** is **mmHg**; a millimetre of mercury.
- **A blood pressure reading is made up of two parts**. The first figure is the systolic blood pressure which is a measure of the pressure while the heart pumps blood as it beats. The second figure is the diastolic blood pressure when the heart is between beats.
- **Blood pressure readings vary** from day to day and fluctuate during the day. A reading of 140 over 90 is considered high (hypertension) and a reading of 90 over 60 is considered **low** (hypotension). A normal reading would lie between these two.

SCALE DRAWING 62
Building plans

- **Floor plans** for buildings show the layout of rooms and measurements.
- **All measurements** are in millimetres.
- **Common symbols and abbreviations** need to be understood.

EXAMPLE: The diagram shows the plan of a kitchen that is 2.8 m by 3.4 m. The cooktop is found about halfway along the left side of the kitchen and the refrigerator is at the top right.

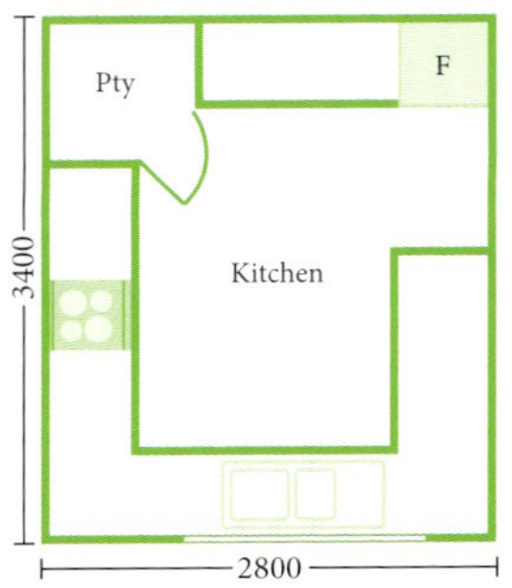

SCALE DRAWING 60
Using scales

- **A plan is said to be drawn to scale** if all measurements are in the same proportions as the original.
- **If a diagram is drawn to scale**, and the scale known, then by careful measurement and calculation all actual measurements can be found.

EXAMPLE: The distance between two points on a map is 3.5 cm. If the scale is 1 : 500 000, what is the actual distance between the points?

Solution: Actual distance = 500 000 × 3.5 cm = 1 750 000 cm = 17 500 m = 17.5 km

EXAMPLE: If a map has a scale of 1:10 000, what distance will 1 cm on the map represent?

Solution: 1 cm represents 10 000 cm or 100 m.

SCALE DRAWING 58
Similar triangles

- **Similar triangles are a special case of similar figures**.
- **Two triangles will be similar** if any of the following occur:
  - corresponding angles are equal
  - corresponding sides are in the same ratio
  - two pairs of corresponding sides are in the same ratio and the included angles are equal
  - the hypotenuse and another side of one right-angled triangle are in the same ratio as the hypotenuse and another side of another right-angled triangle.

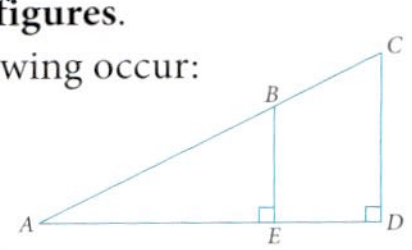

EXAMPLE: Triangles ABE and ACD are similar because they are equiangular. Angle A is common to both triangles and angles AEB and ADC are both 90°.

If two pairs of angles are equal, the third pair must also be equal because the angles of every triangle add to 180°.

INVESTMENT 68
Finding compound interest

- **When calculating compound interest** it is essential that the compounding period and interest rate per period match.
- **The amount of compound interest** earned is the difference between the future value and the present value. $I = FV - PV$

EXAMPLE: Find the amount of interest earned if $20 000 is invested at 6% p.a. compounded monthly for 3 years.

Solution: $PV = \$20\,000$, $r = 0.06 \div 12 = 0.005$, $n = 3 \times 12 = 36$

$FV = PV(1 + r)^n$
$= \$20\,000(1 + 0.005)^{36}$
$= \$23\,933.61$ (nearest cent)

$I = FV - PV$
$= \$23\,933.61 - \$20\,000$
$= \$3933.61$

INVESTMENT 66
Future value

- **In an investment earning compound interest**, the interest is added to the investment so that the amount earning interest changes each time.
- **The final value of an investment is** called the future value (FV). The principal or initial amount is called the present value (PV).
- **The formula $FV = PV(1 + r)^n$** gives the future value of an amount earning compound interest for n periods where r is the interest rate per compounding period expressed as a decimal.

EXAMPLE: Find the future value if $12 000 is invested for 6 years at 4% p.a. compound interest.

Solution: $PV = \$12\,000$, $r = 0.04$, $n = 6$

$FV = PV(1 + r)^n$
$= \$12\,000(1 + 0.04)^6$
$= \$15\,184$ (nearest dollar)

SCALE DRAWING 64
Further applications

- **Measurements from scale drawings** can be used in many applications, such as in painting and furnishing of rooms.
- **The required amounts of materials can be found** and likely costs calculated.
- **Issues such as sustainability** might be more easily addressed by referring to maps and plans.

EXAMPLE: Turf comes in rolls 60 cm wide and 1.75 m long. Determine the best direction to lay the turf on this yard.

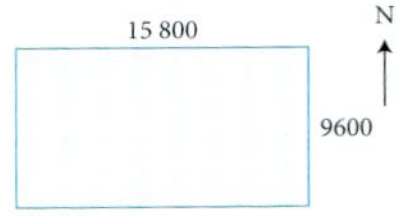

Solution: The turf is 600 mm wide.

Now 9600 ÷ 600 = 16 and 15 800 ÷ 600 = 26.3333…

So having 16 rows running east–west would be best.

INVESTMENT

Finding the present value

69

- **The present value of an investment** is the single amount of money that could be invested now to give the known future value.
- **The formula** $PV = \frac{FV}{(1+r)^n}$ gives the present value when the future value, compound interest rate and period are known. This formula is just a rearrangement of the future value formula.

EXAMPLE: What amount should be invested now, at 6% p.a. interest compounded quarterly, to have \$20 000 at the end of 2 years?

Solution: $FV = \$20\,000$, $r = 0.06 \div 4 = 0.015$, $n = 2 \times 4 = 8$

$PV = \frac{\$20\,000}{(1+0.015)^8}$

$= \$17\,754$ (nearest dollar)

TOPIC: FINANCIAL MATHEMATICS See Excel Year 12 Mathematics Standard 1 pp. 93–94

INVESTMENT

Tables of compounded values

71

- **Often, tables are available** that can be used to find the amounts to which an investment grows.
- **The table gives compounded values of \$1;** multiplying the given factor for the appropriate interest rate and period by the amount involved gives the future value.

Future value of \$1

Period	1%	1.5%	2%
4	1.0406	1.0614	1.0824

EXAMPLE: To find the future value of an investment of \$5000 for 1 year at 6% p.a. interest compounded quarterly, the interest rate per quarter is 1.5% and the number of quarters is 4. From a table of future values of \$1 the factor 1.0614 is found. So the future value is \$5000 × 1.0614 or \$5307.

TOPIC: FINANCIAL MATHEMATICS See Excel Year 12 Mathematics Standard 1 pp. 94–95

INVESTMENT

Solving problems with investments

73

- **Problems might involve comparing** different types of investments or different rates or periods.
- **It might be necessary to determine** whether certain future value targets can be reached and how much more, or less, needs to be invested to achieve those targets.

EXAMPLE: Ed wants to have \$30 000 in 4 years time. If he invests \$20 000 at 9% p.a., compounded monthly, will he have enough? How much more, or less, will he have?

Solution: $r = 0.09 \div 12 = 0.0075$, $n = 4 \times 12 = 48$

$FV = \$20\,000(1 + 0.0075)^{48}$

$= \$28\,628$ (nearest dollar)

As \$30 000 – \$28 628 = \$1372, Ed will fall short by \$1372.

TOPIC: FINANCIAL MATHEMATICS See Excel Year 12 Mathematics Standard 1 pp. 97–98

DEPRECIATION AND LOANS

Depreciation

75

- **Many assets decrease in value over time.** This decrease in value is called depreciation and is a cost to businesses and individuals.
- **The straight-line method is one way** of determining the depreciation of an asset over time. Using this method we assume that the asset decreases in value by a fixed amount each period.
- **The formula for straight-line depreciation** is $S = V_0 - Dn$ where S is the salvage value of the asset after n periods, V_0 is the original value of the asset and D is the amount of depreciation allowed per period.

EXAMPLE: Suppose a machine was purchased for \$35 000 and has been depreciated using the straight-line method with the amount of depreciation allowed each year being \$2750. The value of the machine after 9 years will be \$35 000 – \$2750 × 9 = \$10 250.

TOPIC: FINANCIAL MATHEMATICS See Excel Year 12 Mathematics Standard 1 p. 106

DEPRECIATION AND LOANS

Comparing methods of depreciation

77

- **The graph of the salvage value of an asset,** when the straight-line method of depreciation is used, is a straight line.
- **When using the declining-balance method**, the graph of the value of the asset over time is a curve. It is an example of exponential decay.
- **In the early stages an asset** will be worth more if depreciated under the straight-line method than under the declining-balance method. With the declining-balance method the amount of depreciation will decrease as the value of the asset decreases and so after a while the asset will be worth more under the declining-balance method.

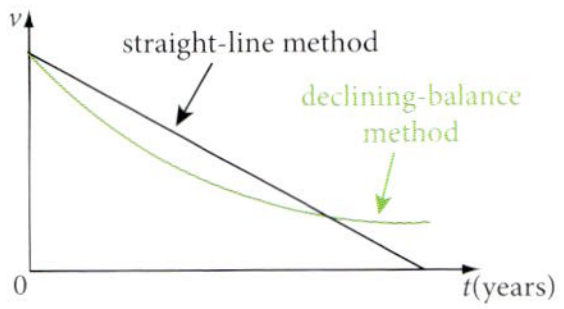

TOPIC: FINANCIAL MATHEMATICS See Excel Year 12 Mathematics Standard 1 pp. 108–110

DEPRECIATION AND LOANS

Reducing-balance loans

79

- **In a reducing-balance loan** the rate of interest charged remains the same, but interest is only paid on the amount owing at a particular time. Because the amount owing decreases as amounts are paid off the loan, the amount of interest also decreases over time.

EXAMPLE: \$10 000 is borrowed at 0.5% per month compound interest. Monthly repayments of \$400 are made. Find the amount of interest charged in the second month.

Solution: Interest for first month = 0.005 × \$10 000 = \$50

Amount owing at end of first month = \$10 000 + \$50 – \$400 = \$9650

Interest for second month = 0.005 × \$9650 = \$48.25

TOPIC: FINANCIAL MATHEMATICS See Excel Year 12 Mathematics Standard 1 pp. 111–114

DEPRECIATION AND LOANS

Published tables

81

- **Financial institutions publish tables** that give the repayments required to repay reducing-balance loans.

EXAMPLE: This table shows the monthly repayments for every \$1000 borrowed on loans over 5 years at different interest rates. Find the total amount repaid if \$80 000 is borrowed over 5 years at 7% p.a.

Monthly repayments per \$1000					
Term (years)	Interest rate				
	6%	6.5%	7%	7.5%	8%
5	19.33	19.57	19.80	20.04	20.28

Solution: Each repayment = 80 × \$19.80 = \$1584

Total repaid = 5 × 12 × \$1584 = \$95 040

TOPIC: FINANCIAL MATHEMATICS See Excel Year 12 Mathematics Standard 1 pp. 114–117

DEPRECIATION AND LOANS

Credit cards

83

- **A credit card is a means of borrowing money** to pay for goods and services.
- **Different credit cards have different fees** and charges and ways of charging interest.
- **Some credit cards have an interest-free period** which means that, provided there is no previous outstanding amount, if the amount owing is paid by the due date no interest will be charged.
- **An outstanding amount is an amount** that should have been paid but has not been paid.

EXAMPLE: Jo has a credit card that has a 55-day interest-free period and an interest rate of 14.6% charged daily on outstanding balances. Jo has no outstanding balance when she uses her credit card to buy clothes costing \$425. If she pays the bill in full by the due date no interest will be paid.

TOPIC: FINANCIAL MATHEMATICS See Excel Year 12 Mathematics Standard 1 pp. 117–121

FURTHER STATISTICAL ANALYSIS

The statistical investigation process

85

- **To properly undertake a statistical investigation**, a certain process should be followed.
- **An issue needs to be identified** and a statistical question posed.
- **A decision must be made on a way to gather information** and then the data collected.
- **Once collected, the data needs to be organised.** It should then be summarised and displayed.
- **The data must be analysed.**
- **The results should be interpreted and communicated**. Valid arguments should support any conclusions.
- **It might be necessary to move back and forth** during the process, making improvements.

TOPIC: STATISTICAL ANALYSIS See Excel Year 12 Mathematics Standard 1 pp. 133–136

INVESTMENT
Practical applications
74

- **If an asset increases in value** over time it is said to appreciate. The future value formula can be used to find the appreciated value.
- **The rate of inflation** can also be determined by considering the present and future values.

EXAMPLE: Find the annual rate of appreciation of a painting that was valued at $1700 eight years ago and that has now doubled in value.

Solution: $\$3400 = \$1700(1 + r)^8$

$(1 + r)^8 = 2$

$1 + r = \sqrt[8]{2}$

$= 1.090\,507\,73\ldots$

$r = 0.09 \quad (2\text{ d.p.})$

The approximate rate of appreciation is 9% p.a.

TOPIC: FINANCIAL MATHEMATICS

See ➲ *Excel* Year 12 Mathematics Standard 1 pp. 97–98

INVESTMENT
Comparing investments
72

- **The benefits or otherwise of different investment strategies** can be explored with the use of technology.
- **Graphs of future values can be used in comparisons**. Such graphs are often examples of exponential growth.

EXAMPLE: This graph shows the value of an investment of $5000 at different interest rates.

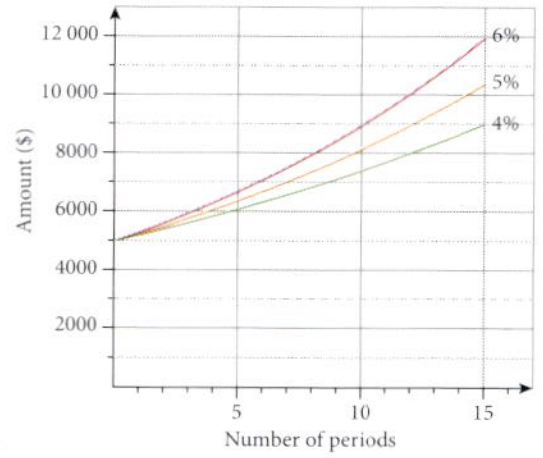

TOPIC: FINANCIAL MATHEMATICS

See ➲ *Excel* Year 12 Mathematics Standard 1 pp. 95–97

INVESTMENT
Finding the rate or compounding period
70

- **The rate of interest can be found** when the present and future values and the period are known, by substituting the given information into the future value formula and solving the resulting equation.
- **The rate must be given in the required form**; for example, as a percentage per annum.
- **The period might be found** using a calculator and a trial-and-error approach.

EXAMPLE: If $9000 accrues to $12 573.26 when invested for 5 years, interest compounded half-yearly, then $\$12\,573.26 = \$9000(1 + r)^{5 \times 2}$.

So $(1 + r)^{10} = 1.397\,028\,88\ldots$ and $1 + r = \sqrt[10]{1.397\,028\,88\ldots} = 1.034$

So $r = 0.034$. Now $0.034 \times 2 = 0.068 = 6.8\%$ p.a.

The annual percentage rate of interest is 6.8% p.a.

TOPIC: FINANCIAL MATHEMATICS

See ➲ *Excel* Year 12 Mathematics Standard 1 pp. 93–94

DEPRECIATION AND LOANS
Tables and reducing-balance loans
80

- **Tables of loan repayments** can be drawn up to show the amount owing at particular time periods. The principal is the amount owing at the beginning of the time period and the interest is calculated on that principal. The interest is added to the principal and then the repayment subtracted to give the amount owing at the end of the period of time. The amount owing at the end of one period is the principal at the beginning of the next period.

EXAMPLE: Suppose $25 000 is borrowed at 5% per period and repayments are $2700.

Period	P	I	$P + I$	$P + I - R$
1	$25 000.00	$1250.00	$26 250.00	$23 550.00
2	$23 550.00	$1177.50	$24 727.50	$22 027.50
3	$22027.50	$1101.38	$23128.88	$20428.88

TOPIC: FINANCIAL MATHEMATICS

See ➲ *Excel* Year 12 Mathematics Standard 1 pp. 114–117

DEPRECIATION AND LOANS
Flat-rate loans
78

- **A flat-rate loan is one where the interest charged is simple interest**. The amount of interest charged is the same over the whole period of the loan.
- **The principal is the amount borrowed**. The total to be repaid is the sum of the interest and principal.
- **The amount of each regular repayment** is the total to be repaid divided by the number of instalments.

EXAMPLE: Suppose a deposit of 15% is paid when a car is bought for $42 000 and simple interest at 7% p.a. is charged on the balance, which is repaid over 5 years with equal monthly instalments. Then the deposit is $0.15 \times \$42\,000$ or $6300 and the balance owing is $\$42\,000 - \$6300 = \$35\,700$. So the interest is $\$35\,700 \times 0.07 \times 5 = \$12\,495$ and the total to repay is $\$35\,700 + \$12\,495 = \$48\,195$. The monthly repayments are $\$48\,195 \div 60 = \$803.25$.

TOPIC: FINANCIAL MATHEMATICS

See ➲ *Excel* Year 12 Mathematics Standard 1 pp. 110–111

DEPRECIATION AND LOANS
Declining-balance depreciation
76

- **In the declining-balance method of depreciation**, the depreciation per period is a fixed percentage of the value of the asset at the end of the previous period.
- **The future value formula**, with a negative value for r, allows us to find the salvage value of an asset that is being depreciated using the declining-balance method. $S = V_0(1 - r)^n$ where S is the salvage value of the asset at the end of n periods, V_0 is the original value of the asset, r is the interest rate per period expressed as a decimal and n is the number of periods.
- **The depreciation after n years** is the difference between the salvage value and the original value.

EXAMPLE: A machine purchased for $16 000 depreciated using the declining-balance method at 12.5% per year will be worth $\$16\,000(1 - 0.125)^4$ or $9379 to the nearest dollar after 4 years.

TOPIC: FINANCIAL MATHEMATICS

See ➲ *Excel* Year 12 Mathematics Standard 1 pp. 106–108

FURTHER STATISTICAL ANALYSIS
Designing questionnaires
86

- **When designing a questionnaire** it is important to be clear about what you are trying to find out and how you will achieve that.
- **Questions should be carefully considered** to make sure they will deliver sensible answers.
- **The target population must be identified** and found.
- **Simple language should be used** in any instructions and questions.
- **Questions should not be ambiguous or biased**.
- **It is important to bear in mind issues** such as privacy and ethnic and cultural sensitivities.

TOPIC: STATISTICAL ANALYSIS

See ➲ *Excel* Year 12 Mathematics Standard 1 pp. 136–137

DEPRECIATION AND LOANS
Credit card calculations
84

- **A cash advance is an amount of money** borrowed as cash using a credit card.
- **Interest calculated on credit cards** will be compound interest and is usually charged daily.
- **Because different amounts are added** to the account at different times, calculations must be made separately for each.

EXAMPLE: Gisele has a credit card that has no interest-free period and a compound interest rate of 0.0495% charged daily from (and including) the date of purchase until (and including) the date of repayment. She uses her credit card to buy jewellery costing $2375 on 18 March. If she pays the credit card bill on 6 April, the total she would need to repay will be $\$2375(1 + 0.000\,495)^{20}$ or $2398.62 to the nearest cent.

TOPIC: FINANCIAL MATHEMATICS

See ➲ *Excel* Year 12 Mathematics Standard 1 pp. 121–123

DEPRECIATION AND LOANS
Working with loans
82

- **The conditions applying to many loans** vary over time.
- **Interest rates for loans** may be fixed or variable.
- **Many lending institutions charge fees** to establish loans and to service loans.
- **There may be stamp duty** to pay on the loan.
- **There could be penalties to be paid**; for example, for early payout of the loan.
- **A graph can show the amount owing** at different times over the period of a loan.
- **A graph can also show the benefits**, or otherwise, of making changes to loans.

TOPIC: FINANCIAL MATHEMATICS

See ➲ *Excel* Year 12 Mathematics Standard 1 pp. 111–117

FURTHER STATISTICAL ANALYSIS

Scatterplots — 87

- **Bivariate data** is data involving two sets of variables.
- **A scatterplot** can be used to show the relationship between two separate sets of numerical data. One set of data is shown on the horizontal axis and the other on the vertical axis.
- **Statistical software or spreadsheet programs** can be used to produce scatterplots.

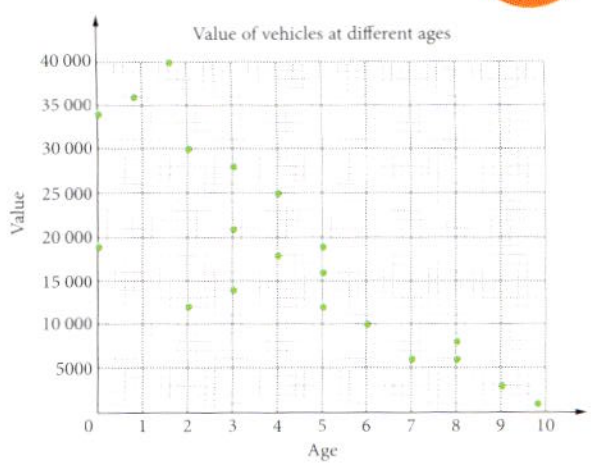

EXAMPLE: This scatterplot was drawn to show the value of some cars at certain ages. The 6-year-old car is valued at \$10 000 and the age of the car that is valued at \$25 000 is 4 years.

FURTHER STATISTICAL ANALYSIS

A line of best fit — 89

- **A line of best fit** is a straight line that can be drawn to approximately show where most points lie.
- **A line of fit can be drawn by eye.** That means simply looking at the scatterplot and trying to draw the line so that it goes roughly through the middle of the set of points. There is no exact answer for such a line of fit.

EXAMPLE: Both of these lines are lines of fit for the same set of data.

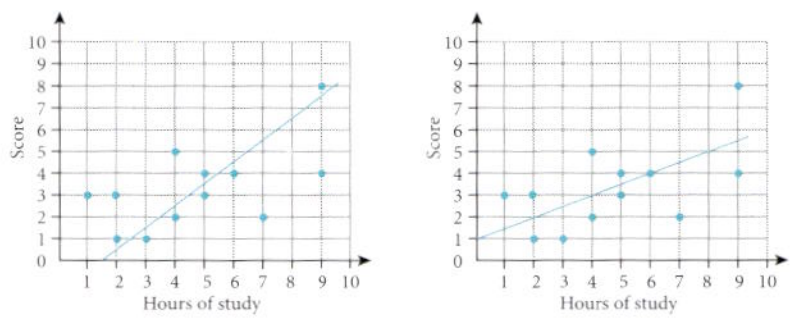

FURTHER STATISTICAL ANALYSIS

The equation of a line of fit — 91

- **The equation of a line of fit** is of the form $y = mx + c$, where m is the gradient and c the vertical intercept.
- **The gradient and vertical intercept** can be interpreted in terms of the data.

EXAMPLE: Suppose the equation $w = 7m + 50$ was found to be a line of fit for a scatterplot showing the weight w, in kilograms, of certain young animals at m months of age. The vertical intercept, 50, would represent the weight at 0 months or the birth weight. The gradient, 7, would represent the weight gained each month.

FURTHER STATISTICAL ANALYSIS

Analysing associations — 93

- **Any of the processes of statistical investigation** can be applied to analyse bivariate data.
- **It is important to consider issues** such as the reliability of the data; for example, consider how large the sample is and whether it is from a reputable source.
- **Look out for any signs of bias.**
- **Ethical considerations** such as privacy and cultural sensitivity should also be taken into account.

NETWORKS AND PATHS

Further definitions — 95

- **A path is a way through a network**, or part of a network, passing through vertices and travelling along edges at most once.
- **A connected network** is one where there is a path between any two vertices.
- **A walk is a set of edges in succession**. The end vertex of one edge must be the start vertex of the following edge.
- **A cycle is a closed walk**. The first and last vertices must be the same, and no other vertices are repeated.
- **The degree of a vertex** is the number of edges that connect to that vertex.

EXAMPLE: For this connected network each vertex has degree 2.

NETWORKS AND PATHS

Weights — 97

- **Some networks have weighted edges**.
- **The weight of an edge** might be a distance, a cost, a time or some other measurement appropriate to the network.

EXAMPLE: This network shows the cost of connecting points to a computer system. The cost to connect A to B is \$618. The cost to connect B to C is \$645 if connected directly or \$975 if B is connected first to E and then to C.

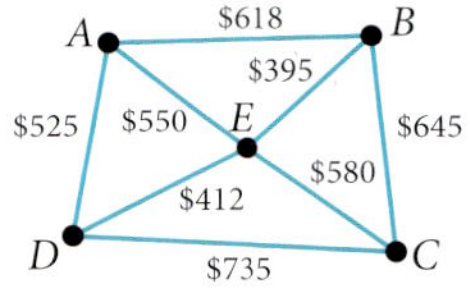

NETWORKS AND PATHS

Trees — 99

- **A tree is an undirected network**, or part of a network, that does not contain a cycle. In a tree, any two vertices are connected by exactly one path.
- **A spanning tree is a tree of an undirected network** that contains every vertex. A network might have many different spanning trees.

NETWORKS AND PATHS

Prim's algorithm — 101

- **Prim's algorithm is one method** of finding a minimum spanning tree.
- **Any vertex can be chosen first**.
- **Connected vertices are added** in turn, depending on the weight of their connecting edges.

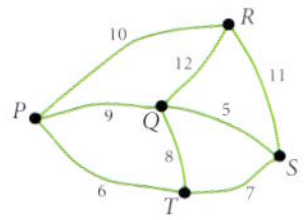

EXAMPLE: Suppose we need to find the minimal spanning tree for this network using Prim's algorithm beginning with vertex P. The vertex that is connected to P with edge of least weight is T, so T is the vertex chosen next. The vertex that is connected to either P or T with edge of least weight is S, so S is chosen next. The vertex that is connected to either P, T or S with edge of least weight is Q, so Q is chosen next and finally R. The minimum spanning tree has weight 28.

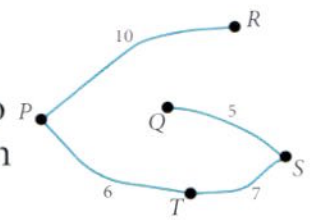

NETWORKS AND PATHS

Shortest paths — 103

- **The shortest path between any two vertices** in a network is the path of minimum weight. It can be found by finding the least cumulative total at each step.
- **The shortest path differs** from the minimum spanning tree because not all vertices must be included.
- **Dijkstra's algorithm is a way** of finding the shortest path from one vertex to another.

EXAMPLE: Dijkstra's algorithm has been used to find the shortest path from A to G in this diagram showing distances in metres. The shortest path is $ADEG$ of length 16 m.

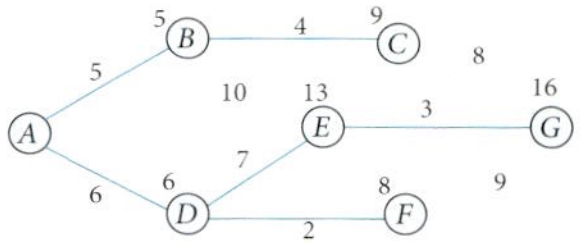

FURTHER STATISTICAL ANALYSIS
Interpolation and extrapolation
92

- **The equation of a line of fit** can be used to make predictions about the data. This is done by substituting values into the equation.
- **Interpolation is when predictions are made** concerning values that lie within the range of the original data.
- **Extrapolation is when predictions are made** concerning values outside the range of the original data. Extreme care must be taken when extrapolating because predictions made for values outside the data range may not be accurate or even sensible.
- **The more data points that are used**, the more accurate the predictions will be and they will be most accurate when the relationship between the variables is strongest.

FURTHER STATISTICAL ANALYSIS
Further lines of fit
90

- **The least-squares regression line** is a special line of fit. The sum of all the squared distances from the line to the actual data points is minimised, which means it is the closest possible line to the data points. (You are not required to determine the least-squares regression line; just be aware that it exists.)
- **The trendline feature in a spreadsheet program** is an example of a line of best fit.

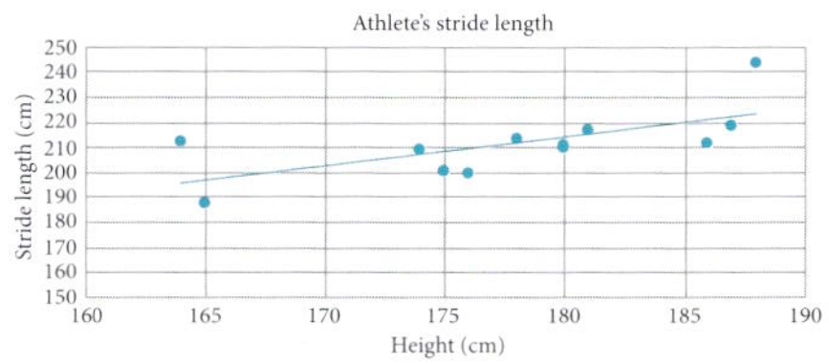

FURTHER STATISTICAL ANALYSIS
Analysing scatterplots
88

- **From a scatterplot** we can easily see whether there appears to be any sort of pattern. This allows us to draw conclusions about any relationship between the two data sets.
- **A pattern is said to be linear** if it forms a straight line (that is, if we can roughly see a line formed by the points).
- **If a pattern is linear,** the direction might be positive (leaning to the right, /) or negative (leaning to the left, \).

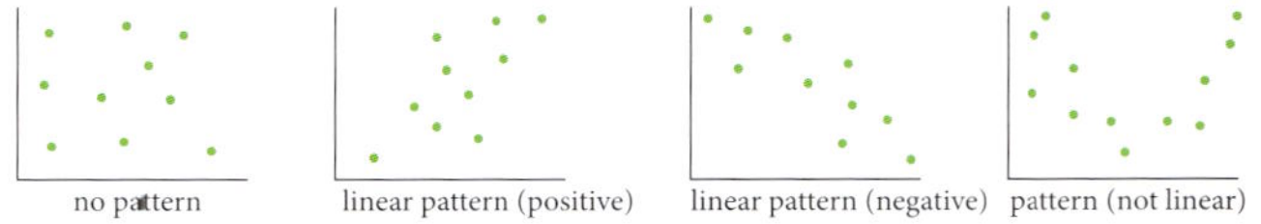

NETWORKS AND PATHS
Tables and networks
98

- **A table can show information** about the connections and weights in a network.
- **Using a table allows** computers to process the information in a network.
- **Constructing a network diagram from a table** might give diagrams that differ in appearance, but they are still basically the same.

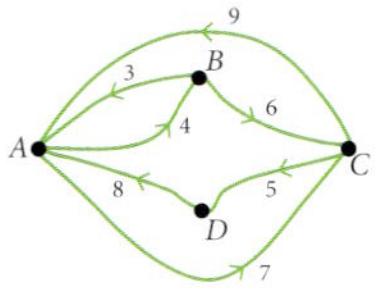

EXAMPLE: This table has been drawn up for the directed network shown above.

From\To	A	B	C	D
A	0	4	7	–
B	3	0	6	–
C	9	–	0	5
D	8	–	–	0

NETWORKS AND PATHS
Directed networks
96

- **A directed network is a network** where the edges must be travelled in a certain direction as indicated by arrows.

EXAMPLE: How many different paths are there from A to D in this network, without completing any cycles? Describe them.

Solution: Two; from A to C to D or from A to B to C and then D

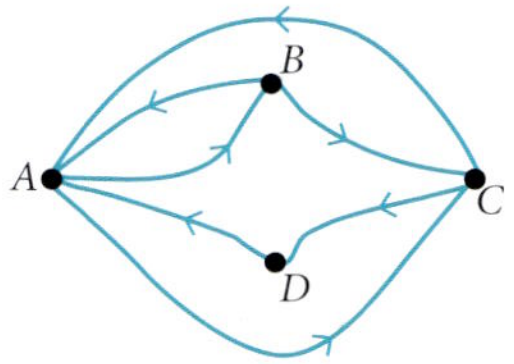

NETWORKS AND PATHS
Network definitions
94

- **A network is a group of objects** that have some sort of connection.
- **A network is made up of vertices**, representing the objects, and connections, which are called edges. Edges do not have to be straight. For example, this network has 5 vertices and 7 edges.

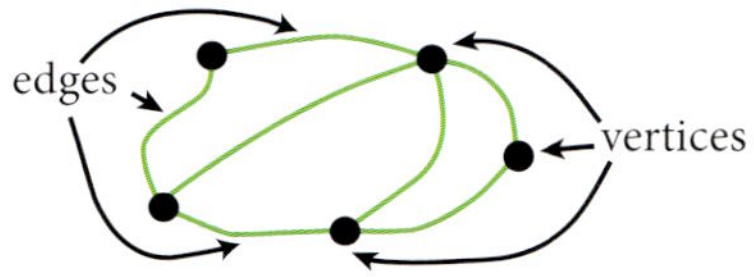

- **A network diagram is a representation** of a network. Diagrams drawn for the same network might differ in appearance but still represent the same connections.

NETWORKS AND PATHS
Applications
104

- **Networks can be used** to solve problems; for example, to minimise the cost of connections of cables or pipes or to plan efficient routes.
- **A network is said to be traversable** if it is possible to move from one vertex to every other vertex moving along each edge once and only once.

EXAMPLE: Find the minimum cost of connecting all the points in this network.

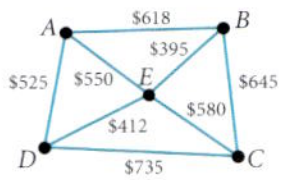

Solution: The minimum spanning tree is shown.

Minimum cost = $(525 + 412 + 395 + 580)
= $1912

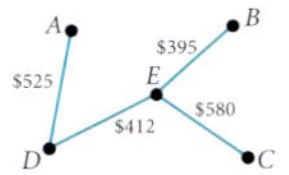

NETWORKS AND PATHS
Kruskal's algorithm
102

- **Kruskal's algorithm** is another method of finding a minimum spanning tree. (Prim's algorithm and Kruskal's algorithm will always give the same result.)
- **The edge with the least weight is chosen first**.
- **Other edges are added** in order of weight, provided they do not form a cycle.

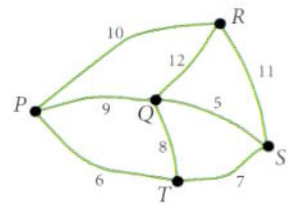

EXAMPLE: Suppose we need to find the minimal spanning tree for this network using Kruskal's algorithm. The first edge chosen is QS, followed by PT and then TS. Neither QT nor PQ can be chosen because a cycle would be formed. The final edge chosen is PR.

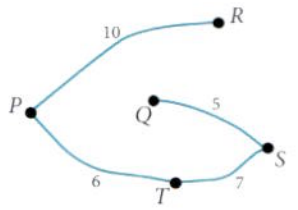

Prim's algorithm and Kruskal's algorithm will always give the same result.

NETWORKS AND PATHS
Minimum spanning trees
100

- **In a spanning tree,** the number of edges will always be one less than the number of vertices.
- **The minimum spanning tree** (or minimal spanning tree) of a network that has weighted edges is the spanning tree for which the total of all the edges is least.

EXAMPLE: The minimum spanning tree for this network will have two edges and those two edges will be the ones with the least weight.

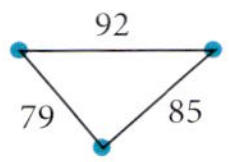